AF752984

Human-Animal Relationships in San and Hunter-Gatherer Cosmology, Volume I

Mathias Guenther

Human-Animal Relationships in San and Hunter-Gatherer Cosmology, Volume I

Therianthropes and Transformation

palgrave
macmillan

Mathias Guenther
Department of Anthropology
Wilfrid Laurier University
Ontario, ON, Canada

ISBN 978-3-030-21184-4 ISBN 978-3-030-21182-0 (eBook)
https://doi.org/10.1007/978-3-030-21182-0

© The Editor(s) (if applicable) and The Author(s), under exclusive licence to Springer Nature Switzerland AG 2020, corrected publication 2020
This work is subject to copyright. All rights are solely and exclusively licensed by the Publisher, whether the whole or part of the material is concerned, specifically the rights of translation, reprinting, reuse of illustrations, recitation, broadcasting, reproduction on microfilms or in any other physical way, and transmission or information storage and retrieval, electronic adaptation, computer software, or by similar or dissimilar methodology now known or hereafter developed.
The use of general descriptive names, registered names, trademarks, service marks, etc. in this publication does not imply, even in the absence of a specific statement, that such names are exempt from the relevant protective laws and regulations and therefore free for general use.
The publisher, the authors and the editors are safe to assume that the advice and information in this book are believed to be true and accurate at the date of publication. Neither the publisher nor the authors or the editors give a warranty, express or implied, with respect to the material contained herein or for any errors or omissions that may have been made. The publisher remains neutral with regard to jurisdictional claims in published maps and institutional affiliations.

Cover illustration: © Woodhouse Collection/Special Collections/University of Pretoria

This Palgrave Macmillan imprint is published by the registered company Springer Nature Switzerland AG
The registered company address is: Gewerbestrasse 11, 6330 Cham, Switzerland

For Daniel and Stephan

The original version of this book was revised. This book was inadvertently published with few errors which has been corrected now. An erratum to this book can be found at https://doi.org/10.1007/978-3-030-21182-0_8.

Preface

This book is written to fill two gaps in anthropology, in two of the discipline's research fields, the one in hunter-gatherer studies, specifically its subfield of Bushman studies, the other in the more recent field of what some refer to as "The Anthropology of Ontology" (Scott 2013) and others have dubbed the "New Animism" (Harvey 2006: xi)—as opposed to its "Old", evolutionary rather than relational, predecessor, *pace* Tylor. These two fields at present exclude each other, in terms of ethnographic substance and theoretical discourse, to the detriment of both. This book sets out to bring the relational ontology paradigm to San studies, and *vice versa*, to the respective research fields' benefit.

This goal is all the more apposite in that hunter-gatherer studies and relational ontology have been linked from the start, back in the 1990s. This is when the "ontological turn", which has since then been taken in socio-cultural anthropology generally and is part of an even wider—posthumanist—turn across Western thinking generally, was first taken in Amazonian studies, among such hunting people as the Achuar, Araweté and Avila Runa, by Philippe Descola, Eduardo Vivieros de Castro and Eduardo Kohn, the three leading voices in Amazonian studies (Costa and Fausto 2010). Through the influence of another leading voice, Tim Ingold, studies of relational ontology were undertaken at around the same time in the Subarctic, from northern Scandinavia, through Siberia (Brightman et al. 2012; Halbmayer 2012) to North America, where ethnologists such as Adrian Tanner, Harvey Feit, Robin Ridington, Colin Scott and Robert Brightman had worked on relational and cosmological aspects of hunter-prey relations even before the 1990s. The influence is evident in these ethnographies of another Subarctic researchers, Irven Hallowell, who a generation before, in an essay on Ojibwa ontology that has since become a foundational article in relational ontology, conceptualized the "non-human person" (or "other-than-human person"), thereby widening the field of social relations—and the concept of both society and of culture—beyond humankind (1960). A similar recasting of "animism as relational epistemology", that acknowledged Hallowell's influence, was the theme 20 years ago of a then seminal and now classic *Current Anthropology* article by Nurit Bird-David (1999: S71), which situated relational ontology among a number of hunter-gatherer-horticulturalists in southern Asia.

Yet, the ontological turn, for all of its paradigm-shifting effects on the study of hunter-gatherers during the last and first decades of the previous and present centuries, all but by-passed the Kalahari, amongst whose hunting-gathering people ethnographers were wont to examine the human-animal relationship not in social, cosmological, mystical fashion but instrumentally and strategically, as a meat-on-the-hoof resource, cherished—more so than plant—for its high caloric yield and thus a key concern of the "foraging mode of production" and its *modus operandi*, "optimal foraging strategy". This cultural-ecological, theoretical-materialist bent in San studies was especially marked and engrained in San studies, with the San, ever since the path-breaking "Man the Hunter" conference in 1966 (and conference volume with the same title edited by the two conference organizers published two years later), and as a result of a large number of high-quality ethnographic writings on the San. The effect of all of this was to render this foraging group one of the two (alongside the Aché) textbook case of the optimal forager, whose "immediate-return" subsistence economy was seen to afford people "affluent" lifeways. When Amazonian and Subarctic hunting became considered in social-relational and cosmological terms rather than instrumental-alimentary ones, in the 1990s, the materialist paradigm continued to inform research in San studies (albeit, not exclusively so, especially through the "Revisionism Debate" this field generated, in terms of political economy and World Systems theory, both paradigms the discursive links of which to relational ontology are no closer than they are to optimal foraging).[1]

I set out in this book to show that San worldview and lifeways are in fact also, at the ontological level, the way people conceive of, perceive and

[1] For elaboration on these points see Guenther (2015: 281–82, 302–9; 2017: 3–4).

experience their interaction with animals, along with other beings of their (preter)natural world, pervaded with relationality and intersubjectivity (and have done so in the past, on the basis of ethnohistorical and archaeological evidence largely on southern San that will be marshalled). In filling this gap in our understanding of San ethnography and culture I will also fill the gap in ontological anthropology, which has excluded these southern hunting people from its neo-animistic purview. Apart from adding new insights to the relational ontology perspective in anthropology, this study, of "San-imism", also underscores the important insight that animism is not some monolithic schema or cosmologico-religious complex but something diverse and multiplex, structurally varied, ecologically and historically contingent. Indeed, as I will also argue, one such included in many and varied animisms of people and cultures of this world are Westerners.

I have recently dealt with these issues in two exploratory articles on relational ontology in the context of San cosmology and lifeways, namely "'Therefore Their Parts Resemble Humans, for They Feel That They Are People': Ontological Flux in San Myth, Cosmology and Belief" (in *Hunter-Gatherer Research* (2015) and "'The Eyes Are No Longer Wild: You Have Taken the Kudu into Your Mind': The Supererogatory Aspects of San Hunting" (in *The South African Archaeological Bulletin* (2017). These articles provided the impetus for this book, with some encouragement from colleagues and friends. It adds to, as well as expands and complements, what is presented, more or less provisionally, in these two articles.

The ethnographic base of this book consists of both my own field work and of ethnographies by other Kalahari anthropologists, as well as of ethnohistorical sources, both published and archival. Given the quantity and variety of all of this source material, most of the contemporary and historical San linguistic groupings of southern Africa are referenced in this book. (See Map 1 for their distribution over southern Africa, and some of their Khoe- and Bantu-speaking neighbours.)

Most of the archival sources consist of unpublished /Xam texts from the Bleek/Lloyd archive. They are referred to by the notational system used by Wilhelm Bleek and Lucy Lloyd that differentiated between interlocutors, informants (by the first, Roman numeral), and by the notebook number and its page number(s); for example L VIII.—4, p. 6365 rev. (Lloyd, /Han‡kasso, notebook 4, page number 6365, back of page). These archival text references can be readily looked up in University of Cape Town's open-access digitalized Bleek/Lloyd archive ("Digital Bleek and Lloyd", lloydbleekcollection.cs.uct.za). The identifying Roman numerals for the other two key narrators are II for //Kabbo, V for Diä!kwain; for the two majn !Kung informants, !Nanni

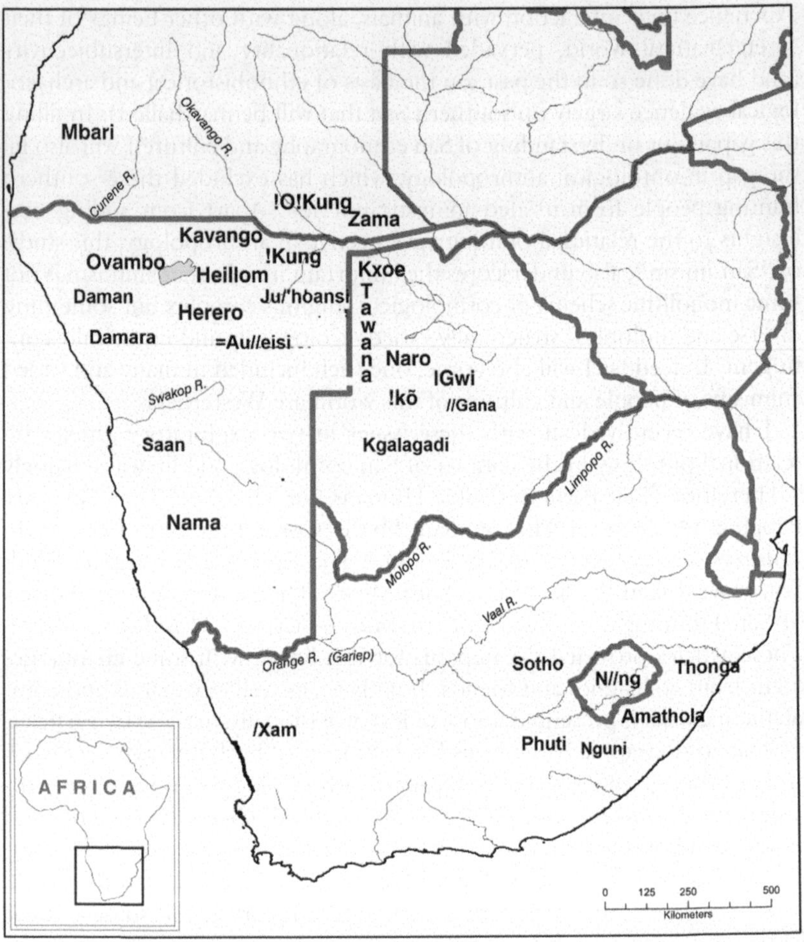

Map 1 Distribution of Khoisan- and Bantu-speaking groupings of southern Africa

and Tamme the identifying numerals are XI and XII. For more information on the /Xam informants, see Bleek and Lloyd (1911: vi–xvii), Deacon and Dowson (1996: 11–43), Guenther (1989: 25–29), Lewis-Williams (2000: 32–33) and—for the most comprehensive account—Bank (2006a).

In addition to these mostly anthropological sources I draw on the writings, rich in quantity and quality, of scholars from a number of other disciplines who have worked in the field of Khoisan studies (many of them drawing on the Bleek/Lloyd archive). These are archaeology, rock art studies and history, as well as folklore and cultural studies, art and literary criticism. The interdisciplinarity of source material has also left its imprint on the content and scope of this book, which, in Vol. II, moves from the San to their Khoe- and Bantu-speaking neighbours in southern Africa, to the Inuit of the eastern Arctic and to the Two Cultures of the West.

Waterloo, ON, Canada Mathias Guenther

References

Bank, Andrew, ed. 2006a. *Bushmen in a Victorian World: The Remarkable Story of the Bleek-Lloyd Collection of Bushman Folklore*. Cape Town: Double Storey Books.

Bird-David, Nurit. 1999. 'Animism' Revisited: Personhood, Environment, and Relational Epistemology. *Current Anthropology* 40 (Supplement): S67–S92.

Bleek, Wilhelm H.I., and Lucy Lloyd. 1911. *Specimens of Bushman Folklore*. London: George Allen & Co.

Brightman, Marc, Vanessa E. Grotti, and Olga Ulturgasheva, eds. 2012. *Animism in Rainforest and Tundra: Personhood, Animals, Plants and Things in Contemporary Amazonia and Siberia*. New York/Oxford: Berghahn Books.

Costa, L., and C. Fausto. 2010. The Return of the Animists: Recent Studies of Amazonian Ontologies. *Religion and Society: Advances in Research* 1: 89–109.

Deacon, Janette, and Thomas A. Dowson, eds. 1996. *Voices from the Past: /Xam Bushmen and the Bleek and Lloyd Collection*. Johannesburg: Wits University Press.

Guenther, Mathias. 1989. *Bushman Folktales: Oral Traditions of the Nharo of Botswana and the /Xam of the Cape*, Studien Zur Kulturkunde No. 93. Stuttgart: Franz Steiner Verlag Wiesbaden.

Guenther, Mathias. 1999. *Tricksters and Trancers Bushman Religion and Society*. Bloomington: Indiana University Press.

Guenther, Mathias. 2015. 'Therefore Their Parts Resemble Humans, for They Feel That They Are People': Ontological Flux in San Myth, Cosmology and Belief. *Hunter-Gatherer Research* 1 (3): 277–315.

Halbmayer, Ernst, ed. 2012. Debating Animism, Perspectivism and the Construction of Ontologies. Special Issue of *Indiana* 29: 9–169.

Hallowell, A. Irving. 1960. Ojibwa Ontology, Behavior, and World View. In *Essays in Honor of Paul Radin*, ed. Stanley Diamond, 19–52. New York: Columbia University Press. (Republished in Stanley Diamond, ed. *Primitive Views of the World: Essays from Culture in History*. New York: Columbia University Press, 1964.)

Harvey, Graham. 2006. *Animism: Respecting the Living World*. New York: Columbia University Press.

Lewis-Williams, David J. 2000. *Stories that Float from Afar: Ancestral Folklore of the San of Southern Africa*. Cape Town: David Philip.

Scott, Michael. 2013. What I'm Reading: The Anthropology of Ontology (Religious Science?) *Journal of the Royal Anthropological Institute* (N.S.) 19: 859–872.

Acknowledgements

In the process of this far-reaching project, I have engaged with the ideas of many scholars, both through their writings and in person. Many of these researchers are recognized in the introduction—the section on interdisciplinarity—and here I again acknowledge the quality of their scholarship and the intellectual stimulation I have derived thereof in my own work, over the years and decades.

I especially acknowledge the many colleagues and friends whose ideas and counsel I have been able to draw on with respect to this project in discussions, both at home in Canada, or at various conference or guest lecture venues in other countries, and in southern Africa. The former, in alphabetical order, are Leila Baracchini, Alan Barnard, Megan Biesele, Laird Christie, Ute Dieckmann, Thorsten Gieser, Jean-Guy Goulet, Erica Hill, Bob Hitchcock, Rockney Jacobsen, Dean Knight, Tihamer Kover, Frédéric Laugrand, Megan Laws, Jenny Lawy, Richard Lee, Chris Low, Andrew Lyons, Harriet Lyons, Junko Maruyama, Bob McKinley, Mark Münzel Elisabeth Marshall Thomas, Ingrid Thurner, Thomas Widlok, Rane Willerslev, Sandra Woolfrey.

Before turning to the long list of colleagues from Africa, I acknowledge three special debts of gratitude. They are to Pieter Jolly and Neil Rusch of the University of Cape Town, for inviting me (in October 2017) to accompany them on a field trip to the Northern Karoo, the home territory of // Kabbo, /Han‡kasso, Diä!kwain and the other /Xam story tellers from whom Wilhelm Bleek and Lucy Lloyd obtained their—and the world's—treasure trove of /Xam myth and lore. Apart from viewing many rock art sites that derive from ancestral /Xam hands and specific features of the landscape that are referred to by the story tellers in one or another tale, this visit afforded me a feel for the landscape—the /Xam *!kau:xu*, or "hunting ground". Another debt to my two hosts and travel companions is that they gave me permission to use some of their images (photographs and rock art tracings).

At the occasion of that trip to South Africa I also visited David Lewis-Williams, who had invited me to Witwatersrand University's Rock Art Research Institute (of which he is the founder), as well as showed me around its exhibition at the Institute's Origins Centre museum. I have admired David's work on San religion, mythology and art ever since I first became acquainted with it when reading and reviewing his path-breaking *Believing and Seeing* almost four decades ago. I was eager to discuss my current project with him at the occasion of my visit (which, as expected, was most fruitful).

The third debt of gratitude is to the late Irene Staehelin, founder and *spiritus rector*, as well as initial funder, through her Swiss-based UBUNTU Foundation, of !Khwa-ttu, a San culture and education centre which, quoting from its mission statement, "celebrates San culture, present and past, for a better future". This NGO has been able to do all that and more, in the 20 years of its operation. I have been fortunate to be associated with !Khwa-ttu on a couple of occasions. One was in 2011 when Irene invited me (along Megan Biesele) to assist in setting up a museum exhibition on /Xam cosmology, around the theme of "The Mantis and the Eland". It was the discussions, planning and research on this project, and head-long delving into the /Xam archive, that spawned this book project. Irene, who died early this year (2019), has drawn a number of other San researchers to !Khwa-ttu, to assist and consult, with the salutary effect of making and keeping them aware of problems and issues about the San people that are more real and urgent than so much that "academics" think and write about in academe's ivory tower.

Colleagues in southern Africa who provided valued input to my project over the past few years are Sam Challis, Josè Manuel de Prada-Samper, Janette Deacon, Janet Hermans, Jeremy Hollmann, Mary Lange, Willemien LeRoux, Al Morris, Alicia Mullen, Richard Northover, John Parkington, Riaan Rifkin, Pippa Skotnes, Andy Smith, Larissa Swan, Carolyn Thorpe, Helize van Vuuren and David Witleson.

Special thanks to David Pearce and David Witleson of the Rock Art Research Institute for their assistance in procuring permissions for some of the rock art images; to Maude Brown of Kuru Development Trust for the same assistance with respect to the contemporary San art; to Pam Schaus for drawing the map; to Stephan Guenther for producing four of the rock art tracings (Fig. 3.18, 4.15, 4.21, vol. I, 4.1, vol. II); to Brittany Ostic and Daniel Guenther for assistance in copy-editing, to Mary Al-Sayed and Linda Braus at Palgrave Macmillan, New York, and Beth McAuley and Lesley-Anne Longo at The Editing Company, Toronto, for their assistance in editing the manuscript and getting it in ship-shape condition for the printer.

Contents

Note on Orthography

All San (and Khoe Khoe) languages include clicks within their inventory of consonants (Guenther 1999: 11). The four that are best defined, phonetically and phonemically, and that appear throughout this book whenever vernacular words, terms and expression are cited are:

1. The dental click (/), produced by placing the tip of the tongue against the back of the upper incisors, creating a sound similar to what we transcribe as "tsk, tsk" (the vocalization used when gently chiding a child).
2. The lateral click (//), produced by placing the sides of the tongue against the sides of the upper row of teeth, creating the sound a rider makes when urging his/her horse on to greater speed.
3. The alveolar click (ǂ), produced with the tongue pressed against the bony projection on the roof of the mouth (alveolus).
4. The cerebral (or alveopalatal) click (!), produced by placing the front of the tongue against the roof of the mouth, behind the alveolus, creating a "cork-popping" sound.

List of Figures

CHAPTER 1

Introduction: "… some subjective identification … which I failed to understand more deeply"

Almost 50 years back, in an article on Namibian! Kung (Ju/'hoan) Giraffe Dance, Lorna Marshall cited an informant's remark about the dance which struck her as enigmatic and puzzling: "We sing it for the tail, it begins at the tail, the way it waves it" (Marshall 1969: 364). The reason she was perplexed was the discrepancy between what the trance dancer's remark suggested—that "the !Kung may think that the sounds or gestures are in some way imitative"—and what was, in her considered opinion based on her ethnographic understanding, actually transpired at the dance—that "the dance steps are definitely not pantomime." Marshall notes that the latter, mimicry of animals, was something the !Kung sometimes practised, with uncanny effect, in ludic dancing or when silently and stealthily approaching a hunted animal, but it was "definitely" not a component of the curing dance, notwithstanding her dancer-informant's comment. She adds the caveat that, "nevertheless, a dancer may occasionally move his head or shoulders almost imperceptibly, in such a way as to suggest the movement of an animal's head," suggesting thereby that there is in fact an element of mimicry, involuntary to the dancer, insinuating itself on her/his body as (s)he dances the Giraffe Dance, and almost imperceptible to the ethnographer.

Thirty years later, in the chapter on the trance dance in her *magnum opus* on Ju/'hoan religion (1999), Marshall cited the same informant's statement on the Giraffe Song, along with another dancer's statement that

© The Author(s) 2020

M. Guenther, *Human-Animal Relationships in San and Hunter-Gatherer Cosmology, Volume I*,
https://doi.org/10.1007/978-3-030-21182-0_1

hints not only at mimicry but also, squarely, at metamorphosis: "When a man dances the Giraffe Dance he becomes a giraffe" (ibid.: 73). Marshall continues to be puzzled about such statements, as "the dance steps are definitely not intentional pantomime" and she rephrases her earlier caveat, more strongly than before: "this suggests some subjective identification on the part of the dancer." To this she adds that it was an aspect of the dance "which I failed to understand more deeply".

The aim of this book is to continue where Lorna Marshall left off: to "understand more deeply" what was intimated and tentatively suggested by the ethnographer—and explicitly stated by her San informants—about the elements of mimesis and metamorphosis in the San curing dance, by placing the same within the wider context of San cosmology in which, as I will demonstrate, these two processes of ontological transformation are prominent.

When Lorna Marshall dealt with these aspects in her two articles, in the 1960s and 1990s, the study of San religion did not readily provide an explanation for the dancer's—or shaman's[1]—ontological transformation. The first article was situated within the *zeitgeist* of anthropological theory of cultural ecology, a cultural-materialist paradigm (Guenther 2007: 371–74, 2017: 3–4) that paid little heed to the symbolic, ritual or mythological domains of San culture (let alone if approached with a phenomenological or ontological slant). As for the second of Marshall's articles, students did in the 1990s consider that domain; however, they did so in the grips very much of the influential "trance hypothesis" which held sway not only in San studies but in the study of shamanism generally, in contexts ethnographic, ethnohistorical and prehistoric (Clottes and Lewis-Williams 1998; Lewis-Williams 2002), not without criticism (Bahn 2003, 2010: 123–31; Dowson 2007). This paradigm led researchers of San religion to focus symbolically and phenomenologically on how the shaman-dancer intensifies and transcends his *human* self, through a psychological experience of an Altered State of Consciousness, rather than being transported out of this self into another, *non-human* being and subject, through a phenomenological experience of ontological transformation (or, as it were, of an Alternate State of Being). This is the focus of the present book—on the modes and limits not of a human being's human conscious-

[1] As before (Guenther 1999: 7), I refer to the San trance dancer by this forever arguably arguable term. See also Lewis-Williams, who concurs (2015b: 62).

ness, through trance, but on those of his or her humanness, through transformation, especially *vis-á-vis* animals.

The now classic study of the San—specifically Ju/'hoan—trance dance that focused on Altered States of Consciousness is Richard Katz's *Boiling Energy* (1982). The work was followed by a sequel of sorts a generation later (with Megan Biesele and Verna St. Denis in 1997), which considers the political implications of the dance and the ever more professionalized dancers. The element of the dancer's transformation, specifically into a lion, was dealt with as well in this study (as will be seen later in this book, which draws on Katz's ethnography); however, its core concern was trance and transcendence (see also Katz 1976) and their psychological and synergistic effects on the trancer-curer and the participants and attendants at the dance. The anthropologist-cum creative therapist and spiritual healer Bradford Keeney's studies on trance curing among the Ju/'hoansi of north-eastern Namibia two decades later are likewise squarely focused on trance and transcendence, in an analytical key that blends anthropological cybernetics with New Age spirituality, through which transcendence and transformation are transposed by the analyst into the loftily elusive domains of God and Creation (1999, 2015, with Hilary Keeney).

Around the same time, in 1999, I published a book—*Tricksters and Trancers Bushman Religion and Society*—in which I considered transformation of a person, a trance dancer, primarily in the psychological, Altered States sense, into a transcended self, and secondarily in the ontological, Alternate Species sense: his—the trance dancer's—or her—menarcheal girl's—transformation into an animal, either a lion or eland (1999: 173–78, 182–92). Having revisited and expanded on this matter in two recent articles (2015, 2017), it constitutes the subject matter of the present book, on San human-animal relations. A title I had considered for the book at one time—and soon rejected for its cumbersomeness—was *Therianthropes and Transformations* as its alliterative echo with the *Tricksters and Trancers* reflects a certain degree of continuity, as well as complementarity. Both books are treatments of the theme of ambiguity. The earlier book considered its pervasive hold on San social organization and values as well as on ritual and belief, with the spotlight on the former cultural domain, *via* the trance dancer/shaman and the initiand, and the latter *via* the trickster protagonist of myth. These social and cultural domains are seen to be linked through feedback loops that sustain the anti-structural makeup of each. In dealing with these linkages not so much conceptually as phenomenologically—that is, via experience and perception rather than cognition—the present project goes deeper: it grounds social-structural and cosmological-mythical ambiguity ontologically rather than conceptually, in terms of the continuity, rather than discreteness, of human and non-human beings.

Connective Cosmology and Mutable Ontology

This continuity, and the state of ontological mutability that derives from it, is embedded within a cosmology of interconnectedness, of beings and states that in so many other cosmologies are discrete—more or less and never consistently, as we will see when looking at one of them (our own Western cosmology, often dubbed "Cartesian" because of its penchant for dichotomization). The South African archaeologist Sven Ouzman has characterized San cosmology with the apt term "connective cosmos" (2008: 219–22), a designation that resonates with South African poet Antje Krog's "interconnected world view" (2009: 184). Like Ouzman (and Low 2014: 351), Krog sees "this interconnectedness with 'the wholeness of life'" (Krog 2009: 184), as something distinctively Khoisan, which some of the other indigenous South African peoples, such as Nguni/Sotho-speakers, "inherited from the First People population … [of] southern Africa" (ibid., quoted in Wessels 2012: 187). In a connective cosmos—"a boundary-less universe whose entangled people and animals move across time and space" (Low ibid.)—ontological boundaries, between species and worlds, are fluid and porous and beings and states are not set each in their respective moulds and modes but interact with and flow into each other.

Ontological mutability is manifested by hybrid beings and by transformation that brings them into and out of being, to varying degrees of explicitness, from incipient stirrings of sympathy or attunedness to the animal, through different forms of mimesis, to complete metamorphosis. Like other researchers before me, I examined trance-induced transformation—"tranceformation"—into a lion, antelope or giraffe, in which transformation can be at its most real to the person undergoing it and most direct and discernible to those who watch it unfolding. In addition I consider transformation outside the trance experience, in such cultural domains and practices as mythology, art, initiation rites, ludic dancing and hunting. These all display transformation in various phases, from incipient to full-blown, fixed, in the form of a hybrid being, or in the process of becoming one, at various phases of perceptibility on the part of the principals or the participants. Such a proneness for transformation and ontological inconstancy is given play in the context of a cosmology within which ontological boundaries between human and non-human are porous. The non-human that is central to this study is the animal (although, as seen in the next chapter, that may also be trees, as well as other plants and "things" of the hunting ground).

While some therianthropes—many of them inhabiting Myth Time—may be constitutionally hybrid, others—some of them featured on rock surfaces—gain their ontological hybridity through transformation, the process that brings such ontologically ambiguous beings into being. The process is elicited either through a being's own volition or through the agency and magical power of another being. This in San mythology is usually a trickster; in other mythologies the agent of transformation may be a sorcerer or witch, such as Ježibaba in the Slavic transformation-tale Rusalka, whose three therianthropic servants—a were-raven, a were-cat and a were-rat[2]—are failed transformation attempts. Or he may be a mad scientist—Dr Frankenstein comes to mind, or Dr Moreau, who, assisted by his chimerical multi-species servant M'ling, populates his dreary island with a dozen-odd like ontological oddities.

Transformation is therianthropic *sui generis*: in its transitional phase between being A and being B, and, as we will see, even as transformed, the latter being, in San mythology, will always contain in its beings the elements of the former, in varying measures. Transforming and transformed beings, as well as the therianthropes of San myth and art, each retain elements of their original being, and through it, autonomy and identity, notwithstanding the inroads of ontological alterity on the integrity of its being. As will be seen, the same can be said of lion- or eland-transformed humans in trance and initiation rituals; indeed, it applies even to hunters. They are all therianthropes of a sort, with the anthro-morphic dose of their being, somatically and mentally, proportionately higher than in their theri-morphic kin-beings of myth, all the while retaining each element of its species autonomy and integrity.

Therianthropes and transformation, the two fundamental manifestations of ambiguity in San cosmology, are thus conceptually and phenom-

[2] They were featured as such in the Met's 2017 production of Dvořák's fairy tale opera "Rusalka", staged by Mary Zimmermann. As playwright and producer a few years earlier of an adaptation of Ovid's "Metamorphoses" (2002), Zimmermann has a professional interest in the mytho-magical and mytho-poeic phenomenon of transformation which is beguilingly evident in her work on Rusalka.

enologically linked: the one, therianthropes, the product of ontological mutability; the other, transformation, its process. These two manifestations of ontological mutability lie at the basis of another conceptual and terminological distinction, the one static, the other dynamic: that between therianthropes—human-animal hybrid beings—and zoomorphs—beings transforming into such (Skotnes 1996: 243), from either a human or an animal form. We encounter both in San myth and art, as the embodiments *par excellence* of ontological mutability. The latter is given the widest scope phenomenologically for humans engaging with their expressive culture, ritual and hunting, as animal-beings and as being-animal.

These two ontological concepts and experiences—and the process that links them, transformation—highlight the non-human beings that hold centre-stage in this study: animals. They are central also to this book's theoretical framework, animism (the "new" version), the core concept of which, "anima" ("soul"), is semantically linked to "animal". Animals are front and centre also in San myth and cosmology. Animal stories are generated through the hunt, which provides an inexhaustible supply of narrative to San story tellers, who, in retelling the hunt and the animals encountered, through exciting or dangerous hunting endeavours or because of uncanny, "counter-intuitive" behaviour on the animal's part rendering it beguiling and "attention-demanding"[3] and transporting it into the realm of legend and myth. The latter domain, the First Order of Myth Time, was inhabited by the First Race, human or humanoid, alongside and interacting with animal-beings, all with a penchant for transformation into one another's kind. Other members of the First Race are animals or hybrid beings that blend human aspects within their animalian ontological makeup, creating multi-species chimeras or, more commonly, human-animal hybrid beings. Such were-beings also appear on rock surfaces, most commonly as antelope-headed men or antelope-legged women, transposing the theme of ontological hybridity from myth and oft-told stories to art and onto frequently encountered, widely seen paintings or engravings (Lewis-Williams 2002, 2015a: 149–72).

Therianthropes from Myth Time may be present also in the Second Order, set in historic or present time, joining their hybrid counterparts from that world, trance-transformed lion-shamans of the Ju/'hoansi and ≠Au//eisi of Namibia and Botswana or greed- and malice-transformed

[3] The terms are Pascal Boyer's (2001) and refer to what according to him constitutes the evolutionary and cognitive basis for religious thought.

Hyena People of the Omaheke Nharo (Naro) of Namibia. Animal identities are mimed and may be partially and fleetingly assumed in ritual or recreational dances, by shamans, young and old women and men at initiation rites and old and young at recreational dances. All of these animal experiences, coupled with sightings and hunts of real animals, and killing and butchering their carcasses, dividing and sharing, cooking and eating their flesh, while, in the process "telling the hunt", give prominence to animal-others in San cosmology and experience.

All of this instantiates and endorses a cosmology of ontological mutability of human and animal, relayed through widely told myths and widely viewed images and embodied, either through mimesis or metamorphosis, by the transforming and transformed shaman and somatically sensed by the girl in her seclusion hut and the hunter when he stalks his poison-wounded prey and closes in on it for the kill. They did so back in the nineteenth century, among the now vanished and vanquished /Xam of the Cape and, it appears, notwithstanding processes of disenchantment in the intervening generations, still in truncated fashion do, among contemporary Kalahari San.

Mimesis and Metamorphosis

These two terms crop up throughout this book, as they are the two modalities through which San experiences ontological transformation. How connected, phenomenologically, are these two processes? The experience of being-changing among the San suggests to me that they are in fact connected, on an experiential continuum, with mimesis, as experienced by San dancers and hunters (as well as some of the were-beings featured in myth and art), a precursor to metamorphosis. While some of the human-animal transformations of people from the real world are less explicit and extravagant than those from Myth Time, they all, to varying degrees, bring about an experience, more or less fleeting, of cross-species blurring of identity and alterity. San cosmology recognizes, in myth, ritual and hunting, this spectrum of ontological transformation, from "playing at" transformation through pantomime to playing it out, experiencing it through the morphing of identities. This process, as seen in most of the following chapters, ranges from incipient to full-blown, partial to complete, ephemeral to lasting, temporary to permanent.

Because transformation lies on so wide a spectrum and assumes so many forms, ontological mutability becomes enrooted and implanted within San cosmology, so much so that, using David Lewis-Williams and

David Pearce's words, "for the San transformations like these are part of everyone's thinking, if not their experience; they are part of life" (2004: 159). Indeed, in his study of the connection between hunting and rock art that included a stint of field work with Namibian Ju/'hoan hunters in 2014, the Finnish artist and art historian Mikko Ijäs found animal transformation, as connected to hunting and trance curing, "were all everyday knowledge to them", so much so that this researcher "found it striking that these connections had not been made previously"[4] (2017: 12).

A "connective cosmos" so premised on ontological mutability as to make "transformations a part of life" raises phenomenological issues and questions for an anthropologist or archaeologist, a folklorist or literary or art scholar, trying to understand the world view, the values and beliefs, the life- and being-in-the-world-ways of the people who subscribe to such a cosmology. Much of the theoretical component of this book (in Vol. II) is concerned with them.

Anthropocentrism and Anthropomorphism

Another conceptual pair that needs to be recognized in the context of this study of human-animal relations, and to be distinguished, is that of "anthropomorphism" and "anthropocentrism". It brings into focus one of the "us"-"them" distinctions that will be dealt with later in the book: the Western and the San (and hunter-gatherer) perspectives on this matter.

The first, anthropocentrism, is deeply engrained in the Western mind the intellectual and religious gatekeepers of which, Cartesian philosophy and science and Christianity, assign a unique status to humans. This leads to two seemingly paradoxical perspectives on animals—the one anthropocentric, the other anthropomorphic, expressed by the English zoologist-philosopher Charles Foster as "humans striding colonially around, describing what they see from six feet above the ground, or about humans pretending that animals wear clothing" (2016: 1). Anthropocentrists either exclude animals altogether from any human sphere or they may assign them to a lower ontological, moral and social stratum on the Great

[4] I suggest some reasons for this elsewhere, prime of them theoretical and analytical blind-sightedness of many San researchers by a perspective deriving from a cultural-materialist paradigm (2017: 1–2). The Danish anthropologist Thea Skaanes, on the basis of recent field work among the Hadza of Tanzania, found a similar bias among researchers of this iconic African hunter-gatherer people, and, resulting from it, a lack of recognition of, and research on, their "religious, ritual and cosmological complexity" (2017: 12).

Chain of Being. We see the latter tendency lexically exemplified in German world view by the use of two sets of words for vital functions, human ones—such as *essen* (eat), *trinken* (drink), *gebären* (give birth), *sterben* (die)—and animal ones (*fressen, saufen, werfen, verrecken*, respectively). Not only are these terms mutually exclusive but the latter are also employed as insult terms by humans for members of their own kind they dislike or look down upon. As for anthropomorphism, in its Western guise, in—if—paying attention to animals and their behaviour, for instance "companion animals" (i.e. pets), the Western tendency is to give these a human cast, morphologically, mentally and behaviourally, thereby obscuring or denying animal's autonomy, identity and integrity.

San world view, and other hunter-gatherers, does not hold so human-exceptionalist a view of animals. Its anthropocentrism is balanced by "theriocentrism", through myth and lore that feature animal protagonists and perspectives, and through a "becoming-animal" ability by humans in certain situations and to varying degrees of awareness. Such "zoomorphing" also tempers San—and hunter-gatherer—anthropomorphism, as such occasional ontological identity and species boundary dissolutions, such experiential "becoming-animal" moments, put a check on humanizing animals through species-solipsistic projection.

Interdisciplinarity in Khoisan Studies

As so many other works in the field of Khoisan studies (Solomon 2009, 2014), this book, too, draws on a number of disciplines other than anthropology. The one most indispensable to this project is archaeology, especially the work of the many, mostly South African, archaeologists[5] who have dealt with aspects of San symbolic and expressive culture, especially rock art. I draw, almost as much, also on the humanities, the prolific writings of folklorists, literary critics, historians, rock art specialists, art historians and art critics, as well as artists, poets and novelists, whose work is defined and inspired by San oral literature, mythology and imagery and

[5] Such as David Lewis-Williams, Janette Deacon, John Parkington, Pieter Jolly, Andy Smith, Sam Challis, Mark McGranaghan, Sven Ouzman, Anne Solomon, Jeremy Hollmann, Aaron Mazel, Francis Thackeray, John Kinahan, Thomas Dowson, Siyaka Mguni, Edward and Cathelijne Eastwood, Geoff Blundell, Frans Prins, Andrew Skinner, David Witleson, Larissa Swan, Alicia Mullen.

who all have made important contributions to the field of Khoisan studies.[6] As noted by one of these scholars, from literary criticism, the late Michael Wessels, in an article-long interpretation of a seminal /Xam narrative—//Kabbo's haunting tale about a hapless member of the Early Race mistaking a lion for a dog—his work on the article[7] required him to "navigate between the hermeneutics of a range of disciplines – folklore, mythography, literary studies, anthropology, rock art studies, Khoisan studies" (2013: 19). That such navigations can be fruitful and enriching for the field of /Xam and Khoisan studies was evidenced in Cape Town in 2011, at a multi-disciplinary conference convened by the South African artist Pippa Skotnes and archaeologist Janette Deacon that celebrated the 100th anniversary of Bleek's and Lloyds *Specimens of Bushman Folklore* (Deacon and Skotnes 2014).[8]

[6]They are, *inter alia*, folklorists (Megan Biesele, Sigrid Schmidt, José Manuel De Prada-Samper, Ansie Hoff), linguists (Roger Hewitt), literary critics (Duncan Brown, Helize van Vuuren, Robert Thornton, Hermann Wittenberg, Michael Wessels, Elana Bregin), historians (John Wright, Nigel Penn, Andrew Bank, Mohamed Adhikari), communication and cultural studies researchers (Keyan Tomaselli, Mary Lange), rock art specialists (Patricia Vinnicombe, Neil Rusch), art critics (Miklós Szalay, Nyasha Mboti, Jessica Stephenson, Leila Baracchini) and artists (Walter Battiss, Pippa Skotnes, Mikko Ijäs), poets (Eugène Marais, Antje Krog, Stephen Watson, Alan James) and novelists (Gideon von Wielligh, Peter J. Schoeman, Elias Canetti, Laurens van der Post, Willemien Le Roux, David Donald). Having just recently read the last-mentioned author's recent novel *Blood's Mist* (Donald 2009) and enjoyed it, I expand on the same as it exemplifies the intellectual rewards of interdisciplinarity: The story, squarely set within /Xam mythology and rock art, as well as colonial history, brings a fiction writer's eloquence and poignancy—along with a enlivening dose of literary licence—to descriptions of a number of ethnographic or ethnohistorical features carefully researched and referenced and annotated by the novelist. These factual features all add to the interest and relevance to this work of fiction. Its relevance to this book specifically are the novel's sections on a girl's inner experiences, of transition and transformation at her menarcheal rite and a shaman's or hunter's experiences of Altered States and of tappings' sympathy and presentiments, respectively. An especially interesting ethnohistorical example for anthropologists is the description by the novelist—who is also a University of Cape Town emeritus psychology professor—of the social chemistry of invidiousness-fuelled incipient tension over inequality and autocratic leadership of multi-band aggregations formed by dispossessed, scattered San people.

[7]See also his *Bushman Letters: Interpreting /Xam Narratives* (2010), a hermeneutical, literary monograph on the /Xam archive, in which Wessels engages explicitly with the sociologist/linguist Roger Hewitt (pp. 121–50; see also Wessels 2009) and the anthropologist Mathias Guenther (pp. 151–74; see also Wessels 2008).

[8]Twenty years earlier, a conference on the same theme—the /Xam Archive—was convened in Cape Town by the South African archaeologists Janette Deacon, John Parkington, David Lewis-Williams and Thomas Dowson, on a somewhat smaller scale (Deacon and Dowson 1996). A similar multi-disciplinary /Xam-focused event was Pippa Skotnes's influ-

What links these copious writings, from a wide range of disciplines, and renders them complementary, is their common focus on a remarkable archive of /Xam oral texts, collected in the 1870s by the German linguist Wilhelm Bleek and his sister-in-law Lucy Lloyd. The Bleek/Lloyd /Xam archive—recently included by UNESCO in its "Memory of the World" register (UNESCO 2012: 328–31) and digitalized by Pippa Skotnes and her co-workers at the University of Cape Town[9]—consists of over 12,000 pages of texts, annotated by the /Xam story tellers and their interlocutors and partially translated. Formally, in terms of disciplinary affiliation, housed in the humanistic disciplines of folklore, literature and linguistics and the more or less exclusive subject matter of most of the writings of these and related disciplines, anthropologists and archaeologists have also drawn extensively on this sheerly inexhaustible font of textual material, in studies on /Xam and San myth and lore, belief and ritual. And, most especially southern San rock art: in deciphering its elusive and enigmatic meaning, the /Xam archive has been regarded as something of a "Rosetta Stone" (ibid.: 331) and its many readings by archaeologists and anthropologists are grounded in this archive. In the process these scholars have also published and annotated hitherto unpublished story texts (Guenther 1989, 1996: 192–93, 179–80; Lewis-Williams 2000, 2018), a scholarly endeavour that extended interdisciplinary cross-fertilization between these two disciplines and the humanities.

My own "navigating" forays, in the context of this book, into both the /Xam texts and into the humanistic literature around them, have been as intellectually challenging as they have been rewarding. (Moreover, it has emboldened me, in a comparative section on San and Western relational ontology in Vol. II, to consider the treatment one or two Western novelists and poets have given to metamorphosis, one of this book's principal

ential and controversial (e.g. Butler 2000, Lyons 2018) *Miscast* exhibition in Cape Town in 1996 (Skotnes 1996). For a critical assessment of the theoretical impact of multi-disciplinarity on Khoisan research, see Solomon (2014). Two multi-disciplinary conferences on San (specifically /Xam) rock art and identity and Khoisan politics of representation were held in 1994 and 1996, respectively, in Johannesburg (Guenther 1995) and Cape Town (Bank 1998) and the South African Cultural Studies journal *Critical Arts* edited by Keyan Tomaselli has published two multi-disciplinary special issues on San, one on their "recuperation" (by popular media) in post-apartheid South Africa (vol. 9, no 2, 1995), the other 20 years later, on San (self)representation (vol. 28, nos 3 and 4, 2015).

[9] The digitalized archive is an open-access resource. A CD of the collection is appended to Skotnes's edited volume *Claims to the Country* (2007).

concerns.) I can only hope that these forays, *qua* anthropologist, into literature, and literary criticism and other humanistic fields, will benefit /Xam and San studies as much as have their forays into my disciplinary domain.

Synopsis of Book

The two volumes of this book are complementary, the first being primarily descriptive in tone and substance, the second discursive. The ethnographic information of Vol. I is presented in anticipation of the arguments of Vol. II, which in turn refers back to the preceding volume, grounding analysis here in description there. Ideally, the two volumes should thus both be read.

However, each volume also to some extent stands on its own, the first as an ethnographic monograph on San cosmology and ontology, the second an anthropological study of ontological ambiguity or, as I refer to it because of the inherent dynamic element of transformation, ontological mutability. It does so in terms of what in the discipline is a standard, tried-and-tested, two-pronged *modus operandi* for anthropological analysis. The one mode is an in-depth study of a certain matter in one culture, the one visited by and known to the writer on the basis of intensive and protracted ethnographic fieldwork that strives towards an understanding of the visited people in terms of their culture. The other is comparison, in an attempt to broaden the understanding gained on the researched matter by the first study. In this book the latter endeavour, dealt with in Vol. II, inherently, through its epistemological operation, refers to the San ethnography in Vol. I; however, in presenting new ethnographic information on other cultures and peoples this part of the book also tells its own story.

Volume I deals with how ontological mutability is manifested, through hybridity and transformation, via the imagination, in myth and lore, conveyed by storytellers as well as, more concretely and starkly, through images produced by past and present-day San artists on rock surfaces or canvas and paper. Also considered is how ontological mutability enters people's awareness not virtually, via the imagination, by means of stories and images, but actually, through experience, in the lived world, specifically the real-life contexts of ritual, play and hunting. Each of these events provides the principals and participants involved in them—trance dancers, intiands, play dancers, hunters—moments at which being-change may be experienced, either mentally ("feeling eland") or bodily ("being eland").

How ontological mutability is experienced, as well as the impacts of this inherently disjunctive and potentially disorienting experience on human and personal identity and integrity, is elaborated on in Vol. II, as that volume's primary concern. This is examined in the context of the San and with reference throughout to the ethnographic information presented in the other volume, in terms of epistemological, experiential and environmental parameters, through which awareness of ontological mutability is conveyed to and through the mind and the body and through being-in-the-world groundedness.

After this discussion, in Vol. II, the ethnographic ground and analytical scope shift and expand, to how other people and cultures think about, perceive and experience ontological mutability. This is done within a loosely comparative framework referenced to the San. It considers three cultural contexts, each broader in scope than the next, expanding the number and kind of factors—structural, acculturational, historical, ecological ones—that impinge on how people in different cultures engage with animals. The first is the Bantu-speaking neighbours of the San with whom some San groups have had contact for centuries, with mutual influences on one another's cosmologies, mythologies and ritual practices and their human-animal aspects. The second comparative context is another hunting society, in another, remote and ecologically radically different part of the world (Inuit of Canada's eastern Arctic).

The third context, the one broadest in scope and vision, is Western cosmology, especially its post-Cartesian, posthumanist take on the human-animal nexus and animals' personhood, being and *umwelt*. All this is quite new and little-charted cosmological territory for anthropocentric, species-solipsistic Westerners and outside their epistemological and ontological mainstream, raising fundamental questions and issues, about species identity and autonomy and more generally, human beings and being human. For the San, and other hunter-gatherers, such matters lie in their intellectual and cosmological mainstream and within well-charted terrain. Thus a study of their view of human-animal relations—of the kind here presented—may provide Westerners, specifically the recent researchers, cognitive ethologists and other Western "anthrozoologists" who have jettisoned the Cartesian perspective, with helpful clues and insights in their new and novel, intellectually recalibrated take on the age-old and universal question of what is human.

The book's conclusion, in Vol. II, discusses critically the impact of the relational ontology paradigm on San studies and considers epistemological and ontological implications of the San (and hunter-gatherer) perception of the human-animal relationship for Western ideas on the same matter.

Outline of Chapters

The book's seven substantive, ethnographic chapters (2, 3, 4, 5, 6 and 7, in Vol. I) are on therianthropes and transformations, two manifestations of ontological mutability that are conceptually and phenomenologically linked. The theoretical and analytical issues and questions they raise, mostly of a phenomenological cast, are dealt with in the four discursive chapters that follow, in Vol. II (Chaps. 2, 3, 4 and 5). The last chapter of Vol. II looks at the "animistic" reading of San cosmology considered in this book comparatively. The conclusion deals critically with the impact of the "ontological turn" on the study of the symbolic, religious and expressive domains both of the San and of hunter-gatherers generally. This raises "further questions and issues", a concluding chapter's staple.

Therianthropic beings are described in their full splendour in Chap. 2 (and then again in Chap. 4, as a motif in San art). The ontological hybridity and species diversity of Myth Time's "Early Race" is evident in the /Xam trickster /Kaggen and his multi-species extended family or band. Therianthropic or chimerical ontological ambiguity may be manifested with dazzling extravagance by some members of the Early Race, or, more commonly, in subtly elusive and mentally beguiling fashion, among them the First Race Elephants and Ostriches. Because myth time and historical time are not exclusive temporal and ontological domains in San mythology and cosmology, First Order therianthropes can also appear in the Second Order, where they may live in remote areas of the veld or hunting ground making occasional contact with human hunters or gatherers, adding grist to the San mill of stories, as well as corroboration to their World of Myth.

Transformation, and its many forms within San cosmology, is the subject matter of the five remaining substantive chapters. The first (Chap. 3) examines these in the context of San myth, in what some San people refer to as the First Order of Existence (or Creation). This inchoate world's most ontologically volatile and slippery denizen was the trickster, in particular the !Kung's /Xué (/Xue) who, at moments of agitation, was wont to undergo transformation upon transformation, at staccato pace, mutating his being the way a vexed and confused chameleon changes colour. Two other transformation-prone beings of the First Order are the rain divinity !Khwa and the lion (whom we will meet again in the chapter on ritual, another arena within which this myth- and symbol-laden, human being-linked animal becomes a daunting and haunting presence).

The therianthropes of myth and their ontologically unstable, at times dazzling transformations are motifs not only of San mythology but also of art, the subject matter of Chap. 4. Therianthropes display ontological ambiguity to the greatest degree—of flamboyance, profusion and diversity—in San rock art, and in a fashion that in some ways is different from how these beings appear in myths. This raises the question, much debated by writers in the field, whether or not the were-beings on the rock panels are in fact the First Race of Myth Time and if not (the general consensus, more or less) what the connection between the two might be. While basically moot, as the question cannot be posed to the artists who painted centuries in the past, it is part of the examination of the therianthrope motif in contemporary San art, the second component of Chap. 4. This art is examined in the context primarily of the Naro and ≠Au//ei artists at the Kuru Art project at the village of Dekar in western Botswana, on the basis of field work I conducted on this project in the mid-1990s. The art of two of the artists—Dada and Qwaa, their artist's pseudonyms—is highlighted as it is informed with transformation, expressing the same in different ways, ranging from flashy to subtle. In examining how transformations are worked out in images by these two present-day artists, and their own commentaries on them, the images created by (pre)historic San artists—some of them painting in colonial times and within similar life situations and a shared world of belief—may perhaps become somewhat more fathomable.

Chapters 5, 6 and 7 deal with transformation as it occurs to humans in the Second Order of Existence, in the contexts, respectively, of ritual, ludic dancing and hunting. Chapter 5 considers humans—of both genders and generations, men and women, girls and boys—in ritual and liminal moments during the trance dance and initiation rite, when, who and what they are undergoes alteration and transition and through it, different forms and degrees transformation. Ludic dances are the topic of Chap. 6; its central focus is on how mimicry of animals, the dances' principal theme, enacted at times with dramatic histrionics and complex and realistic choreography, that includes animal vocalizations and the use of animal paraphernalia that heighten the realism, may evoke ontological identity stirrings in the dancer, akin to transformation.

Another real-time human who is touched by transformation is the hunter, the topic of Chap. 7. In the setting of certain hunting modes, and at certain moments of the hunt a San hunter—as pre-industrial hunters elsewhere in the world—may engage with this prey animal in terms of sympathy and intersubjectivity. Moreover, his use of "hunting medicines" made from animal substances that are rubbed into cuts in his skin, and the employment of animal disguises, especially the wearing of animal skins, too, may contribute towards a hunter's becoming-animal sense. It is also noted that neither all hunting contains these supererogatory elements nor that all hunters feel sympathetically attuned to or ontologically linked with them, caveats that compound people's ambivalence and ambiguity about the human–animal connection.

References

Bahn, Paul. 2003. Chauvet-ism and Shamania: Two Ailments Afflicting Ice Age Art Studies. *The London Magazine*, August/September: 38–46.

———. 2010. *Prehistoric Rock Art Polemics and Progress.* Cambridge: Cambridge University Press.

Bank, Andrew, ed. 1998. *The Proceedings of the Khoisan Identities & Cultural Heritage Conference*, South African Museum, Cape Town, July 12–16, 1997. Cape Town: InfoSource.

Boyer, Pascal. 2001. *Religion Explained: The Evolutionary Origins of Religious Thought.* New York: Basic Books.

Butler, Shelley R. 2000. The Politics of Exhibiting Culture: Legacies and Possibilities. *Museum Anthropology* 23 (3): 74–92.

Clottes, J., and D.J. Lewis-Williams. 1998. *The Shamans of Prehistory: Trance and Magic in the Painted Caves.* New York: Harry Abrams.

Deacon, Janette, and Thomas A. Dowson, eds. 1996. *Voices from the Past: /Xam Bushmen and the Bleek and Lloyd Collection.* Johannesburg: Wits University Press.

Deacon, Janette, and Pippa Skotnes, eds. 2014. *Courage of //Kabbo: Celebrating the 100th Anniversary of the Publication of* Specimens of Bushman Folklore. Cape Town: UCT Press.

Donald, David. 2009. *Blood's Mist.* Auckland Park: Jacana.

Dowson, Thomas A. 2007. Debating Shamanism in Southern African Rock Art: Time to Move on .... *South African Archaeological Bulletin* 62 (183): 49–61.

Foster, Charles. 2016. *Being a Beast: Adventures Across the Species Divide.* New York: Metropolitan Books Henry Holt and Company.

Guenther, Mathias. 1989. *Bushman Folktales: Oral Traditions of the Nharo of Botswana and the /Xam of the Cape*, Studien Zur Kulturkunde No. 93. Stuttgart: Franz Steiner Verlag Wiesbaden.

———. 1995. Contested Images, Contested Texts: The Politics of Representation of the Bushmen of Southern Africa. *Critical Arts* 9 (2): 110–118.

———. 1996. Old Stories/Life Stories: Memory and Dissolution in Contemporary Bushman Folklore. In *Who Says? Essays on Pivotal Issues in Contemporary Storytelling*, ed. Carol L. Birch and Melissa A. Heckler, 177–197. Little Rock: August House Publishers.

———. 1999. *Tricksters and Trancers Bushman Religion and Society*. Bloomington: Indiana University Press.

———. 2007. 'The Return of Myth and Symbolism': Articulation of Foraging, Trance Curing and Storytelling Among the San of the Old Way and Today. *Before Farming: The Archaeology and Anthropology of Hunter-Gatherers* 4, article 4: 1–10. (Online Journal).

———. 2015. 'Therefore Their Parts Resemble Humans, for They Feel That They Are People': Ontological Flux in San Myth, Cosmology and Belief. *Hunter-Gatherer Research* 1 (3): 277–315.

———. 2017. '...The Eyes Are No Longer Wild, You Have Taken the Kudu into Your Mind': The Supererogatory Aspect of San Hunting. *The South African Archaeological Bulletin* 72: 3–16.

Ijäs, Mikko. 2017. Fragments of the Hunt: Persistence Hunting, Tracking and Prehistoric Art. Doctoral Thesis, Aalto University: School of Arts, Design and Architecture, Department of Art.

Katz, Richard. 1976. Education for Transcendence: !Kia-Healing with the Kalahari !Kung. In *Kalahari Hunter-Gatherers: Studies of the !Kung San and Their Neighbours*, ed. Richard B. Lee and Irven DeVore, 281–301. Cambridge, MA: Cambridge University Press.

———. 1982. *Boiling Energy: Community Healing Among the Kalahari Kung*. Harvard: Harvard University Press.

Keeney, Bradford. 1999. *Kalahari Bushman Healers*. Philadelphia: Ringing Rocks Press.

Keeney, Bradford, and Hillary Keeney, eds. 2015. *Way of the Bushman as Told by the Tribal Elders*. Rochester: Bear & Company.

Krog, Antje. 2009. *Begging to Be Black*. Cape Town: Random House Struik.

Lee, Richard B. 2003. *The Dobe Ju/'hoansi*. 3rd ed. Fort Worth/Belmont: Wadsworth.

Lewis-Williams, David J. 2000. *Stories that Float from Afar: Ancestral Folklore of the San of Southern Africa*. Cape Town: David Philip.

———. 2002. *The Mind in the Cave*. London: Thames & Hudson.

———. 2015a. *Myth and Meaning: San-Bushman Folklore in Global Context*. Walnut Creek: Left Coast Press.

———. 2015b. Text and Evidence: Negotiating San Words and Images. *South African Archaeological Bulletin* 70 (201): 53–63.

———. 2018. "Three nineteenth-century Southern African San myths: A study in meaning." *Africa* 88: 138–159.

Lewis-Williams, David J., and David G. Pearce. 2004. *San Spirituality: Roots, Expression, and Social Consequences*. Walnut Creek: Altamira Press.

Low, Chris. 2014. Locating /Xam Beliefs and Practices in a Contemporary KhoeSan Context. In *Courage of //Kabbo: Celebrating the 100th Anniversary of the Publication of* Specimens of Bushman Folklore, ed. Janette Deacon and Pippa Skotnes, 349–361. Cape Town: UCT Press.

Lyons, Andrew P. 2018. The Two Lives of Sara Baartman: Gender, 'Race', Politics and the Historiography of Mis/Representation. *Anthropologica* 60: 327–346.
Marshall, Lorna. 1969. The Medicine Dance of the !Kung Bushmen. *Africa* 39 (4): 347–381.
Marshall, Lorna J. 1999. *Nyae Nyae !Kung Beliefs and Rites*, Peabody Museum Monographs, No. 8. Cambridge, MA: Harvard University.
Ouzman, Svend. 2008. Cosmology of the African San People. In *Encyclopedia of the History of Science, Technology, and Medicine in Non-Western Cultures*, ed. Helaine Selin, vol. 1, 2nd ed., 219–225. New York: Springer Science and Business Media.
Skaanes, Thea. 2017. Cosmology Matters: Power Objects, Rituals, and Meat-Sharing Among the Hadza of Tanzania. Ph.D. Dissertation, Aarhus University.
Skotnes, Pippa. 1996. The Thin Black Line: Diversity and Transformation in the Bleek and Lloyd Collection and the Paintings of the Southern San. In *Voices from the Past: /Xam Bushmen and the Bleek and Lloyd Collection*, ed. Janette Deacon and Thomas A. Dowson, 234–244. Johannesburg: Wits University Press.
———., ed. 2007. *Claims to the Country: The Archives of Lucy Lloyd and Wilhelm Bleek*. Johannesburg/Athens: Jacana Media/Ohio University Press.
Solomon, Anne. 2009. Broken Strings: Interdisciplinarity and /Xam Oral Literature. *Critical Arts* 23 (1): 25–41.
———. 2014. Truths, Representationalism and Disciplinarity in Khoesan Researches. *Critical Arts: South-North Cultural and Media Studies* 28: 710–721.
UNESCO. 2012. *Memory of the World: The Treasures that Record our History Form 1700 BC to the Present Day*. Paris/Glasgow: UNESCO/Harper/Collins.
Wessels, Michael. 2010. *Bushman Letters: Interpreting /Xam Narratives*. Johannesburg: Wits Press.
———. 2012. The Khoisan Origin of the Interconnected World View in Antje Krog's *Begging to Be Black*. *Current Writing: Text and Reception in Southern Africa* 24 (2): 186–197.
———. 2013. Story of a /Xam Bushman Narrative. *Journal of Literary Studies* 29 (3): 1–22.
Zimmerman, Mary. 2002. *Metamorphoses: A Play Based on David R. Slavitt's Translation of* The Metamorphoses of Ovid. Evanston: Northwestern University Press.

CHAPTER 2

Therianthropes

All things (that is living creatures) were once people …
/Han‡kasso (Bleek *1936*: 164)

Xam story teller /Han‡kasso's statement to Lucy Lloyd is as fundamental a postulate about human and animal ontology to San cosmology as is its reverse, Darwinian, counterpart to Western cosmology. Its clearest expression is found in the stories San tell about myth time, the First Order of Existence, whose denizens, for the most part, were therianthropes, that is, animals that "were once people". As such, these people—the Early Race—contained in their being, their First Order human'ness and elements of their Second Order animal'ness. Humans and animals from both myth and historical time are thus blended ontologically, explaining why, as we will see in the second section, we also find therianthropes, of a mythic, Early Race cast, in the Second Order of Existence.

In-between these two tempo-ontological realms, and partaking of elements from both, is a somewhat vaguely defined spirit realm. I will briefly describe its nature and features, before turning to the two other ontological domains and worlds, which are more clearly delineated in San cosmology than their spirit realm, a shreds-and-patches amalgam of indigenous and acquired features and notions.

© The Author(s) 2020

M. Guenther, *Human-Animal Relationships in San and Hunter-Gatherer Cosmology, Volume I*,
https://doi.org/10.1007/978-3-030-21182-0_2

The San Spirit Realm

In San cosmology the First Order of myth and the spirit realm are two different, but interconnected other-worlds, the one located in the mythic past, before the historical and contemporary Second Order that followed from the First. The spirit realm, while preternatural in a number of ways, is also of this world, being located somewhere beyond the hunting ground and accessible, by different pathways, to humans, especially shamans. It may also be situated within the sky, rather than the veld, or somewhere in-between, in lower-sky regions, the spatial-ontological domain within which, according to the German anthropologist of religion Bernhard Streck, preindustrial people tend to locate spirits—"*Götter in Augenhöhe*" (gods at eye-level)—thereby rendering them ontologically ambiguous "supernatural beings whose ancestry is neither unequivocally human nor divine"[1] (2013: 206[2]). A San instance is the G/ui (G/wi), among whom the divine couple *N!adima*'s and *N!adisa*'s realm is just above the visible sky, which the G/ui hold to be a stretch of well-watered veld with an abundance of game (Silberbauer 1965: 51–2). The Ju/'hoansi, reportedly also locate "God's village" in the sky, which shamans reach by climbing to it on "God's ropes" that hang invisibly down from the sky (Keeney and Keeney 2015: 26–32; see also Eastwood and Eastwood 2006: 126–29).

Wherever it is located, on the earth or the sky, or both domains or in-between in a low-sky region close to the earth, the spirit world has its own geography which may contain such mystical features as a special tree or tall termite hill—which may also be the god's dwelling—both of which San may regard as an *axis mundi* access route to the sky (Wagner-Robertz 1976: 541–42; Mguni 2006: 65–7). It includes also more worldly features, veld animals and trees, water holes and river beds, along with grass huts, small and big ones, including a European-style, two-story house, in which lives the trickster god with his spirit-of-the-dead servants. These dwellings may comprise a village, or even a small town with roads and paths, goats and bakkies (pick-up trucks), stores and schools.[3] Another

[1] "*übernatürliche Wesen die weder eindeutig menschlicher, noch göttlicher Abstammung sind*" (my translation).

[2] For an examination of spirits in such terms—as beings "who are their own kind of real"—in the context of Amazonian shamanic cultures, see Kohn (2013: 217); also see Münzel (2013).

[3] Some of the Naro informants from whom I received these accounts of the spirit realm were closely associated with a mission station (of the Gereformeerde Kerk, a branch of the

mode of moving to and through this preternatural domain is along invisible spider-webs, along which the trickster and spirits may move. A variant of these lateral web-strings is vertical "threads" that lead upwards, to a spirit realm located not in the veld but in the sky, the region of the Second Order within which some San situate the spirit domain. The inhabitants of the preternatural spirit world consist of a number of enigmatic god-figures that may or may not be the same spirit beings even though bearing a diversity of other names. The other species of spirit being are the spirits of the dead, along with ghosts, their initial incarnation, who may be encountered hovering near grave sites (Thurner 1983; Marshall 1999: 5–27; Guenther 1999: 61–4; Bennum 2004: 250–6, 272).

Ontologically, these preternatural beings are ambiguous. They are "souls or spirits ... that have left humans whose bodies have died" to Namibian Damara, a black people with Khoisan linguistic and cultural affinities, amongst whom, as reported by Sian Sullivan, there appears to be a more developed notion of spirit and of a spirit domain than the San, their cultural cousins. As such, spirits, in Damara supernaturalism,

> have ontological reality in the present: they are not simply people who lived in the past ... that require worship or regular social or ceremonial commemoration ... [but] are understood more as specific entities that, through pragmatic relationship practices, are greeted and called upon to intervene ... in the present, so as to influence outcomes. (Hannis and Sullivan 2018: 289)

This affords agency to Damara spirits. And as such—"agency-enacting entities with ontological reality" (ibid.: 21)—they constitute another non-human being with whom Khoisan people share their world, along with the animals who, among the Damara, may also be affected by spirits as these

South African Dutch Reformed Church) and their notions of the spirit world and its features were influenced by their exposure, through an active evangelist-missionary, to Christian notions, resulting in a syncretistic amalgam of features of the spirit realm that included the Garden of Eden, "Addam" and "Effa", "Jessu Kriste", the Tower of "Babbel" and a hell-fire in a dark, remote region of the Kalahari veld around which skulk shivering, cowering the spirit avatars of "bad people" munching on flies. I have described these notions in some detail elsewhere (Guenther 1989: 41–50; 1997: 203–11; 1999: 117–18). For a less acculturated depiction of the domain of the gods and spirits of the dead (among Ju/'hoansi), see Marshall (1999: 5–27); also see Thurner (1983: 304–5). For an account of the San spirit world that focuses on the aspect of termite hills—as God's house in the spirit world and "symbolic wombs" of fat and spiritual potency within San cosmology, see Mguni (2006: 65, 2015).

may direct their human-elicited interventionist agency towards animals (especially lions).

A pan-Khoisan mythic-spirit being which is found in both the spirit world and the mythic world is the trickster. A protagonist and culture hero in the latter domain and a divinity—Keeper of Medicines and Spirit Keeper of Animals—in the former, we find him in each of these two netherworlds. Myths and stories tell of his roaming through myth time as well as today's veld and farms, and shamans seek him out, either in his spirit place or whenever he contacts them (as well as initiands) within their ritual and liminal space around the dance fire (or menarcheal hut). The domain, as we will see presently, within which the trickster is most prominent and most fully profiled is in the First Order of Existence, the World of Myth, the inchoate nature of which he, and his many other First Race fellows, recapitulate in their being.

The First Order and the Early Race

> The dead and the living, men, beasts and gods, and rivers and sun, and moon, dance through the regions of myths in a burlesque ballet of Priapus, where everything may be anything, where nature has no laws and imagination no limit. (Lang 1901: 17)

> No certain form on any was imprest;
> All were confus'd, and each disturb'd the rest,
> For hot and cold were in one body fixt,
> And soft and hard and light and heavy mixt. (Ovid 1815: 23)

While the first passage by Andrew Lang refers primarily to Greek mythology, this Victorian "Old Animist" (*cum* "Primal Monotheist") also references the Bushmen—specifically the /Xam—in his characterization of the "regions of myth".[4] Lang's depiction of the Myth World of the Ancient

[4] As "a professional man of letters with primarily literary interests [who] was widely read in folklore and related fields of anthropology" (Lowie 1948: 119), Lang had read all the then available literature on Khoisan folklore, consisting of Bleek (1864, 1875) and Orpen (1874), the former a summary of hitherto collected /Xam folklore texts, the latter a short article on folklore texts from one Maluti /Xam informant. Lang drew on this recent ethnographic information in his two-volume *magnum opus Myth, Ritual and Religion* (1887, 1901) in his own discussion of "primitive religion", including animism. Wilhelm Bleek's characterization of "the Bushman mind" as holding "an especially fertile genius for myth formulation"

Greeks captures the flavour also of the San's First Order, with the exception, perhaps, that in the /Xam—though not the Ju/'hoan or Nharo—instance Priapus' burlesque ballet was more restrained.[5] I will return to Lang—and his contemporary and compatriot Tylor—again briefly in Vol. II (conclusion), which will note the difference between the New Animism and the Old, into whose discourse Lang brought elements of San myth and lore.

I have described First Order, or Myth Time, and its inhabitants, the Early Race, in some detail elsewhere (Guenther 1999: 66–80), as a Primal Time and Place of inchoateness, ambiguity and indeterminacy, inhabited by beings who led ontologically, culturally and socially unstable existences. Having recently re-read Ovid's *Metamorphoses* (in preparation for the present project), I was struck how the poet's depiction in Book the First of the Graeco-Roman version of Primal Time resonates with that of the San. San Creation Time had the same central characteristic, of pervasive and ceaseless ontological flux, with amorphous states and forms, "all confus'd, and each disturb'd the rest." Its effect on Creation Time's beings was that "every creature ended up being changed into all possible forms, including the form of each creature because they were all changing equally," as some Ju/'hoan elders put it to Bradford and Hillary Keeney. "It was a world," the San informants continued, dwelling further on the point about flux, "with nothing but changing forms. It was always changing." (Keeney and Keeney 2015: 61).

I will return in the next chapter to the element of transformation that pervades the states and beings of the First Order of Existence and created many of the therianthropic members of the Early Race (as well as their descendants in the Second Order). They consisted of, on the one hand,

(1874: 11) resonates with Lang's postulate of a "universality of the mythopoeic mental condition" (1901: 16) and peaked his interest in the then available literature on Bushman orature. Regarding animism, Lang engaged critically with Tylor's take on this pivotal concept (belief in souls), agreeing to disagree, in conciliatory debate, with "my friend Mr. E. B. Tylor" (1901: 17). With Tylor he shared the more general notion of "animation of nature", which Lang, again, saw manifested especially fulsomely in Bushman mythology. This to him was an exemplar of "the savage condition of the mind … in which all things, animate or inanimate, human, animal, vegetable or inorganic, seem on the same level of life, passion and reason" (ibid.: 39).

[5] The /Xam storytellers' restraint here may have been in deference to the Prussian and Victorian sense of propriety of their interlocutors, one a theologian's son, the other a rector's daughter (Guenther 1996: 90–91; Wittenberg 2012). This editorial intervention was evidently continued by Bleek's daughter Dorothea (Bennum 2004: 336).

humans and regular animals, both of them rather obscure and secondary beings of Myth Time and low-profile characters in the stories who make cameo appearances. In addition to such ontologically "regular" animals, actual, real plants, too, were found in the First Order, as the Early Race, just like the San today, hunted and gathered. Trees, too, could be just that, "just trees"; others could be transformed humans.

Yet, for all their apparent and most often also evident ontological discreteness, like all the other "living creatures" of the First Order its human and animal members, too, were ontologically ambiguous. This ambiguity lays on a scale from "mere" animals and full humans to "were" animals and partial humans, the former scale the least, the latter the most ambiguous. The latter two were therianthropes, the principal denizens of the mythic First Order of Existence, who blended ontologically with Myth Time's regular humans and animals (such as they were).

That the humans and animals of the First Order were ontologically unstable, and their human and animalian attributes were at odds with each other, is evident from a number of /Xam tales to be examined below and in the next chapter, as well as in a ≠Khomani tale that L. F. Maingard collected in the 1930s in the context of linguistic research he conducted amongst these Khoisan people (1937).[6] The story features a "wolf" (i.e. hyena) and a springbok doe whom wolf is courting, with much ardour and persistence. His wooing discourse, directed to—and rebuffed by—the reluctant doe, is replete with ontological ambiguity and ambivalence. This is the bone of contention in their intense exchange, fuelling both the wolf's ardour and the doe's reluctance. The latter—who is herself ambiguous, referring to herself as human, as "a woman", albeit one with "hairs", but referred to by the wolf as "springbok doe"—is wary of the wolf's entreaties and of his repeated avowals as to his humanness:

> I feel I am a man, I do. I feel I am a man. I can speak to you, I can. My heart, I speak to you, you springbok doe. (McCall 1970: 2)

Her wariness unrelenting, she persists in rejecting him:

[6] Daniel McCall built his much-cited pioneer study on "the equivalence of hunting and mating in Bushman thought" around this symbolically pregnant ≠Khomani story (1970: 1–2).

> "No, I cannot marry a wolf ... this shall not happen ... I shall not say yes, definitely, I shall not, I the springbok doe. Indeed, I refuse ... I the springbok doe can have nothing to do with you."

Until he exclaims, frustrated and defeated:

> "You reject me definitely. It cannot be so, it cannot be so. I am a man and I shall run before you. You say how can I be a man? Listen, I am a man, a complete man. I shall find you out and chase you. I shall run and fetch you. I shall fetch you, yes, I shall fetch you. I shall do so. I the Wolf, for I am a wolf more than a man. I call out to you: 'Turn back!'"

His animalian identity comes to the fore at the end of the tale, in an animalian being whose ontological constitution is not fixed and who, when everything is said and done, turns out to be "a wolf more than a man", notwithstanding his attestations two sentences prior, to being "a man, a complete man".

The First Order humans—"an earlier race of people who preceded the Bushmen in their country" (Bleek 1875: 14)—were vaguely sketched humans who, while ontologically fluid and prone to social mingling with therianthropes through intermarriage and out-adoption, were somewhat more stable in their ontological make-up than were the therianthropes. They were less so as social beings, however, afflicted as they were with moral turpitude, bordering at times on monstrosity (as we will see in Vol. II, Chap. 3) and with limited intellectual capacity evidenced, for example, by a First Order man who, nagged by his wife, cut off his ears to get a smooth pate, in an attempt to assuage his spouse for whom her husband's baldness was the issue of contention. First Order humans had a chronic propensity for foolishness, such as a man mistaking a young lion for a dog and raising the latter as a hunting dog, only to be eaten by the "dog" when out on a hunt (Lewis-Williams 2000: 192–204; Wessels 2013). Another propensity was for moral transgression, that could lead to First Order humans being transformed into frogs or other animals (along with their possessions that reverted back to their natural state), the punishment meted out to girls for menarcheal infractions and hunters for killing a "rain animal" (Hewitt 1986: 75–88). The animals hunted, killed and eaten by the Early Race of humans (as of and therianthropes) were the First Order's regular animals, for instance the veld's and mountains' quaggas, hunted for meat by long-nosed and striped mice-Early Race people (Lewis-Williams 2015a: 134–6).

For the most part: the Early Race—the therianthropes—sometimes also hunted, killed, butchered and ate one another, as though the animal component in a First Race were-being took over its being, leading its fellows, even members of their own extended family, to view him or her as animal, prey and meat. A case in point is the pan-Khoisan myth-motif of the antelope-wife that appears in many versions and variants[7] (McCall 1970; Schmidt 2013a: 128–32). Having been married and moved to her husband's people's place, the latter come to change their perception of their newly acquired sister-in-law (who, in the context of the most frequent version of the tale, is a Quagga girl, married to a Jackal). This starts when one of them comes to the ever more nagging, inescapable realization that their newly wed "Young-He-Dog" brother or son has "married meat". Having convinced the initially incredulous and reluctant husband of this fact, he realizes that what his Jackal kinfolk tell him is true and joins them in killing and eating the animal-wife (Guenther 1989: 94–5, 98–100, 1999: 69–70, 75, 153; Biesele 1993: 2, 197). (We see from this story and its recurrent trope that the wooed springbok doe's wariness in the above-featured ≠Khomani tale was not unwarranted.)

Indeed, an Early Race hunter may actually butcher himself to roast his own meat and gorge himself on it out in the veld or to feed his family back at his home camp (Schmidt 1995: 89–91, 213–5, 2013a: 48–51, 128–32; Biesele 1993: 171–3; Guenther 2015: 283). Such an act of "autovenery" and "autophagy" can be seen as the radical, solipsistic culmination of "endoanthropophagy"—"eating not just other people but your own"—which Thea Skaanes finds in Hadza cosmology and extends metaphorically to their eating of eland flesh (2017: 193–5). It attests dramatically to the ontological ambiguity of the First Race. They were not only both human and animal, but also both hunter and prey and eaters and eaten, of meat that was simultaneously their own flesh and that of a game animal (Guenther 2015: 306–7).

It was such hybrid beings as these, beings, "confus'd and disturb'd", morally muddled and ontologically conflated, who were the centre-fold exemplars of the First Race. They are the salient protagonists and antagonists of San myth and lore, none more so than the myth- and real time-shuttling trickster.

[7] In 1995, I obtained an especially vivid and expansive Naro version from D'Kar's best story teller, Qhomatcã. The tale is presented and discussed in Vol. II (Chap. 3).

As suggested above, the Early Race, including certain trickster figures, may themselves have been products of transformation. A case in point are the Springbok People, a "tribe" of the /Xam-ka !ei,[8] as the /Xam referred to Early Race. They are mentioned parenthetically by /Han‡kasso, in his version of a widely told Mantis tale (Schmidt 2013b: 579–80), about how Mantis rescued the child of one of his sisters, a small springbok, who was kidnapped by a she-elephant while he was babysitting his little nephew.[9] Being preoccupied with another task, Mantis was inattentive to his charge, resulting in the little springbok's abduction. Inserted into /Han‡kasso's masterful telling of this epic tale about Mantis's springbok nephew is the following explanatory comment to Lucy Lloyd:

> We who are Bushmen, were once springbucks [*sic*], and the Mantis shot us, and we really cried (like a little child). Then the Mantis said, we should become a person, become people, because we really cried. (L VIII.—4, p. 6365 rev.)

Much the same cosmogonic origin for the Bushmen is revealed from a lengthy creation story Lloyd obtained from her !Kung informant Tamme, whose home was north-eastern South-West Africa. Tamme sets his myth at an early time in the First Order of Existence—when "there were no Bushmen"—at which the shape-shifting trickster-culture hero /Xue roamed the world. Its inhabitants then were "bucks"; the story teller indicates no species (in line with rock artists and engravers when they depict the common "generic buck" motif). These pre-human antelopes told / Xue (who, in the story is also Hare) "that they are Bushmen". (As to the question on who made the bucks, which Lloyd posed to Tamme, "my informant thinks that the bucks probably 'made themselves'".) Acting on this ontological cue, /Xue-Hare then made actual Bushmen, as well as "mere" bucks, for the Bushmen to trap and "cook in a pot to eat",

[8] There were several phonetic transcriptions and glosses for this term (see Guenther 2014: 198–9). In the Keeneys' "Dictionary of the Ju/'hoan Religion" that appears as an appendix in their *Way of the Bushman* (2015: 208–15) the entry for "the original ancestors" is *g// aon='ansi*. The Ju/'hoan term for "first creation" is *G=aing=aig=ani* (ibid.: 210).

[9] In kidnapping the infant springbok the Early Race she-elephant disposition for nurturing—a "maternal instinct"—that /Xam myth and cosmology appears to have attributed to elephants. This, suggest Neil Rusch and John Parkington, that elephants, a frequent motif in Northern Cape rock engravings, were "particularly powerful models for behaviours and aspirations of growing women" (2010: 120). "/Xam and other San", they continue, "felt a common purpose with elephants—and none more so than women gatherers" (ibid.).

separating out these early therianthropes human and animal sides. It was in the context also of this creative act that /Xue-Hare began to realize the ontological split in his dual being: "I am /Xue, and am a Bushman and will lie in a hut", he insists, addressing his Hare-self, whereas "thou art the Hare, lay thyself on the ground." The Hare's indignant response reflects the ontological indeterminacy and *Nichteigenständigkeit* (lack of autonomy) (Thurner 1983: 236–37) of /Xue and is indicative of the state of ontological ambiguity of the First Race generally:

> I am not a mere Hare, but am a Bushman Hare, and am /Xue; for thou alone art not /Xue, for we two are /Xue, and are Bushmen. (Bleek 1934/35: 263)

We will return to /Xue again, who, as San myths' most transformation-prone member of the First Order, will make repeated appearances in the upcoming exploration of further aspects of that Order's state of ontological mutability.

Returning to /Han‡kasso's narrative, the enigmatic comment by this /Xam storyteller about the Bushmen's springbok ancestry through trickster Mantis's agency was offered to Lloyd as an enigmatic footnote by the story teller, to a story replete with ontological hybridity. Its protagonist Mantis is a human-insect-bird being—"a little green thing" that "looks like a long thin Locust" (van Vuuren 2016: 6[10]), at times spreading its feathered wings and fly off, after a scrape or misadventure. His wife, who does not appear in this tale as she does in so many others, is Dassie.[11] The principal female character is Mantis's sister, who, wife-like, berates him volubly for his dereliction of duty at the end of the story. Unlike Mantis's other sister, Blue Crane, who is absent from this tale, the sister here featured is of indeterminate ontological provenance; her springbok child suggests that she is either ontologico-consanguinally or socio-affinally,

[10] A description of Mantis given to Gideon von Wielligh by one of the /Xam descendants in the Northern Cape in the 1920s.

[11] The Afrikaans term for the rock hyrax (*Procavia capensis*, order *Hyracoidea*). Also known as "rock rabbit" because it vaguely resembles the latter animal's size and shape, the animal is anatomically and phylogenetically related to elephants, conferring to it, in a western zoologist's eyes, a superficial morphological and taxonomic hybridity that resonates with its ontological ambiguity in /Xam myth. Etymologically, the taxonomists' term for the animal's Order is itself rooted in a species-confounding term: *hurax*, the ancient Greeks' word for the creature, means "shrewmouse".

springbok-linked. Mantis's nephew, the Little Springbok, has human speech, as, at the beginning of the story, he is engaged in animated prattle with his granduncle Mantis, in a tone different from the hectoring he always gets from his own grandson, a precocious Ichneumon[12] Lad, whose mother is a Porcupine woman and whose father's family are Meerkat and Lion People.[13] All in all, Mantis's ontologically diverse extended family, spanning four generations and including bilateral affines, all of it the size and kin composition of a San band, encompasses close to a dozen different animal species, including one monster, on Mantis's Dassie wife's side, the infamous ogre //Khwai-hemm (a.k.a. All-Devourer), whom we will meet again in Vol. II (McGranaghan 2014: 8–10). Also included are what appear to be real humans, rather than were-humans, perhaps two or three of them. One of these, Mantis's adopted son-in-law /Kwammang-a, Porcupine's husband and Ichneumon's father, who, in addition to his lion and meerkat kin, has a bat-eared fox aunt—yet more ontological entanglement. Adding to it is that /Kwammang-a's name is the same as the Rainbow's, a lexical link that may derive from ontological affinity (Thurner 1983: 154–5).

What underscores the ontological distinctiveness of /Kaggen and each of the members of his multi-species family is that each of them is individuated with respect to his or her social persona and moral character traits. These, as perceptively noted by Ingrid Thurner, are not only clearly delineated in the /Xam ***kukumi*** but are also displayed consistently throughout the tales that feature them, underscoring the autonomy of each of these ontologically distinctive characters. Fickle, foolish, peevish Mantis-/Kaggen's Dassie wife !Kauru (or /Huntu!katt!atten) is a "calm cross-check to /Kaggen's impetuous character" (1983: 148). Sensible and caring towards her wayward husband, she tries to steer /Kaggen away from the madcap (mis)adventures he is prone to get himself entangled in. Of similar disposition—"sensible and resolute ... who takes charge when others flounder, directing the course of events when the others around her fall short" (ibid.: 151)—is the couple's adopted daughter, the porcupine !Xo. Son-in-law /Kwammang-a, human (more or less) and foil to his insectoid, erratic father-in-law, is "quiet and balanced, relaxed, self-controlled,

[12] Egyptian mongoose (*Herpestes ichneumon*).

[13] For a complete overview of Mantis's ontologically diverse extended family, over several generations see Bleek (1923: v–vi), Thurner (1983: 135–36, 148–61), Guenther (1999: 67–8), Lewis-Williams and Pearce (2004: 112–25) and Lewis-Williams (2015a: 82–5).

placating, both friendly and determined" (ibid.: 152), as opposed to grandson Ichneumon, characterized by Thurner as "loquacious, cheeky, impertinent, impudent and precocious, however, not altogether unlikable: the eternal irksome know-it-all who's always in the right" (ibid.).

The antagonists in /Han‡kasso's tale, the elephants, themselves display some of the confounding ontological ambiguity that is characteristic for so many of the First Race. This characteristic makes the latter beings so difficult to grasp mentally or describe, on the part the San story teller's part, to the anthropologist or folklorist (Guenther 2015: 284–85), tying them up, in an attempt to do so, in contradictions and double negatives such as the statement that Mantis "neither *is* nor *is not* a praying mantis" (Lewis-Williams 1980: 20, emphases in original) or "*both* a praying mantis and not a praying mantis" (Barnard 2013: 62, emphasis in original). The elephants in /Han‡kasso's story appear very much as elephants, yet, they are also not such, having a number of human traits that make them human (but also not). The antagonists' two distinctive traits, a trunk and massive size, are dwelled on in the story, especially at its climactic ending when Mantis escapes out of the she-elephant's stomach through her trunk, after he has entered her stomach, full-sized, through the elephant's navel, to rescue the little springbok, whom she had just before swallowed. Having emerged from her body, he sprouts wings, in Mantis-trickster fashion, and flies off clutching his little nephew and boasting and jeering at his adversaries. Other elephantine traits are suggested in the tale: large ears (through one of which the elephant had swallowed Mantis's nephew), large feet, with which the elephant repeatedly threatens to stamp Mantis to death, and moving about in herds, including at night (as is their, Second Order, habit). Yet, that said, they are also human: one of the ploys Mantis considered was to enter her body through the palm of one of her hands,[14] a plan he abandoned when, twice over, she "said that clapping her hands [she] would break him" (L VIII.—4, pp. 6364–5). The First Race, apparently orthograde Elephant People live in houses at settlements, and, like all of the therianthropes of myth, they have speech. Speech, a trait that links Early Race beings like elephants to humans, also differentiates them as some of the First Race were-people used special clicks (a trait, as will be discussed at a later point, that enlivens not only beings from the mythic

[14] In Dorothea Bleek's composite version of the tale Mantis's point of entry is under her fingernail, to which Elephant replies, in human anatomical fashion, "that she would pick him out" (Bleek 1923: 43).

past but also story tellers' narrative performance about them). There were also First People whose idiolectical speech contained no clicks at all, such as Ostrich (whom we will meet presently).

A number of members of the Early Race display this sort of confounding ontological ambiguity, in which their humanness is obscured by their animalness or *vice versa*, as in the case of the "wolf" met above, who, having repeatedly insisted on being human then asserts that he is more a wolf than a man. Either the one or the other—animalness or humanness—is foregrounded at certain times or moments in the lives and affairs of the Early Race and the stories about them, in kaleidoscopic fashion. An example, as seen in a /Xam narrative excerpt, is the Lion as a prominent—and ontologically ambiguous—animal in the First Order of Existence as in the Second:

> The lion goes, for he goes above in the heaven. Therefore he does stand firm above in the sky. He is a lion who talks, he eats people, he talks, he is a man, he is a lion, his hand's are a man's. (Bennum 2004: 360)

Such ontological indeterminacy and inconstancy could have dire consequences for people of the Early Race, as when one of them, his animal side at the fore, is perceived as such, as prey and meat, and, as seen above, killed, butchered or eaten by his or her fellows, or through self-immolation. Jeremy Hollmann puts the point aptly:

> The combination of human and animal characteristics made Early Race society inherently unstable; misunderstandings and disasters were common and occurred because of continual tensions between human rules and behaviour on the one hand, and animal instincts on the other. (2005: 332)

The ontological ambiguity and mutability of the Early Race placed them on "the borderline between amenable and unamenable nature" (1986: 136), as noted by Roger Hewitt, referring specifically to the /Xam trickster /Kaggen/Mantis, that Race's most chimerical and volatile member.

That borderline was all the more precarious as the distinction between First Order's were-animals and mere animals was sometimes fuzzy. And, given this ontological indeterminacy, it is not always clear to an Early Race human (or therianthrope) just what the being he or she encounters in the veld is. A case in point is a man and his people who, when out in the veld, come across some "ducks who were at first people" and who were in the water washing themselves. The man and his fellows could not decide amongst themselves if these "things" were people or animals, and went towards them to find out, and hunt and eat them in case they were animals. This surmise was confirmed when the beings, whom they had approached in the water, "were sitting on them and biting … and pecking them and biting them", until the people ran off. The ducks then "again became people and went on living". As they had before, as people—but also as ducks, of which, as the Ju/'hoan story teller explains, "there are some kinds and then there are other kinds" (Biesele 2009: 25).

Much of the same confusion is experienced by a group of Early Race human hunters in a Naro story, when, out in the veld, they come across a group of enigmatic beings that look—and don't—like ostriches. I collected this dramatic story among the Naro of Ghanzi from the masterful story teller !Khuma//ka, who dwells, in the second part of his drawn-out tale, on its Rabelaisian elements, on how the Early Race of men discovered not only women but also carnal knowledge (Guenther 1989: 61–3). The first half of the story deals with the ontological ambiguity of the ostrich beings. It begins with a group of men out on a hunt led by Xau "Kiki, one of several trickster figures of Naro myth and lore. The men had hitherto all lived without women and were thus perplexed when they caught sight of a group of ostrich-like, grass-skirt clad creatures in the distance. Never having seen such beings before they wondered what to make of them. Xau "Kiki exclaimed:

> "What kind of things are those?!" He did not recognize them as being women. He looked again and tried to figure out just what kind of creatures they might be: "What kind of thing are they? Are they, perhaps, women? Are they some ostriches? What are they?" He kept asking himself questions like this because the men at that time were alone [i.e. lived on their own, without women]. (Guenther 1989: 62)

Well, women they turned out to be who let themselves be taken by the men to their hitherto men-only camp. Having elicited strange new stirrings in the men's nether regions—this is where the story takes its Rabelaisian turn—the women took charge and taught the men—with much bungling on their part—*ars armandi*'s ins and outs. Having started on a note of ontological (as well as gender) ambiguity, the story ends up assigning species—and gender—status unequivocally[15] to the Myth Time's ostrich-women, as well as, in the process, seemingly moving from myth to real time.

Turning from the Kalahari in the north to the Karoo in the south, into the Myth World once again of the /Xam, we note that the ostrich there, too, is an ontologically ambiguous being. This emerges from a lengthy tale //Kabbo told Wilhelm Bleek in 1872 (Bleek 1923: 28–30) which, in the context of an engaging Mantis-trickster tale, juxtaposes a real ostrich with a magical one and tries to sort out just how they differ. Underlying its plot is a discourse, silently in Mantis's own mind as well as vocally with his grandson Ichneumon, on whether or not the Magic Bird is in fact a "real Ostrich" or a "different kind of Ostrich". It is a story in which members of the Early Race probe the ontological tumult of their world. Having told this story, //Kabbo adds his own take on the matter, in the form of a prolonged natural history-style account on "ostriches and Bushmen" (Bleek 1875: 8).

Like any ostrich the "fabulous bird" */Káken/ká /ka /áuĩ* lays eggs. However, her eggs are different from those of any other ostrich: they are magic eggs that, as described by Wilhelm Bleek in a plot summary in his report to the Cape governor on his "Bushman researches" (1875: 8), "adhere in the most comical way to the mouth of the Mantis, as well as the whole load of eggs to his back, whence they cannot be removed until all the eggs are humbly carried back by the robber to the magic bird's nest". All of this puzzles and bewilders Mantis, who is put through this frustrating and painful experience in his encounter with the Magic Bird, leading him into a state of perplexed ontological musing around the question: "Is this an ostrich?" His puzzlement had started right from the outset of his weird encounter. When he first espied the bird, sitting on her eggs, and shot an arrow at her the arrow returned, which he managed to dodge, saying to himself "Is this an Ostrich? She does not run away" (Bleek 1923: 28). His perplexed state mounted when the quasi-ostrich then started to talk to him drawing him into her magic eggs' sticky machinations, throughout which he kept repeating the same question to himself, "Is this an Ostrich?" Having returned home, with eggs sticking to his mouth and back and haplessly relating his strange misadventure to his grandson Ichneumon, his musings and misgivings on the Ostrich question were affirmed in the usual know-it-all reproving tone: "O Mantis ... an Ostrich

[15] In other Kalahari ostrich stories, the ostrich character's ontological and gender ambiguity is retained (Guenther 1989: 55; Eastwood and Eastwood 2006: 108).

does not talk, for the Ostrich is afraid of us. The Magic Bird talks, because she is a different kind of Ostrich, the kind of Ostrich which talks."

When at the end of the story Mantis meets one such, a "real Ostrich", he is reassured and relieved as his precarious understanding of an equally precarious world without "certain form on any impres't" seems once again to be on track: when he approaches the bird it is concealed behind a bush and it runs away, as a non-human bird would and should do when a "Human" approaches it. "A real Ostrich always acts like this," he said to himself, "it runs away. It lies among the little Driedoorn bushes. Its eggs are like this; they are white, they are beautiful." He handles the eggs in Bushman fashion, savouring the whipped-up yolk of one of them relieved that they don't do their un-ostrich magical sticking-to-mouth shtick.

> He took an egg and put it down. He took out his brush, put it into the yolk and beat it up in haste. He was beating it up with his finger-tips, lest his hands should again stick fast to the brush handle. Then he gently put the tip of the brush to his mouth; he pulled it away and felt his tongue, He did not want his tongue to stick fast again. He whisked up the egg again and finished off the yolk. (Bleek 1923: 29)

That savoury task done, he does another Bushman thing: packs up two more eggs, with his hands, to which, yet again, they did not "stick fast", places them in his carrying net and heads home to share the eggs out with his family.

The encounter with the real ostrich in the second half of the story and his routine Bushman hunter-gatherer-like dealings with its eggs is juxtaposed to the "different kind of Ostrich" in the story's first section, the doings of which fall in line with the chaotically inchoate First Order's ontological, cultural and social topsy-turvydom.

Yet, even then, when all seems clear, it is not. Mantis's last thought, as he hurried along to get home before the sun set, is "Ah, I must have been mistaken (in thinking) that I could not take hold of the eggs." The Magic and the real Ostrich are about to become conflated again, perhaps the former—"the kind of Ostrich which talks"—was not a "different Ostrich" after all. Once again, we have arrived at the same point that is inherent in all of the San stories about the First Order and its therianthropic inhabitants: that they are neither animals nor not animals, nor are they humans nor not humans, both surrendering and retaining one another' identity and being. I will return to this elusive aspect of ontological ambiguity again later (in both this and the next volume), in part because it is at the heart of what I understand to be San cosmology and because it is the key question also of the perspectivist approach of the New Animism paradigm, the applicability of which to San relational ontology is examined in the final chapter of Vol. II.

In addition to Mantis and his multi-species band, and the Springbok, the Ostrich and the Elephant people, there are many other such ontologically conflated animal-people roaming the hunting ground of the First Order. Just about all of the phyla are represented, as well as all of the letters of the alphabet, from aardvark to zebra. We see this in the "animals" entry in folklorist Sigrid Schmidt's latest compendium,[16] in the recent update of her comprehensive *Catalogue of Khoisan Folklore of Southern Africa* (2013a, vol. 1: 98). This lists the following species, 106 in number: Aardvark, Agama, Ant, Anteater, Antlion, Baboon, Bat, Bear, Bee, Beetle, Bird, Buffalo, Butterfly, Crocodile, Dassie, Dikpens [Corn Cricket], Dog, Donkey, Dove, Duck, Duiker, Dung Beetle, Eagle, Eland, Elephant, Fish Flamingo, Fly, Fox, Frog, Giraffe, Goat, Goose, Grasshoppers, Guinea Fowl, Hammerhead [a stork-like shore bird], Hare, Hartebeest, Hen, Herons, Hippopotamus, Honey Badger, Honey guide Horse, Hyena, Ichneumon, Jackal, Klipspringer, Kori Bustard, Kudu, Lamb, Leopard, Lion, Lizard, Locust, Louse, Lynx, Mantis, Mongoose, Monkey, Mosquitos, Mouse, Oryx [Gemsbok], Ostrich, Owl, Ox, Porcupine, Proteles [Aardwolf or Civet Hyena], Python, Rabbit, Rhinoceros, Roan antelope, Sandgrouse, Shark, Sheep, Silver Jackal, Snake, Springbok, Springhare, Springbok, Suricates [Meerkats], Tick, Toad, Tortoise, Turkey, Vulture, Warthog, Wasp, Wildebeest, Wild pig, Woodpecker, Zebra. The list includes the following "mythical animals" (two of them animal hybrids): Flying Lion, Giant Bird, */has* [protective spirit in the form of a jackal or hare], Lioncow, Liondog, Rain animal, Sun-Sheep, Great Watersnake.

This is not a complete list as one of its important entries ("Bird") contains some obscure songbirds with such onomatopoeic names as */kain-/kain*, *=kainjatara* or *!kwãi-!kwãi*. A point about the last avian member of the First Race germane to the theme of this book is that, in tune with his world's ontological mutability, this bird—"who was once a man" (Bleek 1923: 45)—also "in appearance somewhat resembles a 'duiker'" (Lloyd

[16] When years ago I counted up the animal characters and beings that appear in Khoisan myths, I came up with a total of 73 species (Guenther 1990: 246).

1889: 6), as Lucy Lloyd was told by /Han‡kasso, from whom she received this Mantis story about this rather sinister human-bird-antelope chimera.

In terms of the importance of animals as actors in the First Order, a rough and ready quantitative gauge for this is their relative frequency in the stories. In applying this quantitative measure to the /Xam corpus Janette Deacon concluded that "apart from people, those that appear most often are the lion, the mantis, the moon, jackal, hare, ichneumon, eland, baboon, hyena and quagga/zebra, in rough order of frequency" (1994: 242). While this frequency-based list of animal characters does also reflect to some degree their cosmological-mythological and narrative-aesthetic salience, it leaves out animals that may make only one or two appearances in the myths, yet hold great cosmological portent. Qualitative criteria, on the symbolic and cosmological significance of a character's appearance in a story, may thus be a more significant measure of that character's mythological stature. Two examples that come to mind are the Anteater and the Kori Bustard among the /Xam and Ju/'hoansi, respectively, both playing pivotal creative role when the First Order became the Second (Guenther 1999: 68–69; Bennum 2004: 49–51).

Regarding the last, what Sigrid Schmidt refers to as "*die grosse* Wende" ("the great turning-around") in San cosmology (1995: 152), we must mention yet another salient member of the Early Race who is on Deacon's list (holding third-last place), the baboon. Baboons, too, "were once people" as related in a lengthy narrative to Lucy Lloyd by Diä!kwain in the First Order; "their parts smelled like people ... and resemble humans," for they feel they are people. They danced the *=gebbi-gu* dance and sang its song "sounding like Bushman women"; in fact, they taught the same song and dance to humans. And they spoke; indeed, to this day, unlike the other First Race beings, "the baboon still speaks, he sounds like a man"—and, "still understands like a man" (Hollmann 2005: 24–6). It is the trait that has been retained in baboons of the Second Order as a residue of their First Order humanness. This cosmogonic notion is entertained also by Kalahari San, such as the ‡Au//eisi Ju/'hoansi, who considered baboons, "unlike other species, all of which were considered by the Ju/'hoan to be animal-people", to be "people-people"—so much so that they were one of the few species people did not hunt for food (Suzman 2017: 153).

Therianthropic Beings of the Second Order of Creation

> Then began the time when the power of transformation vanished. Forms became fixed. Diversities of the world of appearances derive from this time. Yet, all that is visible within nature is linked to the primal Bushman kindred [*alte Buschmannverwandschaft*]. The animals of present times are the Bushmen of the Early Race. Man is not descended from the ape, but the ape is a human, who no longer looks like such, because the time of arbitrary transformations is past. The mountains and plains, too, are descended from the primal race of Bushmen, as are the waterholes and streams, the game animals and the birds. (Heinrich Vedder 1937: 421 my translation)

A number of therianthropic beings of Myth Time have carried select features of their primal human side over into their later, Second Order animal being, into which they were transformed when the Second Order of Existence took over from the First. Beings and identities became ontologically more fixed, discrete and distinctive. Yet, for all these differences, the beings of the Second Order have retained their connection to Primal Time and the Early Race. The Second Order, of historical and present time, can thus not be bracketed out when dealing with the therianthropic beings of the First as it is not in San cosmology a vanished time of yore and myth, a nether-never-land.

In this respect San myth time is similar to the realm of Faiërie or Elfenland of Celtic or Scandinavian folklore, a preternatural realm that is accessible to humans—through various portals in liminal places such as swamps or beaches—or layers of fur coats in a wardrobe in a closed-off upstairs room in an old English country house—and at liminal times such as Midsummer Nights. In accessing the mythic netherworld, humans also have congress with that world's beings (as do the fairies and elves with them, "for seven years …"). As argued by David Lewis-Williams and David Pearce, cracks and crevices through the rock surface that San painters and engravers used as canvases for their paintings of trance dances, elands and therianthropes "were reified visions of the spirit world executed on a diaphanous, rupestral veil, … a mediator between this world and the spirit world that lay behind it" (2004: 180). As a "surface of cosmological mediation, and interface between the material and spiritual world" (ibid.) it allowed humans access to the other-world of spirit and myth beings. As we will see in Chap. 7, which examines the articulation of myth and experience and the “reality” of religion and cosmology and its ontological aspects, San ritual and storytelling allowed for the same thing.

The opposite is the case as well: mythological and spirit beings, too, can gain access to the Second Order. One of them who does so frequently is, as noted above, the trickster who, *qua* Animal Protector (Hoff 2011; Guenther 2015: 295) and God (Guenther 1999: 109–15) has important dealings with Second Order animals and humans. His presence at trance dances and initiation rites, and on the hunting ground monitoring the actions of hunters and animals, brings this myth-spirit being squarely within the purview of today’s humans. Transformation-prone, he embodies ontological mutability; he thus fits into the next two chapters, on transformation, where we meet up with him again in all of his chimerical splendour. As we will see later on, his human associate, to whom he is ontologically linked, the shaman, too, moves into the spirit and myth worlds when he is in trance and on one of his outer-body outings. Both trickster and shaman shuffle back and forth between the First and Second Orders, injecting into the latter some of the ontological mutability of the former.

The interface between myth and present time is a matter not only of beings from both entering one another’s realm. As we saw in the case of the baboons, these animals today have carried at least two of their Baboon People traits—human language and understanding—into the Second Order where these human traits reside in today’s baboons. Other First Order were-animals have done the same, bequeathed residual element of their original therianthropic humanity to their animalian descendants. Examples are elephants among the Ju/’hoansi and Naro, whose flesh smells human because it still is partially human (Guenther 2015: 283) and hares and quaggas among the /Xam.

Questioned on the latter by Wilhelm Bleek, on “baboons (and quaggas) having formerly been people”, Diä!kwain, arguably the most cerebral and reflective of the /Xam story tellers and the one most given to narratives of a “natural history” cast, along with as asides and footnotes these generated, provided a commentary that problematized this ambiguous aspect of the ontological constitution of these two animals:

> The quagga and the baboon, they have no ‘pens’; they do not resemble other things, for they verily possess a *cloaca maxima* [‘Dickdarm’ (D.)], of which they make a stomach; while they feel they are not game. For they were people. Therefore they always resemble us who are people, for they feel they were people. Therefore their organs still smell of people although they are game. (L.V. 23. 5881 rev.—5882 rev.)

In dealing with the ontological complexities of these two animal species *vis-á-vis* humans, Diä!kwain draws on each of the three explanatory registers of /Xam cosmology, empirical-physiological ones (smell and organs, i.e. large intestine, or *Dickdarm*), mystical ("feel like people") and mythological ones ("they were once people"). In so doing, he points to yet another link and bridge, in terms epistemological and phenomenological, between the First Order of myth time with the Second Order of historical and present time.

San cosmology may also view the retention of primal human traits in terms that are less discrete, conveying a diffuse, vague human'ness to animals in general. We see this in Naro animal nomenclature, which, in the context of a pan-San "connective cosmos", can be seen as linked also to their cosmology. The very designation "animal" in Naro—*kx'o /ua* and *n/ ie /ua*, for "game animal" and "carnivorous animal",[17] respectively—might, in addition to its intrinsic anthropomorphization, also connote a mythic-somatic link of today's animals to their First Race antecedents. The suffix "*/ua*" means "child" in Naro, and is normally applied only to humans, thus bringing animals into the human fold semantically (Guenther 1988: 197, 1999: 74). A mystical link also to Myth Time forebears is suggested by the linguistically related neighbouring G/wi and G//ana in whose stories about that time animal protagonists' names typically have the suffix */koa* (child) attached to their names (such as *Zooba/kkoagu*, "Hare Child" or *=Goo/koagu*, Springhare Child). This suffix, as noted by Jiro Tanaka, "refers to a mythical original state of the animal" (1996: 24). Applied to today's animals, it evokes an ontological linkage with humans on not only semantic but also mythic grounds. Is the */ua* 'ness of today's animals, perhaps, a residue of their primary human'ness, as is the giraffes' and elephants, baboon's and quagga's human-scented flesh, the former mystically conveyed through language, the latter through the body?

In addition to these mystic-somatic and semantic linkages of today's animals and humans that suggest a connection to their human and faunal Early Race antecedents, today's veld is said by some San to be inhabited, in its remote regions, by therianthropic beings that are much like those of the mythic past. Some believe that they derive from the First Order, even though they now exist outside it (much the same way that certain trees in the /Xam's karoo veld may be transformed humans). Among the contemporary descendants of the /Xam of the Northern Cape, these mythic the-

[17] See Barnard (1985: 37) for further Naro glosses for the term "animal".

rianthropic beings may be animal guardians, such as that of the porcupine, a sighting of which by an informant was described to Ansie Hoff during her folklore research among the people:

> There was a porcupine guardian at Springbokpan. He looked like a porcupine, but walked upright on feet, had thin quills and was hairy, He guarded the porcupines at their hole. He whistled through his beard and hit his stick to drive them toward their holes. (Hoff 2011: 41)

Noting that such "guardians", of the Second Order, like the "People of the Early Race", were bipedal and used tools and language, Hoff ventures the opinion that today's "therianthropes who acted as the guardians seem to have been survivals of the First Order and were consequently not limited to the mythical past" (ibid.: 47). "This included /Kaggen", she continues, who, along "with other guardians continued to live on as supernatural beings … after the inhabitants of this world … were transformed into the animals of today" (ibid.).

The Kalahari Ju/'hoansi, too, talk of therianthropic beings of a mythic kind who they believe inhabit remote regions of the Kalahari. In their "seminars" about animal ethology with a number of Ju/'hoansi hunters, Nicholas Blurton Jones and Melvin Konner were told about a sexually predatory "giant mythical baboon" roaming the veld, whom women are warned about when walking in the bush alone (1976: 343); however, given the secular-scientific discursive framing of their discussions (Guenther 2015: 293–94), the topic is not elaborated on by the two authors.

Lorna Marshall gives a more informative account of such mythical beings in Ju/'hoan folklore, both of them remote anthropo-antilopean veld dwellers, the Gemsbok People and the Wildebeest People, whose therianthropic state of being resonates with the /Xam stories. Like the /Xam Early Race—or of the Naro (Guenther 2015: 284–85)—they are both animal and not animal. Moreover, their not-animality does not render them human people but Gemsbok People.

> According to !Kung lore, people who have the form of gemsbok live in the south of the Nyae Nyae area. The Gemsbok People look exactly like animal gemsbok (*!gwe*), we were told. The similarity of their feet was especially mentioned; one cannot distinguish the spoor of one from the other. Gemsbok People can talk. They are reticent, however; and do not approach humans. Their place is far to the south; but they wander, feeding in the veld

> as animals do, and one could encounter them anywhere. The Old Old People could recognize them and knew not to hunt them, but present-day hunters might mistake them for animal gemsbok. If a hunter did kill one, however inadvertently, he would die. (Marshall 1999: 245)

We learn from another source that "such accidents are rare, for if a group of gemsbok people were out grazing, and notice that a Bushman hunter is drawing his bow at them, they will say to him: 'Do not shoot us, we are not gemsbok' and the hunter will naturally leave them alone" (Thomas 1958: 147).

Ju/'hoan stories tell of humans having visited the Gemsbok People and been treated hospitably by them, despite their aloofness to humans. They were given water and meat, as well as ostrich-eggshell beads. Their hosts, who made these by boring holes in the beads with their sharp horns, also taught the humans the bead-making craft. Indeed, it was they, the *!Xoosi*, who were the originators of the ancient Bushman craft of ostrich-eggshell beadwork (Biesele 1993: 95), the trail of which to mythic time archaeologists can follow into the earliest horizons of historical time (Orton 2008). Such tales are set within the Second Order and feature the human characters as humans; however, the protagonists, in appearance and actions are one with those from the First Order, suggesting that the Gemsbok People may be Early Race, who have strayed over, in their migratory wanderings, into the world of "animal gemsbok" and human people. They attest, along with tricksters who roam the veld or attend rituals, to the continued presence of the First Order, even after the appearance of the Second, which did not eclipse it but manifested it as an alternate world.

Marshall came across another such ontologically ambiguous therianthropic folk, the Wildebeest People. The Nyae Nyae Ju/'hoansi locate the same far to the south, perhaps the Ghanzi veld or the central Kalahari, and know little about them. Like the Gemsbok People, "they look like the animals ... [with] wildebeest horns and feet and their spoor is indistinguishable from animal wildebeest spoor" (Marshall 1999: 246). And yet: following a wildebeest spoor a hunter may be led far into the veld to an old man sitting under a shade tree. Here the wildebeest spoor ends, revealing to him with anxious shock that what he followed was not in fact a wildebeest spoor but of a Wildebeest Person. Realizing this fact as he moves along—"I thought this was a wildebeest spoor. Now I find it is people's spoor"—!Gai Goro turns around as he does not "want to kill people like myself". He heads for home, passes the shade tree—and finds the old man gone! When he got back home to "his own people" he told them about his disorienting experience when out hunting in the veld where "animal wildebeest" tracks became a person's footprints, through some mytho-mystical goings-on that bewildered the hunter and his people (and added grist and substance to their store of hunting stories):

> Oh, I nearly killed myself. I was following the spoor of a wildebeest. I thought it was an animal wildebeest. All the time it was the spoor of a person. (Marshall 1999: 246)

We will return to therianthropes again in Chap. 4 when exploring the other major domain of San expressive culture, art, both ancestral rock art and contemporary studio art, to explore how it depicts these intriguing beings. Also considered is what these depictions of ontological ambiguity might have meant to the artists and to what extent and in what way they might have drawn on myth's Early Race when the painters or engravers created their brand of therianthropes.

Ontological and Temporal Indeterminacy in the Two Orders of Existence

Such stories and beliefs indicate that the therianthropes of the First Order of Existence, and the ontological ambiguity that attaches to them and obscures the human animal divide, are beings and states not only from the mythic past of yore that has become eclipsed when the Second Order succeeded it. Some of these beings have remained, as has the ontological indeterminacy that inheres in their being.

More than a few San stories—Naro *hua*, Ju/'hoan *n=oahnsi*, /Xam *kukummi*—underscore ontological indeterminacy with temporal indeterminacy. The latter comes about when the two worlds of being, the First Order of Existence, set in virtual myth time, and the Second, set in actual real time, merge. Examples of such "trans-cosmological" or "trans-realm" tales (Lewis-Williams 2015a: 143) are //Kabbo's story "The First /Xam Man Brings Home a Young Lion" (Wessels 2013) and Diä!kwain's and /Han‡kasso's stories, respectively, "The Sending of the Crows" (Bennum 2004: 287–88) and "The Story of the Young Man of the Early Race who Warned his Family of the Approach of a Koranna Commando" (ibid.: 302–4). Michael Wessels, in his examination of //Kabbo's *kumm*, multi-discipline-based in the context of both the "pre-colonial imaginary" and "historical actuality" (Wessels 2013: 5), notes that the story can be—and has been—interpreted as both a "legend, set in history and featuring the first /Xam man" and a "myth, featuring one of the First Men [of the Early Race]". Wessel's discussion of the "meeting and merging of myth and history" (2013: 8–10; see also Guenther 1989: 152) reveals a continuum of actual historical time and present with virtual, mythological timelessness and (n)ever-ever'ness. Behind it he sees a specific actual event—a none-too-bright man, say, trying to raise a lion cub and training it to be a hunting dog and being killed by the same in the process—which, through countless retellings, over years, decades and generations, becomes ever more legendized and mythologized, to enter timeless, decontextualized, non-contingent myth-time and space. The temporal indeterminacy, around a hunting dog, a historical, Second Order animal species, is underscored by the ontological ambiguity of the lion, which may be seen as both a dangerous lion, of either the First or the Second Order hunting ground or a malevolent shaman-lion, threatening the Early Race as much as present-day people.

Diä!kwain's story comingles sheep and a Boer commando raid with a group of (First Order?) Bushmen out in the hunting ground, whom the (Second Order?) Boers kill. The hunter's womenfolk back home send crows out to find them when they fail to return back home. In the process of their errand, the crows acquire their Second Order species-markings. /Han‡kasso's *kumm* likewise features that most salient and deplored and deplorable feature of San history, the commando raid—this one by Koranna[18] rather than Boers. The narrative process from fact, through legend, to myth is explained by Neil Bennum as follows:

> /Han‡kasso's story was set simultaneously in the time of the First at Sitting People [the Early Race], when it was possible to disguise yourself by altering your shape at will, and the world brought about by the arrival of the farmers of European descent ...Koranna "war parties", commandos, had been visiting themselves on the /Xam-ka !ei for no more than a generation but in the oral tradition of the /Xam-ka !ei their violence was already located in the mythic past. The mythic past was not necessarily an *ancient* past, as

[18] In another story of the raid genre by this narrator the myth time–real time conflation of Koranna raiders is even more explicit: the mythological provenance of one of the Early Race characters featured in the *kumm* is a bat-eared fox who, the narrator states, "was once a person [and] who felt he was once one of the Early Race" (L. VIII. 18: 7593–95).

> Diä!kwain's story of the sending of the crows confirms. The /Xam people conflated time and place and landscape defined their culture. (2004: 304)

And in the course of this conflation, "genocide and dispossession were leaking into myths because they were killing the past" (ibid.). The last is a topic to which I will return again in a later chapter (Chap. 4 in Vol. II), on recent and contemporary /Xam and San "disenchantment" which was generated and fuelled by this dismal process, of "killing the past". The context of this discussion is an examination of the extent to which the aspect of San cosmology here considered—the evident inherence of myth beings and states from Myth Time in today's world—makes an impact on how people actually regard themselves, the landscape they inhabit and the animals they encounter, hunt and eat.

We now turn to transformation, which, as San cosmology and cosmogony's animating ontological force, pervades San myth and art. It's the dynamic that brings into and out of being the vast array of First Order therianthropes found in the people's expressive culture and collective imagination. As well, it rouses zoomorphic stirrings amongst the humans and non-humans of the Second Order, with moments of consummation when "feeling eland", or lion, may become "being eland", or lion.

References

Barnard, Alan. 1985. *A Nharo Word List.* Occasional Publications No. 2. Durban: University of Natal, Department of African Studies.

———. 2013. Cognitive and Social Aspects of Language Origins. In *New Perspectives on the Origin of Language*, ed. Claire Lefebre, Bernard Comrie, and Henri Cohen, 53–71. Amsterdam: John Benjamins Publishing Company.

Bennum, Neil. 2004. *The Broken String: The Last Words of an Extinct People.* London: Viking.

Biesele, Megan. 1993. *Women Like Meat: The Folklore and Foraging Ideology of the Kalahari Ju/'hoansi.* Johannesburg/Bloomington: Witwatersrand University Press/Indiana University Press.

———., ed. 2009. *Ju/'hoan Folktales: Transcriptions and English Translations: A Literacy Primer for Youth and Adults of the Ju/'hoan Community.* San Francisco: Trafford Publishing.

Bleek, Wilhelm H.I. 1864. *Reynard the Fox in South Africa; or Hottentot Fables and Tales.* London: Trübener and Co.

———. 1874. Remarks by Dr. Bleek. In Orpen, Joseph M. 1874. A Glimpse into the Mythology of the Maluti Bushmen. *Cape Monthly Magazine* [N.S.] 9: 9–13.

———. 1875. *A Brief Account of Bushman Folk-Lore and Other Texts*. Cape Town/ London/Leipzig: J. C. Juta/Trübner & Co./F. A. Brockhaus.
Bleek, Dorothea. 1923. *The Mantis and His Friends: Bushman Folklore*. Cape Town: T. Maskew Miller.
Bleek, Dorothea. 1934/53. !kun Mythology. *Zeitschrift für Eingeborenen-Sprachen* XXV (4): 261–283.
———. 1936. Special Speech of Animals and Moon Used by the /Xam Bushmen. *Bantu Studies* 10: 161–199.
Blurton Jones, Nicholas, and Melvin Konner. 1976. !Kung Knowledge of Animal Behavior (*or: The Proper Study of Mankind is Animals*). In *Kalahari Hunter-Gatherers: Studies of the !Kung San and Their Neighbours*, ed. Richard B. Lee and Irven DeVore, 325–348. Cambridge, MA: Cambridge University Press.
Deacon, Janette. 1994. Rock Engravings and the Folklore of Bleek and Lloyd's / Xam San Informants. In *Contested Images: Diversity in Southern African Rock Art Research*, ed. Thomas A. Dowson and David Lewis-Williams, 238–256. Johannesburg: Witwatersrand University Press.
Eastwood, Edward B., and Cathelijne Eastwood. 2006. *Capturing the Spoor: An Exploration of Southern African Rock Art*. Cape Town: David Philip.
Guenther, Mathias. 1988. Animals in Bushman Thought, Myth and Art. In *Property, Power and Ideology in Hunting-Gathering Societies*, ed. James Woodburn, Tim Ingold, and David Riches, 192–202. Oxford: Berg.
———. 1989. *Bushman Folktales: Oral Traditions of the Nharo of Botswana and the /Xam of the Cape*, Studien Zur Kulturkunde No. 93. Stuttgart: Franz Steiner Verlag Wiesbaden.
———. 1990. Convergent and Divergent Themes in Bushman Myth and Art. In *Die Vielfalt der Kultu: Ethnologische Aspekte von Verwandschaft, Kunst, und Weltauffassung*, ed. Karl-Heinz Kohl, Heinzarnold Muszinski, and Ivo Strecker, 237–254. Berlin: Dietrich Reimer Verlag.
———. 1996. Old Stories/Life Stories: Memory and Dissolution in Contemporary Bushman Folklore. In *Who Says? Essays on Pivotal Issues in Contemporary Storytelling*, ed. Carol L. Birch and Melissa A. Heckler, 177–197. Little Rock: August House Publishers.
———. 1997. Jesus Christ as Trickster in the Religion of Contemporary Bushmne. In *The Games of Gods and Men: Essays in Play and Performance*, ed. Klaus-Peter Koepping, 203–230. Hamburg: Lit Verlag.
———. 1999. *Tricksters and Trancers Bushman Religion and Society*. Bloomington: Indiana University Press.
———. 2014. Dreams and Stories. In *Courage of //Kabbo: Celebrating the 100th Anniversary of the Publication of* Specimens of Bushman Folklore, ed. Janette Deacon and Pippa Skotnes, 195–210. Cape Town: UCT Press.
———. 2015. 'Therefore Their Parts Resemble Humans, for They Feel That They Are People': Ontological Flux in San Myth, Cosmology and Belief. *Hunter-Gatherer Research* 1 (3): 277–315.

Hannis, Michael, and Sian Sullivan. 2018. Relationality, Reciprocity, and Flourishing in an African Landscape. In *That All May Flourish: Comparative Religious Environmental Ethica*, ed. Laura Hartman, 279–296. Oxford: Oxford University Press.

Hewitt, Roger. 1986. *Structure, Meaning and Ritual in the Narratives of the Southern San*. Hamburg: Helmut Buske Verlag.

Hoff, Ansie. 2011. Guardians of Nature Among the /Xam San: An Exploratory Study. *South African Archaeological Bulletin* 66: 41–50.

Hollmann, Jeremy C., ed. 2005. *Customs and Beliefs of the /Xam Bushmen*. Johannesburg/Philadelphia: Wits University Press/Ringing Rock Press.

Keeney, Bradford, and Hillary Keeney., eds. 2015. *Way of the Bushman as Told by the Tribal Elders*. Rochester: Bear & Company.

Kohn, Eduardo. 2013. *How Forests Think: Towards an Anthropology of Nature Beyond the Human*. Berkeley: University of California Press.

Lang, Andrew. 1887. *Myth, Ritual and Religion*. Vol. II. London: Longman, Green and Co.

———. 1901. *Myth, Ritual and Religion*. Vol. I. 2nd ed. London: Longman, Green and Co. (Both Available as ebooks Through Project Gutenberg, Released November 12, 2009).

Lewis-Williams, David J. 1980. Remarks on Southern San Religion and Art. *Religion in Southern Africa* 1: 19–32.

———. 2000. *Stories that Float from Afar: Ancestral Folklore of the San of Southern Africa*. Cape Town: David Philip.

———. 2015a. *Myth and Meaning: San-Bushman Folklore in Global Context*. Walnut Creek: Left Coast Press.

———. 2015b. Text and Evidence: Negotiating San Words and Images. *South African Archaeological Bulletin* 70 (201): 53–63.

Lewis-Williams, David J., and David G. Pearce. 2004. *San Spirituality: Roots, Expression, and Social Consequences*. Walnut Creek: Altamira Press.

Lloyd, Lucy. 1889. *A Short Account of Further Bushman Material Collected. Third Report Concerning Bushman Researches, Presented to Both Houses of Parliament of the Cape of Good Hope by Command of His Excellency the Governor*. London: David Nutt.

Lowie, Robert H. 1948. *Primitive Religion*. New York: Grosset & Dunlap.

Maingard, L.F. 1937. The ≠Khomani Dialect, Its Morphology and Other Characteristics. In *Bushmen of the Southern Kalahari*, ed. John D. Reinhallt Jones and Clement M. Doke, 237–276. Johannesburg: University of the Witwatersrand Press.

Marshall, Lorna J. 1999. *Nyae Nyae !Kung Beliefs and Rites*, Peabody Museum Monographs, No. 8. Cambridge, MA: Harvard University.

McCall, Daniel F. 1970. Wolf Courts Girl: The Equivalence of Hunting and Mating in Bushman Thought. *Ohio University Papers in International Studies, Africa Series*, No. 7. Athens: Ohio University, Center for International Studies.

McGranaghan, Mark. 2014. 'He Who Is a Devourer of Things': Monstrosity and the Construction of Difference in /Xam Bushman Oral Literature. *Folklore* 125: 1–21.

Mguni, Siyakha. 2006. Iconography of Termites' Nests and Termites: Symbolic Nuances of Foundlings in Southern African Rock Art. *Cambridge Archaeological Journal* 16 (1): 53–71.

Münzel, Mark. 2013. Warum verlassen die Geister die Insel? In *Wege im Garten der Ethnologie: Zwischen dort und hier: Festschrift für Maria Susanna Cipolleti*, ed. Harald Grauer, 35–49. Sankt Augustin: Akademia.

Orpen, Joseph M. 1874. A Glimpse into the Mythology of the Maluti Bushmen. *Cape Monthly Magazine* [N.S.] 9: 1–13.

Orton, Jayson. 2008. Later Stone Age Ostrich Eggshell Bead Manufacture in the Northern Cap, South Africa. *Journal of Archaeological Science* 35: 1765–1775.

Ovid. 1815. *Metamorphoses: Translated into English Verse Under the Direction of Sir Samuel Garth by John Dryden, Alexander Pope, Joseph Addison, William Congreve and Other Eminent Hands.* Book the First. London: R. McDermott & D. D. Arden.

Rusch, Neil, and John Parkington. 2010. *San Rock Engravings: Marking the Karoo Landscape.* Cape Town: Random House Struik.

Schmidt, Sigrid. 1995. *Als die Tiere noch Menschen waren: Urzeit- und Trickstergeschichten der Damara und Nama in Namibia*, Afrika Erzählt. Vol. 3. Cologne: Rüdiger Köppe Verlag.

———. 2013a. *A Catalogue of Khoisan Folktales of Southern Africa*, 2 vols. Research in Khoisan Studies 28.1&2. Cologne: Rüdiger Köppe Verlag.

———. 2013b. *South African /Xam Bushman Traditions and Their Relationship to Further Khoisan Folklore*, Research in Khoisan Studies, 31. Cologne: Rüdiger Köppe Verlag.

Silberbauer, George B. 1965. *Report to the Bechuanaland on the Bushman Survey.* Gaberones: Bechuanaland Government.

Skaanes, Thea. 2017. Cosmology Matters: Power Objects, Rituals, and Meat-Sharing Among the Hadza of Tanzania. Ph.D. Dissertation, Aarhus University.

Suzman, James. 2017. *Affluence Without Abundance: The Disappearing World of the Bushmen.* New York: Bloomsbury.

Tanaka, Jiro. 1996. The World of Animals Viewed by the San. *African Studies Monographs*, Supplementary Issue, 22: 11–28.

Thomas, Elizabeth M. 1958. *The Harmless People.* New York: Random House.

Thurner, Ingrid. 1983. *Die transzendenten und mythischen Wesen der San (Buschmänner) Eine religionsethnologische Analyse historischer Quellen.* Wien: Föhrenau.

Van Vuuren, Helize. 2016. *A Necklace of Springbok Ears: /Xam Orality and South African Literature.* Stellenbosch: Sun Press.

Vedder, Heinrich. 1937. Die Buschmänner Südwestafrikas und ihre Weltanschauung. *South African Journal of Science* 34: 416–436.

Wagner-Robertz, Dagmar. 1976. Schamanismus bei den Hai//om in Südwestafrika. *Anthropos* 71: 533–554.

Wessels, Michael. 2013. Story of a /Xam Bushman Narrative. *Journal of Literary Studies* 29 (3): 1–22.

Wittenberg, Hermann. 2012. Wilhelm Bleek and the Khoisan Imagination: A Study of Censorship, Genocide and Colonial Science. *Journal of Southern African Studies* 38: 667–679.

CHAPTER 3

Transformation in Myth

and the snake came out of the water and raising his head, and looked warily and suspiciously round, and then he glided out of his snake's skin and walked.
Qing (Orpen *1874*: 7)

When the sun rose, /Xue was !naxane*; the birds ate /Xue; /Xue was* !naxane. *The sun set, and /Xue was /Xue; and lay down and slept. … And the sun rose and he was another kind of* !naxane, *which is a tree. And the night fell and /Xue was not a tree … The sun rose, and /Xue was a* dui *[fruit]; and the sun set, and /Xue was an Omuherero, and lay down; and the sun rose, and /Xue was /Xue ….*
!Nanni (Bleek and Lloyd *1911*: 405)

He [!Khwa] resembled a bull, while he felt that (he) was the Rain's body.
/Han‡kasso (Bleek and Lloyd *1911*: 193)

They [lions] also are people. Therefore, they talk, they are also people. They talk, they are also people. They talk, they eat their fellows … They are people who are different … they talk.
//Kabbo (II.16: 1551)

And they take a blanket and wrap it into the trunk of an elephant. And they take another blanket and shape them into the ears of an elephant. In the old time they had big plates on which to save their food.

© The Author(s) 2020

M. Guenther, *Human-Animal Relationships in San and Hunter-Gatherer Cosmology, Volume I*,
https://doi.org/10.1007/978-3-030-21182-0_3

And they took two of those dishes and put the child's foot on it. And he starts walking and pulling the trees down, just breaking the trees ...

Qhomatcã, Naro story teller, myth excerpt (author's field notes, 31 May 1995)

Therianthropes, the topic of the previous chapter, are the manifestations *par excellence* of ontological ambiguity, while transformation, this and the remaining chapters' central topic, manifests ontological mutability. It is the unceasing leaven for change and chaos in the world inhabited by these were-creatures that not only allowed for the existence of amorphous, ambiguous, multi-species beings but also for their constant formation, re-formations and transformation, randomly and arbitrarily, keeping or jettisoning their respective forms, or interchanging them.

Both, as previously noted, may actually be manifestations of ontological mutability, transformation the process of mutability and therianthropes its product (unless they were such constitutionally, *sui generis*, as many of the Early Race folk indeed were). The two being-states may at times be interrelated: transformation, in its metamorphic unfolding, creates a flow of therianthropes or zoomorphs, which, even when the flow ceases and the morphed being is formed, that being, in San cosmology, is rarely an ordinary lion, eland, louse or tree but something a good deal more ontologically ambiguous and tentative. Having each undergone his or her ontological transformation, the outcomes are such hybrid beings as a *!giten* (shaman)-Lion, a girl-Eland, a Mantis-louse, a bow-player- or hunter-Tree. When / Xue—the First Order's chimerical *ueber*-therianthrope who was once a Hare Person—was in the process of excising Hare,[1] his alter-ego, from his being and transforming the same into a "mere" hare, Hare, as we saw in the previous chapter, berated him for attempting such a futile exercise: he was unable to shed Hare from his Bushman /Xue being. The one instance where an Early Race member's transformation into an animal is ontologically clear cut and conclusive is when the /Xam rain divinity !Khwa changes people into frogs (or himself into rain or a pond, "because its body goes into it").

Transformation is an integral dynamic of the First Order that affects all of its beings and states, animate and inanimate ones alike. The human members of the Early Race, tricksters as well as any of the others, could metamorphose

[1] Hare is widespread in Khoisan myth and lore, as a trickster and culture hero and, in particular, as the being from the First Order who brought death to those of the Second Order (Guenther 1999: 126–45, Schmidt 2013a, vol 1: 134).

into animals—for instance, a snake, as in the case of the Maluti San[2] culture hero Qwancigutshwaa, who, as seen above, in an excerpt from a lengthy myth told by Qing to Orpen, ended his snake transformation when he "glided" out of the animal's skin, to walk, upright. In addition to animals, they morphed into plants, stellar bodies, meteorological phenomena (such as wind or rain) and natural objects, conveying both human shape and personhood to the transformed being or thing. As seen in the excerpt from Qhomatcã long First Order story, transformation was not always "from the inside out", through spontaneous metamorphosis undergone by a First Race human, but could also come about *vice versa*, "from the outside in", by means of mimetic props and actions done to or by the subject that would end up in full-scale transformation (a point to which I will return again later). In addition to First Order humans, First Order animals, in particular the lion, likewise, could transform into humans or into other animal species. A transformation may be ontologically ambiguous or unambiguous, complete or incomplete, transient or permanent, ephemeral or lasting, reversible or irreversible, voluntary or involuntary, transitive or intransitive, into a single being (such as a bird) or multiple beings (a flock of birds), all of them giving the theme of instability, the widest ontological and experiential scope. As well as narrative scope: stories about transformation abound within San oral literature. As seen in later chapters, they are supplemented and complemented in San art and dance, which depict therianthropes with riotous abandon and enact animal transformations mimetically, indeed, may even embody the experience through moments of metamorphosis.

To convey the scope of transformation in the Khoisan Myth World, I once again draw on Sigrid Schmidt's *Catalogue of Khoisan Folktales of Southern Africa*, citing her index entry on "Transformation" (2013a, vol. 1: 183). Over 60 transformations are listed, most of them ontological, with a few "sociological" ones, such as from adults to children or older people or "fugitives into old couple hard of hearing" or "into persons in church". Trickster figures transform into rocks and stone, crocodile or hartebeest, women into lioness, lynx or porcupine and lions to persons, girls into trees or reeds and elephants into girls; dogs into ants or lions and lions into men, fruit, or stars; "self" into elephant or bear, bird or ant, antelope or duiker. In addition to this ontological volatility in Khoisan myth and lore amongst humans and animals, other quasi-beings, too, may

[2] The ethnonym of the Maluti "Mountain Bushmen" is obscure. It seems that some of them referred to themselves as themselves *N//ŋ* (on *N//ng*, meaning "Eland" or "People of the Eland" or "Eland's People") (Barnard 1992: 90–1). Jolly refers to them as *//Xegwi*, an ethnonym also applied to southern San further to the north-east, around Lake Chrissie (2015: 27).

undergo ontological transformation, such as clothes and shoes into dogs, wild dogs or vultures, or drops of blood into a mouse or serpent, or an egg to thick bush, river or sand desert and a tooth or an eye into mountain or evening star, respectively. The transformation theme is especially prominent in the /Xam (and !Kung) corpus: in the "keywords" index of the digitalized Bleek and Lloyd archive produced by Pippa Skotnes and Eustacia Riley in the 1990s, there are 175 entries for "transformation"!

Of all of these manifestations we find of transformation, the ones most striking and most consequential for the lives and affairs of the Early Race and the state of being of their other-world, as well as their descendants in the Second Order world, are those pertaining to the trickster, the rain divinity !Khwa and lions. These three beings are the pivotal generators of transformation in the First Order, not only of themselves but also of the beings and states around them. Each displays transformation in different ways, as seen in this chapter, and, as explored in subsequent chapters, with different reactions and emotions, from the principals and their fellow-beings—as well as from the story tellers that tell the tales of their many and varied and weird transformations, and from their audience.

Trickster: Shape-Shifter *Par Excellence*

> Sometimes he appears as a louse that is biting the man. ... He knows it is not a louse, but it is the Mantis.... (Diä!kwain (Bleek 1932: 236))

> The sun set, and /Xue was /Xue, and lay upon the ground, and slept ... And the sun rose, and /Xue awoke and ... and stood up, and saw the sun – a little sun – and was *//gui*, and was a tree.

> And his wife saw the *//gui*, and went to the *//gui*, and went to take hold of the *//gui* fruit, and the *//gui* vanished; and /Xue was a fly. And his wife laid herself upon the earth, and cried about the *//gui*, and died. And /Xue was a fly, and settled upon the grass. And his wife lay down upon the earth, and cried about the *//gui*. (!Nanni (Bleek and Lloyd 1911: 497))

Tricksters everywhere are shape-shifters; none, however, as extravagantly so as /Xue (/Xué), of the !Kung of north-eastern Namibia, the quintessential exemplar of ontological transformation. Of 175 transformation entries in the Bleek/Lloyd archive, 51 are by this figure, Khoisan mythology's most unrestrained shape-shifter. /Xue is only one of perhaps two dozen or so other trickster figures in San myth and lore, with different names and personas, who hold in their bag of magical tricks and feats the ability to transform (Guenther 1989: 115–51, 1999; Schmidt 2001, 2013a, vol. 2, *passim*). They do so more or less frequently and usually spontaneously, at a whim, usually to perform some antic or creative act and, in addition to transforming themselves, they transform other beings or things, either as a trickster'eque prank or, *qua* culture hero, as an act of creation.

While most embellished in the /Xue tales, Bleek and Lloyd collected from their three !Kung informants the transformation that plays itself out also in the /Xam corpus, pervading the Myth Time of these southern San people and sustaining through constant change and metamorphosis, its constitutional state of ontological mutability. Much of this churning emanates from /Kaggen (or Mantis, which, as we saw above, he is both and neither), especially in the surreal myths about this kaleidoscopic figure. Some of these, as suggested elsewhere, are linked, in substance and form, to dreams and dreaming, a connection that confers the fantastical oneiric qualities of altered and alternate states of consciousness experienced by the dreamer in the here-and-now to the other-world of myth (Guenther 2014). Along with the rain divinity !Khwa, /Kaggen is the most transformation-prone member of the Early Race of /Xam, whose different manifestations span different phyla, from louse, through snake, vulture and hare (Lewis-Williams 1980: 21), to his favourite avatars, antelopes, especially eland and hartebeest, both of them, according to Diä!kwain, "things of the Mantis" (Bleek 1923: 10). Another transformation is into wind—among the Kalahari Naro, a whirlwind—which, as a fluid, evanescent meteorological phenomenon, is consistent with his ontologically inconstant state.

In a story that has cognates across Khoisan folklore (Schmidt 2013a, vol. 2: 382–4, 2013b: 46–8), we find /Kaggen lying out in the veld, transformed into a hartebeest and feigning death (Bleek and Lloyd 1911: 2–16). (In the other Khoisan versions of this tale, trickster turns himself into a variety of other antelope species, such as hartebeest, springbok, kudu, gemsbok and generic "buck", attesting, once again, to this being's ontological mutability (Schmidt 2013b: 48).) /Kaggen allows himself, in this doubly-deceiving trickster's guise, to be found and butchered by a group of nubile girls (whom, as evident from most of the other tale variants, he lusts after). Butchered into his different hartebeest parts, and shouldered by the girls for the trek back to camp, /Kaggen tricks and frightens the maidens, who drop their hartebeest meat load and run away, heading for the safety of their home camp. Reassembling his body parts, back not to the hartebeest he had transformed himself into but to his prior human form—thereby reversing his previous transformation in what amounts to an act of double transformation—/Kaggen chases after the girls. The narrator, presumably //Kabbo (Schmidt 2013a, vol. 2: 47), dwells on the process whereby /Kaggen reassembles his body parts, all the while chasing after the frightened maidens (Bleek and Lloyd 1911: 9–11; see also Guenther 1996: 90–1, 1999: 103). The process of rearticulation, back into his human form, body part after body part, from the head down to the legs, provides a graphic account of the anatomical workings and mechanics of bodily transformation:

> The flesh of the Mantis sprang together, it quickly joined itself to the lower part of Mantis's back. The head of Mantis quickly joined (itself) upon the top of the neck of the Mantis. The neck of the Mantis quickly joined (itself) upon the upper part of the Mantis's spine. The upper part of Mantis's spine joined itself to the Mantis's back. The thigh of the Mantis sprang forward (like a frog), it joined itself to the Mantis's back. The chest of the Mantis ran forward, it joined itself to the upper part of the Mantis's spine. The shoulder blade of the Mantis ran forward, it joined itself on to the ribs of the Mantis. The other shoulder blade of the Mantis ran forward, while it felt that the ribs of the Mantis had joined themselves on, when they raced. (Bleek and Lloyd 1911: 11)

We see how, in this process of mutational rearticulation, Mantis's body parts, these—lower back, neck, spine, thigh, chest, shoulder blades—themselves appear to have a degree of autonomy and agency, each joining up with the rest only after its own spell of "racing ahead" in the complete body's—Mantis's—lusty chase after the girls. This madcap chase of Mantis's re-articulating bones suggests, conceptually, that ontological discreteness—such as it is, given his chimerical make-up—applied not only to Mantis but also to a certain extent even to his individual body parts, each of which has a mind of its own. This sort of agency and motility, along with, perhaps, ontological discreteness, applies as well to Mantis's objects—his bow, quiver, sandals, cloak—which may take off on their own volition, like Goethe's sorcerer's apprentice, to be called to heel by their bemused master—more or less successfully as Mantis's "children" may ignore their master's at times contradictory summons and, displaying their master's adeptness at mutability, as well as motility, transform themselves back to the beings they were, shrubs, trees or antelopes. Practically, the rearticulation of Mantis's human-antelope body suggests that the story teller, //Kabbo, who was also a hunter, knew about the anatomy of his antelope prey (Guenther 2015: 286). In addition to story teller and hunter //Kabbo was also a shaman, all of it reflecting the convergence we find in San culture and cosmology of shaman, hunter and story teller (a matter to be discussed in Chap. 7).

Like /Kaggen, /Xue, his !Kung counterpart, is very much a trickster, of the prototypal San cast[3] (Guenther 2002; see also 1999: 95–125), that is, ontologically and morally ambiguous, numinous and ridiculous, quixotic and mercurial, transformation-prone. As both the protagonist of myths, *qua* trickster, creator and culture hero, and divinity and Spirit Protector of Animals, /Xue inhabits the three domains of myth, spirits and humans (who, in all three worlds, both fear and mock him). This multiplicity of being renders /Xue constitutionally ambiguous. Like /Kaggen and other San tricksters, /Xue is also a member of a multi-generation kin group, one less localized than /Kaggen's and less integrated and more loosely tied to their way-ward trickster band member: he has a wife (and a tension-fraught marriage) and a father-in-law, children, one of them a grown-up son, a mother and father and a younger brother. Appearances of these figures in /Xue tales link myth protagonist to a social, human group and moral community—albeit tenuously—rendering

[3] Given /Xue's paradigmatic—"classic San"—trickster make-up, it occasions no surprise that Dorothea Bleek, who did field work on the !Kung of northern Namibia and Angola in the 1920s, would link this being of myth and ritual to the widespread trickster-divinity Huwe who is found, in various guises and varying names, in the mythology and religion of numerous other Kalahari San groups (1927: 121–3; see also Thurner 1983: 230–98); indeed, Lloyd's two young !Kung informants !Nanni and Tamme told Lloyd that "our Huwe is our */xue* and he is a Bushmen". Tamme, by way of explanation, described the same as "a person who works many things, we call his name by that of Huwe, we call his name by that of */xue*" (Bleek 1934/35: 271; also see Bleek 1927: 121–3; Thurner ibid.: 230–98). Moreover, Bleek includes also the linguistically divergent Naro within the Huwe complex of northern !Kung myth and ritual, linking to the same the Central Naro supernatural being Hise (Bleek 1928: 24; Guenther 1999: 61), who, in turn is known also among the Ju/'hoansi !Kung, as Hishe (Marshall 1999: 5, 7). Bleek who was intimately familiar with /Xam mythology and its central protagonist /Kaggen—having published a book, *Mantis and his Friends*, on this beguiling trickster being based on her father's and aunt's folkloric researches (1923)—saw convergences even with this being, about a thousand kilometres to the south of the Kalahari, noting that "in some points he [Huwe] resembles */kaggen*, the Mantis, of */kham* Bushman mythology" (1927: 123). Restating the same observation, Bennum describes /Xue as "infuriating, unpredictable, powerfully and quintessentially *human*", traits that "recall /Kaggen of the /Xam people" (2004: 319, author's emphasis).

him “a Bushman” to whom story teller and audience can relate (Bleek 1934/35: 276). /Xue’s family group—like /Kaggen’s—is very much like //Kabbo described his own family group back at the Bitterpits to Bleek and Lloyd:

> My wife was there; I was there, my son was there; my son’s wife was there, carrying a little child on her back; my daughter was there, while she also carried a little child; my daughter’s husband was there. We were like this. (Bennum 2004: 19)

As seen in Chap. 2, //Kaggen, the other trickster in the Bleek/Lloyd archive, too, heads this sort of “Bushman band” family group, aligning him, for all his ontological and moral ambiguity, with humans.

What distinguishes /Xue from all of the other San trickster beings is an irrepressible trigger-sharp readiness for transformation. Dorothea Bleek comments on “the ordinary set of endless transformations” (Bleek 1934/35: 283) /Xue is apt to undergo in his scrapes and antics, to escape from dangerous situations these brought him into. The Austrian anthropologist Ingrid Thurner, in her meticulously drawn portrait of this San trickster figure (1983: 230–42), notes that “it is no exaggeration to say that /Xue can transform himself into anything conceivable” (ibid.: 236)—from “Bushman’s eye water”, into which he transformed himself and drops to the ground in order to hide from a Herero man (Bleek and Lloyd 1911: 411), through ant and locust, buffalo and elephant, to moon and (or) sun (or both—the story is ambiguous on this sidereal transformation of /Xue).

His transformation may occur in rapid-fire succession and with staggering ontological diversity, as in the following summary of a /Xue story by the !Kung story teller (and illustrator) !Nanni, in which he undergoes as many as 17 transformations, from buffalo to ant:

> In this story /Xue becomes an ant, a butterfly, a fly, water, a buffalo, a snake, a tsan, a //xuonna and a bird. /Xue chases his wife in a variety of forms and eventually dies. His wife picks him up and puts him inside her kaross, but drops him when he becomes a //xunna. She goes on into her father’s country ... /Xue also becomes a !naxane [see figs. 2.2 & 2.3], an owl, a fly, a bee, a hyena and, finally, an ant. As a bee he bites his father. /Xue performs a doctoring ceremony and his father rises up and chases /Xue. /Xue and his father fight. Eventually, /Xue settles on his father’s head and he dies. /Xue’s

|χ̥ụé e !kã̃ń-a; tá e !kań tséma ụḿm; ta |kúä e |χ̥ụé.

|χ̥ụé is a !kã̃ń-a, a littlo food-bearing tree, for he is not |χ̥ụé.

!nanni, *May 19th*, 1880.

Fig. 3.1 /Xue transformed into a *!kan-a* fruit-bearing tree. (Bleek and Lloyd 1911: 309)

> father takes him off his head and sets /Xue down on the ground, where he becomes an ant. (XI & XII -5.: 9395–9403[4])

/Xue is especially fond of transforming into trees and shrubs, the names of half a dozen of which are mentioned by her !Kung story tellers. Some of them were drawn with pencil on paper, provided by Lloyd to her young !Kung narrators, in a couple of instances as illustrations for a certain story. One of them, drawn by !Nanni (Fig. 3.1)—who drew most of 42 images of /Xue in the Bleek/Lloyd archive, most of them transformation scenes—depicts /Xue as a fruit-bearing tree, his arms transforming into branches heavily laden down with *!kan-a* berries. The drawing depicts a seemingly startled or perplexed /Xue looking at one of his transforming limbs, suggesting that his transformation has just begun and in the process of unfold-

[4]File no. BC_151_AZ_113 in digitalized Bleek/Lloyd archive.

ing. Another, similar depiction, again by !Nanni, shows /Xue at a more advanced phase in his metamorphosis (into a !naxane plant), lying on (or in) the ground, tuber-like, with bulbous fruit-bearing shoots radiating from his body (Fig. 3.2). Figure 3.3 depicts his transformation into the same plant, here depicted as a fully formed plant, without any recognizable human traits, which /Xue, now a plant, has poured into this—for the time being—ontologically distinct and discrete being.

Two of the drawings, by the same story teller/illustrator, of /Xue's tree metamorphosis, depict this scene with striking imagery and imagination, along with dream-like surrealism. They depict the trickster's morphing from opposite vantage points, the one /Xue/tree's canopy, the other his/

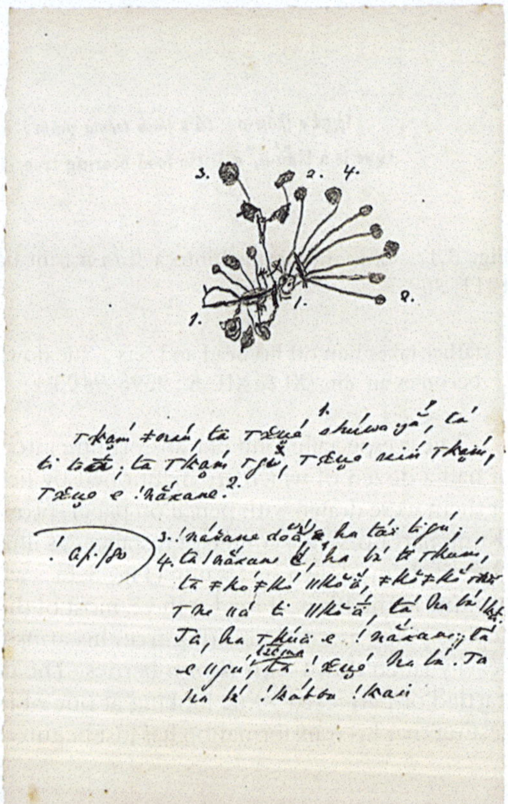

Fig. 3.2 /Xue transforming into a *!naxane* plant. (Image NLLB_D_!n029), ("Digital Bleek and Lloyd")

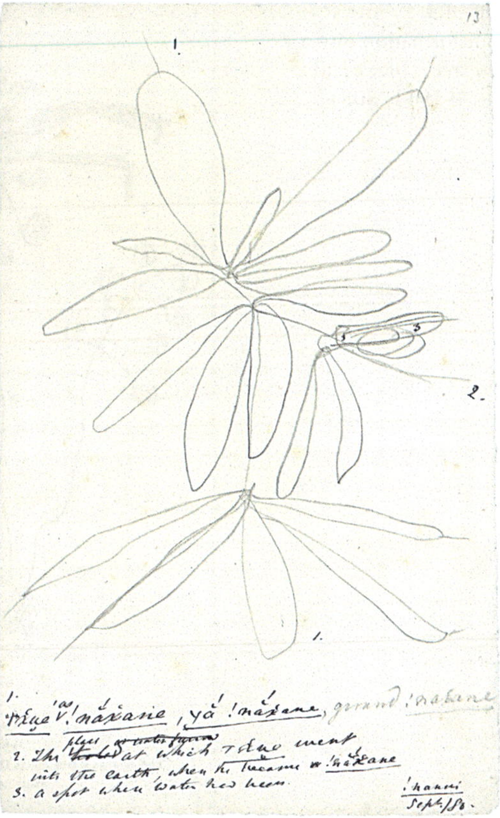

Fig. 3.3 /Xue as a fully transformed *!naxane* plant. (Bleek and Lloyd 1911: 292)

its roots. Once again, both appear to depict transformation dynamically, as a process that is unfolding, in the one case with speed and seemingly wild abandon. The one (Fig. 3.4), in which the trickster morphs into a *//gui* tree—another fruit-bearing tree the berries of which people "do not put into a pot but eat raw" (Bleek and Lloyd 1911: 407)—places /Xue in amongst the tree's upper branches, his body, on a lateral plane, forming a network of branches that enfolds and—in trickster fashion—camouflages his body. Fruit sprout on tendril-like twigs from his torso and head, weighing his head down towards the ground, within reach of unwary foragers. !Nanni's other drawing places /Xue at the bottom of another fruit tree (Fig. 3.5), his limbs and body entangled within the trees' meshwork of

Fig. 3.4 /Xue's transformation into a //*gui* tree. (Bleek and Lloyd 1911: 308)

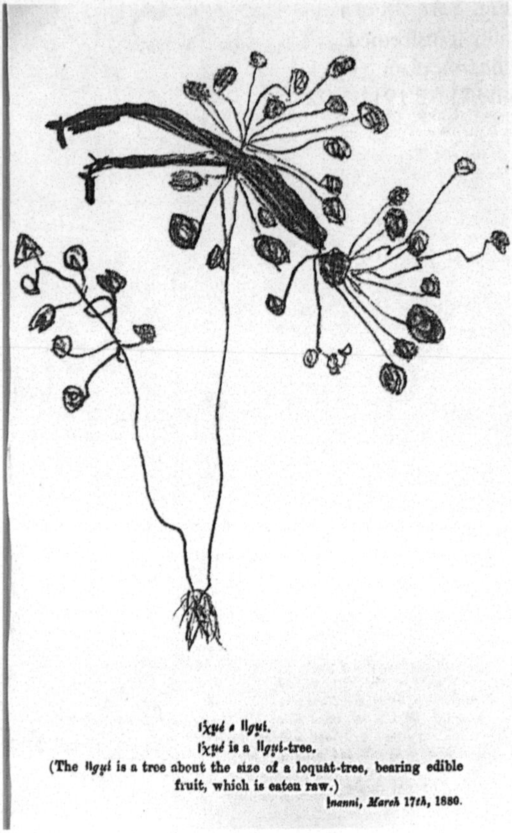

roots and rootlets, his body and head the plant's bulbs. One of them, /Xue's head, sprouts two thin stems that shoot upwards, sprouting thin branches, thorns and fruit.

It is tempting project onto these two drawings a trickster-prankster narrative line. In the first drawing, /Xue's fruit-laden head tilts down from the tree's top, close to the ground—and within plucking range, perhaps, of unwary people the Early Race to approach and be startled by yet another one of his antic transformations, followed, perhaps, by a bird or butterfly metamorphosis? And likewise, with respect to the second drawing: are the fruit-bearing shoots meant to lure unsuspecting beings to himself and give them a good fright?

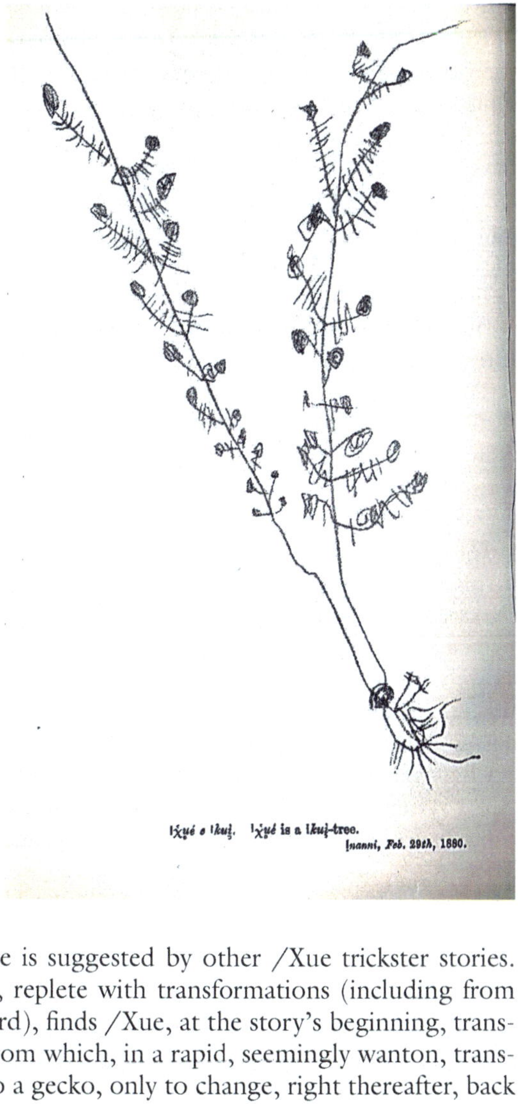

Fig. 3.5 /Xue's transformation into a *!kui* tree. (Bleek and Lloyd 1911: 209)

This sort of a story line is suggested by other /Xue trickster stories. One of them, a long tale, replete with transformations (including from death back to life, as a lizard), finds /Xue, at the story's beginning, transformed into a waterhole from which, in a rapid, seemingly wanton, transformation, he mutates into a gecko, only to change, right thereafter, back again into "the water" (Bleek and Lloyd 1911: 408). This he does in order to lure some nearby wood pigeons to himself, which he wanted to catch and eat. When they landed beside the water hole to drink, he became "large grass, which is (called) reeds" enabling him to catch one of the birds, through a trickster'esquely cumbersome and drôle hunting procedure: "large grass, [who] was /Xue, took hold of the wood pigeon and bit the wood pigeon's bill, and the wood pigeon cried out" (ibid.: 409). After plucking its feathers, prior to roasting and eating the bird, /Xue dons some of the wood pigeon's feathers. It is a mimetic action that appears to have triggered his partial transformation into the bird. This progression, from mimesis to metamorphosis, that we at times come across in San myth and lore (as well as, as I will suggest in a later chapter, in ritual and hunting), is intimated later on in the story where we see /Xue—and/or his father /Xue–//nu'ù, the plot is unclear at this point in the story (as it is at others)—covered with "many, many wood pigeon feathers" (Bleek and Lloyd 1911: 413), suggesting a process of avian transformation.

Animals are /Xue favourite avatars, starting with insects—flies and butterflies—into which he transforms himself either singly—as one fly—or collectively—as a swarm of butterflies. He transforms himself in like fashion—singly or collectively—into a bird or into "many birds". We have come across some of his other vertebrate transformations, into gecko, lizard, hare, monkey (baboon?) and elephant. There are also human transformations, one, into a screaming baby, wanting, seemingly as part of a trickster'esque prank, to be picked up by the humans around him, and, two, a couple of ethnically specified people from the Second Order of existence, one an Omuherero, the other a Makoba. (The latter enigmatic transformation attests—as some of the /Xam therianthropic tales discussed in the previous chapter—to the porousness of the temporal boundary between the two orders of creation: at /Xue's time, in the First Order, when /Xue went about transforming himself, there were as yet no ethnically distinct black groups.) Like other San tricksters, he also undergoes sidereal transformation. This, in /Xue's case, is ambiguous, seemingly merging two opposite heavenly bodies: "the Moon, but [it] was red, very red like the sun" (Bleek 1934/35: 268). Another sidereal aspect about /Xue's transformations is that they sometimes occur in 12-hour intervals, that have him sleeping, as /Xue, during the night and transform into another being, tree or plant in the morning (Thurner 1983: 230). Finally, as we saw in the previous chapter, /Xue transforms himself, or tries to do so, by excising the Hare component from his seemingly dual being, so that he becomes the "human /Xue", who "will lie in a hut", as opposed to his Hare alter-ego, a mere hare who "lays itself on the ground".

Another transformation, one arguably ontological as it pertains to an "unknown country from whose bourn no traveller returns", is from alive to dead and back again, in one case because he was killed by his alter-ego and *Doppelgänger* Hare. In another myth his return back to life results in another transformation of his being: into "the Moon … for /Xue does not die altogether. The Moon is like /Xue" (Bleek 1934/35: 270).

One long tale (Bleek 1934/35: 266–70), told to Lloyd in several segments by !Nanni and Tamme, which the latter had been told by his paternal grandfather, features /Xue and his son and other members of the Early Race (referred to in the tale as "the people"). He is presented in the classic trickster mode, as prankster and creator, blending these two personas: after first making Herero people—four of them, "holding guns"—alongside his son who made Bushmen, holding quivers and arrows and wearing monkey skins, /Xue, later in the tale, makes rain ("sky rain"), and then, having unmade it by blowing on a buck's horn and singing, he followed this creative act by making fire. This quickly escalates to a veld fire that "lighted the grass and burnt, burnt, burnt, burnt, burnt (during many months)". Both creative acts frightened the people and made them run away. Their shouts of fear and dismay intermingle with /Xue's laughter and jeering, as well as attempts to reassure the people (Bleek 1934/35: 268–70).

Interspersed within these antics are multiple transformations, some of them in rapid succession. The first two of the ten transformations undergone by this master of metamorphosis are into a *ssao* plant and a monkey—into which, in a classic instance yet again of mimesis leading to metamorphosis, he became transformed after first simulating the animal through an elaborate monkey-skin costume: "an apron of monkey's head (the bones being left in the skin) … a back apron of (four) monkey's feet, … a monkey's tail around his head" (Bleek 1934/35: 266). Next, in roaming through the country looking for his father, /Xue appears to have transformed into the sun (the story is unclear on this point, as noted by Bleek) and soon thereafter, when people approach to look at him as such—as "*/xue*'s fire"—he hides from them by becoming a */naxane* fruit. This, upon being taken by the people to eat, turned into "many butterflies", spooking the people who, once again, "ran away". To coax them back he turned into a baby—a small "*/xue* child"—laying himself on the ground and crying to be picked up, by wary people, who decided to leave him be. His next transformation was into a stellar body that resembled the moon, but was red, "very red like the Sun". This transformation so scared the people that they "cried, hiding their faces on the ground for fear of */xue*", becoming "small like little children" who, fearing */xue*, "walked softly in the grass to a distance", all the while crying: "My mother, */xue* is killing us, oh dear! My mother, */xue* is killing us, oh dear! My mother, our hands pain us from (from the grass), */xue* is killing us, oh dear! Killing us outright, oh dear!" (Bleek 1934/35: 268). He thereupon metamorphoses into a fly, followed right away, when hearing and fearing approaching people, into a "a butterfly, large and black" (moth?). An injury to his eye caused by a blade of grass that pierced it as he hid from the people brought on his re-transformation, back to /Xue, asking the women to treat his eye by pouring milk into it. The women's fumbling caused blisters in his eye, leading him to make rain, to pour down upon the fearful women, followed by fire, which blazed into a grass fire, once again frightening the women.

One of the women with whom /Xue has seemingly chronic antagonistic relations is his wife, who is herself transformation-prone. One lengthy story by Tamme, that features the two pan-San antagonists, Moon and Hare, over the portentous existential matter of death life and resurrection (Guenther 1999: 126–45) weaves /Xue into the narrative, at times conflating him with the moon, into which being he transforms himself (after first morphing into a "little snake"). The Moon's ferocious fight with his wife—"who took a stick [a knobkerrie] and beat him till he screamed" as he bit her in her thumb ("breaking the bone of it") leading her, in turn, to cut his throat and killing him (Bleek 1934/35: 276–7)—is recapitulated in a fierce spousal altercation between /Xue and his wife. It is brought on by /Xue's twisting his child's neck during so that it died in the morning, for which dastardly act /Xue's wife "burns him with a firestick" (ibid.: 283). In another story, /Xue plays tricks on his wife, vexing her by appearing as a *//gui* tree, replete with fruit, only to vanish into thin air when she attempts to pluck and eat one of the *//gui* fruits (right away, to reappear as a fly, adding yet another rapid, ontologically radical transformation). Vexed and frustrated, his wife "laid herself on the earth, and cried about the *//gui*, and died" (ibid. 1911: 407), while "/Xue was a fly, and settled upon the grass".

Another tale of spousal conflict is presented by Tamme—the narrator who told Lloyd the story of the Moon's (/Xue's?) fight with his wife—in pictorial form, as an annotated pencil drawing (Fig. 3.6) (He accompanied a number of his stories with drawings, at Lloyd's urging.) Moving from bottom left of the cartoon to top right, the narrative shows /Xue transformed into an elephant. "His wife sees him, and is afraid", the commentary continues. "She prepares food.

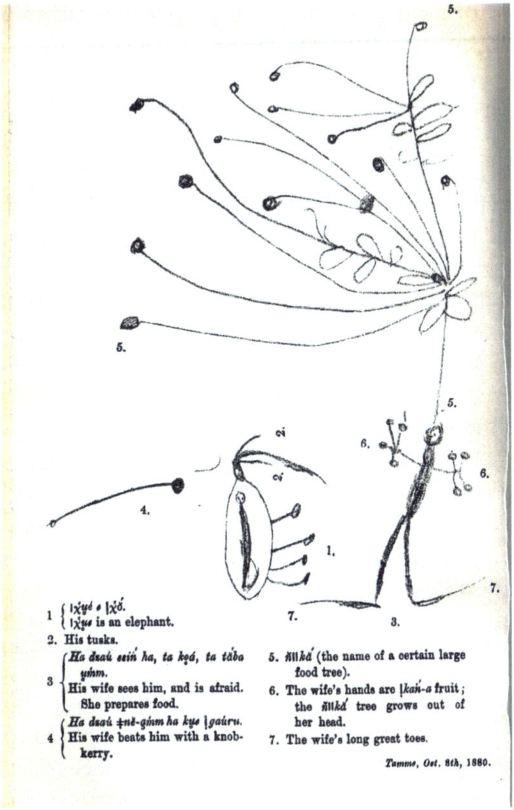

Fig. 3.6 Tamme's cartoon of /Xue and his wife transforming during a spousal dispute. (Bleek and Lloyd 1911: 325)

His wife beats him with a knobkerrie". Moving on, we see a *n//ka* fruit tree, growing out of the top of /Xue's wife's head, whose fingers, splayed on stretched-out, branch-like arms, sprout */kan-a* fruit and feet elongate into tapered, root-like appendages. She appears to be in the process of transformation into a tree (or two types of trees, as the one on her head is a different species from the one sprouting from her hands). It is a process much like the one undergone by three young /Xam men, when their illicit glance at—and from—a maiden in her seclusion hut transforms them into trees. As we will see (in Chap. 5), the latter arboreal transformation was brought on as a result of a transgression, as opposed to the /Xue instance, which was generated by a state of agitation and spousal strife, much the same way that agitation in other situations, such as nearness of a predator or other dangers will trigger ontological transformation in this volatile, high-tension protean being.

We might ask what, given his protean nature manifested through repeated transformations, /Xue might look like "when /Xue was /Xue", as /Xue rather than /Xue transformed. Is he a bland, neutral generic being—perhaps of human or humanoid body plan—with an ontological malleability into multiple forms? Or is he an amalgam of beings and ontological parts, which he absorbed through his multiple transformations that have made him into a chimera? Unsurprisingly, given his slippery ontological make-up, he appears, *qua* /Xue, in both forms. Of the handful of drawings by !Nanni and Tamme of /Xue that depict the trickster s more or less untransformed, his appearance is blandly human, usually standing erect or striding—"a grown-up Bushman", as !Nanni informed Lucy Lloyd (who, in one of his drawings, is seen smoking a Boer pipe!).

However, /Xue's self-depiction, as /Xue, is usually more ontologically entangled, his humanoid features blending with floral or faunal ones. The one description we have of /Xue, by the story teller Tamme (who also produced some of the /Xue drawings), at the end of his lengthy myth about Moon, Hare and /Xue, provides a vivid example:

> And */xue*'s stomach is big, but his body is small, his feet are small, and his legs are small, and his hands are not large, but small; and his head is small, and is not like other people's heads, but like a mouse's head; and his stomach only is big and black, and his body is yellow. And he lies and is (so big, showing about four feet), and is like an elephant. (Bleek 1934/35: 280–1)

It also suggests that /Xue's unceasing and unrestrained transformations have formed this San trickster figure, as /Xue, untransformed, into an ontologically heterogeneous being, whose diminutive, bi-coloured, microcephalic, big-bellied, small-limbed human body shares traits with that of a mouse and of an elephant. Through these unceasing transformations he becomes a serial chimera, his being matched accordingly at the moments at which he is himself, /Xue.

In this chimerical make-up he is not unlike his fellow-trickster beings, Hise and Huwe, as well as //Gana (or //Gauwa), another one of the protagonist-divinity San being to whom /Xue is mystically linked, via his connection to the ghost-, spirit- and dream-like *//gaŋa*, spirit beings, into both of which /Xue will also on occasion transform himself (XI & XII-5: 9351–8; Bleek 1934/35: 281–3; see also Bleek 1927: 123). The latter, *//gaŋa*, may be either a ghost or a spirit of the dead or, as explained to Lloyd with mystical incoherence by Tamme, *//ŋaŋa* may be a dream. A dream may, in turn, also be a transformation of /Xue so that the trickster of myth and the domain of the spirits may enter also the domain of people's dreaming. As a dream, the ethereal personification of a *//ŋaŋa*, /Kue "resembles a Bushman", with its own distinctive apparel and dwelling place, both of faunal provenance. This is how Tamme described this dream-transformed being:

> The dream's hut is little, resembling a bird's nest. For the dream wears a little apron like Bushmen, a little thing, it wears a little apron. It wears a backdress of mouse skin. It wears a little apron of ? ? [indecipherable]. The dream's body is little, it resembles a Bushman. (ibid.: 282)

So mutable a being is /Xue that he transforms himself into something as evanescent, amorphous and surreal as a dream, through the agency, acquired through a transformation, of equally insubstantial dream and spirit beings, *//ŋaŋa*. Dreams are in San cosmology another arena for transformation, drawing to it beings from the ontologically fluid world of myth, especially its most mutable denizen, the trickster (as well as ontologically diffuse beings from the spirit realm). I found this mutability-reinforcing linkage, between dreams and myths, also in my own field work among the Naro, some of whose many and varied, ontologically ambiguous and fluid descriptions of their trickster figure were reportedly based on dreams people had, or dream-like visions, especially on the part of shamans (Guenther 1999: 98, 100). In her field work among the same people two generations previous, Dorothea Bleek found much the same thing, dream-like visions of Hise—whom elsewhere she referred to as a "dream Bushman" (1923:v)—that were ontologically conflated and mutable:

> On one of the nights of the /gi dance (the men's [initiation] dance), a supernatural being called Hiʃe approaches the dancers, circles round them and is driven away by the medicine men. The older men had seen this being. One said it was like a woman in appearance, another that it was not a person at all, but a creature about three feet high with a flat head, red eyes underneath the head, a black body, wings and claws. Some old Auen [Ju/'hoansi] said two beings came, male and female, looking like lions, but walking upright. They were followed by children like baboons. (Bleek 1928: 24)

The linkage of transformation, which is played out to its fullest extent in myths, to dreams confers some of the surrealism that characterizes dreaming to the inchoateness and ontological mutability of Myth Time (Guenther 2014), in a synergistic two-way fashion that reinforces the element of transformation and its pivotal place in /Xam (and San) cosmology. This also explains why /Xue has the ontological mutability that he does. Given the location, and place of encounter by humans—dreamers and story tellers—of /Xue within the domain of both myth and dreams, one might expect this being to be ontologically kaleidoscopic and slippery, with some parts human, some animal, some plant and trees, some meteorological and sidereal, some spirit-like. As a result, in part, of his constant transformations, /Xue becomes the ultimate chimera—"a being composed of attributes of extremely diverse origins" (Descola[5] / Simay 2012: 11)—and the most striking manifestation of ontological mutability of San cosmology.

/Xue's chimerical being brings up an intriguing speculation for a student of the origins or religion: that this San trickster is to San cosmology what, perhaps, the "Sorcerer" on the Trois-Frères cave wall was to the thought world of the Magdalenians, of Europe's Upper Palaeolithic. Rane Willerslev, in a "thought-experiment" in the manner of the Victorian armchair anthropologist on the "primordial source of the animistic soul" (2011: 7), describes this oft-described and discussed figure of Palaeolithic shamanism as "a hybrid creature, undoubtedly male … every part of his anatomy belong[ing] to some animal: an antlered head, a wolf's ears, eyes of an owl, a cat's face, the long beard of a reindeer, paws of a bear, … the tail of a horse" (ibid.: 15). Religious scholars given to speculations about the origins of myth and belief might be tempted to consolidate their surmises about this enigmatic, iconic figure of Palaeolithic shamanistic religion with reference to the San case. This would inject into the chimerical Sorcerer's imaged stillness—as a pictorial still—the dynamism of transformation of /Xue of San myth and story. This, as noted earlier, is

[5] Descola places the chimera not within the animistic ontological schema but within the one he refers to as "analogistic" (which "make compatible and connect qualities, properties, attributes, elements of the world that are very different in origin, in order to weave them together into a meaningful whole" (interview quotes, Simay 2012: 11)). This take on the chimera is problematic in the context of /Xam mythology: the fact that, while combining physical and behavioural traits from different beings in a complex fashion (of "ontological entanglement"), these beings, as members of the Early Race, all share the same origin and, as fellow-denizens of Primal Time, also the same ontological space. These characteristics would keep San chimeras within Descola's animism schema.

something narrative is able to convey so much more effectively than image (although, as we will discover later in the next chapter, some of the San images, too, attempt to, and succeed at, conveying transformation as a process).

!Khwa: The Rain That Transforms

> Then the people one after another go out and fly up into the sky, the cold wind blows them up into it. Then they keep coming out of it, floating down and falling into the pond, where they become frogs.
>
> Meanwhile the karosses become springbok which lie down and roll, thereby shaking out (the water from their skins), while the sticks and branches (of the hut) become bushes; then the arrows (or reeds) just stand about, and so do the quivers. The skins of which people have made the quivers turn into springbok, as the quivers stand about there, they get ears; meanwhile the rain turns altogether into a pond, because its body goes into it. (/Han‡kasso (Bleek 1933: 299))

The figure of !Khwa appears in about 16 narratives in the /Xam corpus, that range from obscure sketches to expansive stories, such as the ones by /Han‡kasso and !Kweiten ta //ken that appear in *Specimens*[6] (Bleek and Lloyd 1911: 192–205). While not nearly as prominent and profiled a mythological and spirit being as /Kaggen, nor as ubiquitous—under different names and with different personas—in Khoisan myth and lore as the latter, !Khwa is nevertheless very much of a presence in the /Xam First Order (as well as the Second, which, as we will see in Chap. 5, he frequently enters as well, specifically at a girl's menarcheal rites). Indeed, amongst twentieth-century descendants of the /Xam, as revealed by Ansie Hoff's recent folkloristic field work in the Northern Cape (2011a, b; also see Schmidt 2013b: 199–201), !Khwa, especially in his bovine and serpentine incarnations, is more widely known than the now almost forgotten /Kaggen.

What makes this enigmatic being fit into this world, and what links him to /Kaggen and the other members of the Early Race, is his mutability. While this ontological quality is displayed in more curtailed contexts and manifested with far less exorbitance than by the trickster and the other Early Race therianthropes, it is deep and abiding within his being,

[6] The two variants are perceptively juxtaposed, with respect to substance and style, by Roger Hewitt (1986: 80–5).

containing mythic and spirit components, as well as meteorological and faunal ones. !Khwa's—like /Kaggen's—mutability is both intransitive, that is inherent within his being, and transitive, dispensed by !Khwa to other beings. The latter beings, in the !Khwa stories, are ***human*** members of the Early Race, rather than therianthropes, who rarely appear in the !Khwa narratives, a plot detail that contributes to projecting these stories out of the world of myth into the present world, within which some of them, about menarcheal transgression, are ambiguously situated.

The animals that are transformed into by !Khwa are virtually always frogs, who, ranine through and through, hop away at the story's end, into a pool or water hole. This is the domain of !Khwa, or even, in one of his transformations, his "body". It is the domain also of his "rain children" (frogs, tadpoles, crabs, turtles, fish, snakes), as well as of the spirits of shamans that enter! Khwa's domain upon their death (Solomon 2007: 152–4; Hewitt 1986: 75–80; Lewis-Williams and Challis 2011: 160–4). The girl's family members all share the same fate (with the exception, in one story featuring a resourceful girl that had taken the care and caution to attended to !Khwa in the proper fashion, by rubbing *buchu* on his bovine forehead, being thereby spared their band member's fate). The transgressing girl's and her family members' frog-transformations are always final, irreversible and total: the girl, her father, "likewise his wife", each "altogether became a frog", using! Kweiten ta //ken's and Diä!kwain's phrasing (Bleek and Lloyd 1911: 203; Lewis-Williams 2000: 278, respectively). Moreover, compounding the state of ontological mutability, the transformation, of humans into frogs, is usually accompanied by transformation of their material, "cultural" possessions—arrows, digging sticks, quivers, karosses—back into their animate "natural" form, as twigs, shrubs, trees and antelopes, to enroot themselves in amongst the reeds or roll on the grass to dry themselves off from their rain's embrace, before bounding off into the veld. This sort of absolute transformation—humans and objects "absorbed into pure nature" (Hewitt 1986: 76)—is quite unlike the transformations wreaked by the trickster, which may be ephemeral and reversible and are almost always incomplete, keeping portions of the transformed beings fist being after its transformation into the second. The same applies to the trickster's own, self-transformations, as opposed to !Khwa, when transforming himself into one of his manifold rain-associated beings—"the Rain", "rainbull" (which may be an eland), whirlwind, a pond. These are neither equivocal nor tentative: "it [the Rain] turns altogether in a pond", notes /Han‡kasso, "because its body goes into it" (Hollmann 2005: 137; see also Bennum 2004: 213).

Moreover, there is none of the trickster's spur-of-the-moment wantonness in !Khwa's transformations, which are inflicted on the human not on a whim but as decisions and actions that are deliberate. And deserved, on the part of the transformed human: they are brought on by a moral transgression, usually by a "disobedient maiden", as well as, on one occasion, a careless hunter or musical bow player (whose remarkable transformation experiences into trees will be explored in the next chapter). August and austere, !Khwa can also be "a dangerous force and a mighty and fearsome figure of death: the southern San equivalent of the Kalahari 'lesser god'" (Solomon 2007: 152), once again differing from /Kaggen and his trickster ilk, towards whom people's feelings range from ridicule and outrage to reverence and awe (Guenther 1999: 109–15). Consistent with his ontological ambiguity, there is also moral ambiguity and ambivalence in people's feelings about this enigmatic being. In addition to dreading !Khwa, people also may have tender feelings towards him, generally "attributing friendly characteristics to the Bull [and] referring lovingly to it in the diminutive form" (Schmidt 2013b: 200). A girl or young mother of the Early Race or its /Xam human descendants may anoint !Khwa with *buchu* as he come trotting towards her, attracted by her womanly scent.

!Khwa's most common incarnations are as a bull and as rain. "He resembled a bull, while he felt that (he) was the Rain's body", /Han≠kasso explained to Lucy Lloyd in an aside to his story "A Woman of the Early Race and the Rain Bull" (Bleek and Lloyd 1911: 193; Lewis-Williams and Pearce 2004a: 141), a narrative that alternately refers to !Khwa as "the Rain"—with a fragrantly scented breath—and as a bull, trotting along, tail lowered. In describing !Khwa as a creature that "was capable of taking many forms", Roger Hewitt qualifies this description in terms of / Han‡kasso's insight into this being's ambiguous ontological state, noting that "strictly speaking … his body was water even though his shape was most frequently said to resemble that of a bull ox" (1986: 78). As rain, !Khwa may manifest himself in rain's various forms, as a heavy, threatening, thunder- and hail-spewing "he-rain"—"the lethal storm incarnate" (Solomon 2009: 33)—or, less commonly, a soft, sustaining, steady "she-rain" (Hewitt 1986: 77; Schmidt 2013b: 199–201). Another of !Khwa's rain manifestations, one diffuse and evanescent, is as clouds or mist ("the Rain's breath"). Yet more insubstantial is his being and presence as "an unspecified force only described by its actions" (Schmidt ibid.: 200), actions that are "aquatic" in some way or the other (Lewis-Williams and Pearce 2004a: 137–57, 2004b: 214). Further underscoring !Khwa's close connection to water we note that the term for water, in /Xam, is *!khwa*.

A cognate to !Khwa's incarnation as rain bull, one that links this mythic being to /Kaggen, may be his embodiment in antelope form, especially as an eland.[7] This is how !Khwa appears in /Han‡kasso's haunting story about an Early Race hunter shooting a grazing eland he came across with a poison arrow, unaware that the antelope was "the Rain" (Lewis-Williams 2000: 222–3). The man left the wounded rain-eland, which succumbed to the wound, and returned to his home camp, where he lay down to sleep, without performing any of the elaborate eland ritual /Xam hunters and their band members are expected to perform on such an occasion (Hewitt 1986: 86). This was a serious transgression, one that likely led to what transpired the next day: after announcing in the morning to his people that he had "shot the rain", the hunter and some of the men set out to "follow the Rain's footprints", through mist-laden terrain. Reaching the expired eland lying on the ground, the men built a fire and butchered the eland and proceeded to roast the meat on the fire.

The ontological ambiguity and amorphousness of !Kaggen-rain-eland becomes evident with striking physicality at this point in the story. As they roasted the meat:

> the meat kept vanishing, being burnt up in the fire. ... They went on roasting, and all the meat vanished from the fire. When they wanted to take out the meat, they turned over the ashes, looking for it, but all the meat had been burnt up in the fire. The fire burnt out; the fire died down. (Lewis-Williams 2000: 222)

It is as though the water in the eland's "Rain's body", that constitutes also the flesh of this ontologically mutable myth-being, had evaporated into thin air, transforming back into !Khwa's evanescent form as mist or vapour, in the process of dousing the fire (Hewitt 1986: 86). Spooked, the men left the rest of the eland carcass, heading home. However, the "Rain shut them in" and transformed them into frogs. The same fate awaited the hap-

[7] Here I note parenthetically—expanding the above-mentioned semantic-symbolic chain between rain and !Khwa beyond the /Xam—that the word, in Naro, for eland—*dui*—is the same as for rain.

less men's band members, when they came looking for them the next day. And at the site, where before "he had been on the hunting ground" (ibid.: 223), !Khwa-Rain then made—transformed himself into (?)—a pool, the new habitat and *Umwelt* for the frog-transformed hunters and their family.

Another aspect of !Khwa's being, consistent with his multiple metaphorical and metonymic connections to water (Lewis-Williams and Pearce 2004a: 137–58), is his transformation into a snake (Lewis-Williams and Pearce ibid.: 138–40; Solomon 2007: 151–5; Schmidt 2013b: 201; Challis et al. 2013). While this notion, and others like them, that mystically link rain to an aquatic snake, is quite widespread amongst contemporary Khoisan, including /Xam descendants (Schmidt ibid.: 198–9, 202–4; Solomon ibid.: 154; Lange and Dyll-Myklebust 2015: 5), there appears to be only one, somewhat shaky, piece of evidence for a !Khwa-informed rain snake. It is the contradictory commentary of two San informants about an image they were shown (Fig. 3.7) depicted on copy J. M. Orpen had

Fig. 3.7 Orpen's Drakensberg rock art copies. (*Cape Monthly Magazine*, 1874, foldout)

produced of some San rock art he had come across in the Drakensberg and had attached to an article on Maluti San mythology he had recently (1874) published in the *Cape Monthly Magazine* as a foldout appendix (Bank 2006a: 208–9). The image in question (upper right panel) depicts four men who appear to lead a hippo-like herbivore by a line that is attached to the animal's nose. Below these figures stands another hippo-like herbivore, approached by two running men holding spears.

The two San commentators were Diä!kwain and the Maluti Bushman Qing (!King), questioned, respectively by Wilhelm Bleek and Joseph Orpen about the image. Their different answers reflect the ontological ambiguity of !Khwa. Diä!kwain's[8] interpretation of the image is as follows: "we see here a water thing, or water cow, which, in the lower part is discovered by a Bushman, behind whom a Bushwoman stands. The Bushman then beckons others to come and help him" (Banks ibid.). This San informant sees the picture as a depiction of the initial phase of a protracted of a /Xam ritual, by real hunters and shamans, in actual time. We should note that his reading of the image was, according to Andrew Bank, heavily redacted by Bleek, in order to bring it in line with the /Xam myth narrative about hunter-shamans catching a "water-cow", as part of a /Xam land-generating ritual (2006a: 308). We are thus at a several-times remove from the original rock painting: copied by Stow, shown to and commented on by Qin (in translation), likewise by Diä!kwain (or /Han‡kasso), whose commentary was edited by Bleek.

Qin's exegetical commentary, which is closer to the source, is a less matter of fact than that of Bleek's /Xam informants, "invoking a mythical past and present" (Solomon 2007: 151). While, applying to it the same mythical-mystical rain-magic, land-restoring interpretive template as did the /Xam commentators, Qing sees the figures not as above-ground, real-life hunter-shamans but—"unquestionably" to Solomon[9]—as spirits of departed shamans, living in !Khwa's under-water domain, in which they carry out, as spirit beings, their rain ritual (ibid.: 153–4). Solomon's clue for this myth-informed interpretation is Qing's "most startling" statement about the picture: that the large, seemingly terrestrial, quadrupedal herbi-

[8]According to Lewis-Williams and Pearce the author of this commentary is /Han‡kasso rather than Diä!kwain (2004a: 140). Bleek seems to have been unclear on this point, noting that he received the information on the picture not from any specific San individual but "from the Bushmen who are now staying with me" (Bank 2006a: 307).

[9]For an alternative interpretation, in terms of trance and potency, see Lewis-Williams and Pearce (2004a: 138–43).

vore being captured by the spirit-shamans is a snake! This baffled Orpen, as well it should have given his limited familiarity with Sam myth and cosmology. However, in the context of the same, this sort of ontological conflation makes eminent sense and I am in accord with Solomon's statement that "its explanation lies in a deeper understanding of San thought and cosmologies" (p. 152), specifically, I would submit, in ontological mutability, the conceptual and phenomenological underpinnings of the same and key to its understanding. Indeed, not only does such a cosmology allow for a conflation of herbivore and snake, in the imagination of members of the culture that subscribe to it, but for even more ontologically entangled beings in the Myth World: horned, "eared", or buck-headed snakes, a motif not uncommon on San rock art panels (e.g. Rusch 2016: 9–13; see also Lewis-Williams 1981: 89), especially in the Brandberg of Namibia with its abundant antelope- and giraffe-headed snake representations (Viereck 1962: 57, 1962: 61, 65; Rudner 1956/57: 14; Pager 1989: 212, 315, 352). Like /Kaggen and his trickster counterparts, this spirit- and god-like shape-shifter, too, is a chimera, exhibiting in its fullest measure the First Order's state of ontological inchoateness.

The being has remained such in the spiritual beliefs of the Khoesan peoples who live in the Northern Cape today some of whom trace ancestral connections with the /Xam. Prominent in their folklore and beliefs is a being they refer to by such terms as "Water Snake", "WaterBull/Cow", "Water People" or "Water Person", which folklorist Ansie Hoff traces back to /Xam myth and belief about like incarnations of !Khwa (2011a: 44–5, 2011b). While the connection to !Khwa in today's stories and beliefs is oblique, what has remained about the modern version of the Water Snake/Bull is its ontological inconstancy. Mary Lange, in her study of the same being—"a water deity located in the Orange/!Garib river, namely the Water Snake"—amongst Khoe people around Upington on the Orange's north bank was struck by its "transformational character", dwelled on repeatedly and volubly by story tellers, such as Maria (Mokkie) Malo:

> He can of course change himself into anything; that is how the story goes! Into a bleating lamb, or a child sitting over there—your own child, or your sister, and you might wonder what she is doing at the river … It's anything. Do you understand now? It's a person … or a hat … or someone calling you from the river. (Lange 2015: 5; see also Lange 2011)

Returning to the nineteenth-century/Xam and !Khwa and his snake association, we should note that, alongside frogs, snakes, as well as their reptilian tortoise cousins, were the "rain's things", which, as noted by / Han‡kasso, "the rain put aside for its food" and which, as a result, "the Bushmen fear greatly" (Hollmann 2005: 145). Tadpoles, too, were such rain's animals, depicted in striking therianthropic guise on some Drakensberg rock panels (Thorp 2013). These creatures—the mid-phase in the "ontogeny of the frog" (Lewis-Williams and Pearce 2004b: 214)—exemplify metamorphosis, symbolically, visibly and physiologically, powerfully underscoring the aura of transformation that informs !Khwa and his aquatic faunal acolytes and ripples through his watery domain (Lewis-Williams and Challis 2011: 162–4).

As we will explore in Chap. 4, this domain, with its extensive suite of symbolic, mythic, ontological and cosmological signifiers of ontological flux, includes also the water- and pool-associated "new maiden" and her menarcheal rite, which is replete with transition and transformation (Thorp 2013, 2015; see also Guenther 1999: 173–8). What attracts !Khwa to the girl, and young women generally, is her scent, which she transfers to him—*qua* Rain Bull—by rubbing *buchu* on his bovine body, after he has cantered after her in his bull incarnation and surrounded her with mist. A transgression on the girl's part—such as being out in the veld seeking food, rather than staying put in her menstrual hut—may in the process trigger her own transformation, as in one of //Kabbo's version of the "disobedient maiden" story in which "the Rain", through a whirlwind, has the girl carried high up into the sky, along with her springbok-skin kaross and digging stick, to float down again, into a pond, where she became a frog and her two possessions each, a springbok and a tree (L II.–37. 3336 rev.–3337 rev.; see also Hewitt 1986: 8). "She was henceforth a frog", notes //Kabbo. Moreover, this ranine transformation also rendered the girl one of the "rain's things", "altogether", in all her body and being, having previously, as girl, been such merely in spirit and through mystical, ritual and symbolic associations (Thorp 2015: 181–2).

We see here yet another linkage between !Khwa and /Kaggen, as well as with his many trickster counterparts in Khoisan mythology. Both beings are attracted to, and lay a measure of proprietorial claim over, nubile girls and young women, especially in the context of their initiation into womanhood. Symbolically—both metaphorically and metonymically—and phenomenologically, this connects women to rain and potency, to fat and meat, through a being—the eland—central and pivotal in San cosmology. The predilection for women (and through it, connectedness to key issues of San cosmology) of these two mythological-mystical beings, along with an abiding proneness for transformation, makes these two figures of /Xam and San myth time and the spirit world kindred beings, notwithstanding certain notable differences, in being and deportment and in the spiritual and ritual role each plays in people's lives (Hewitt 1986: 131–7, 198–201; see also Solomon 2007: 154).

Let us turn from these two mythical and spirit beings to the other transformation-prone being of San myth, ones of an altogether different ontological cast, the lion.

Lion: "They Are People who Are Different"

That is what mother used to tell us about the lion's doings ... about the things which the lion does when he wants to kill us ... he makes things happen which we do not understand .... (Diä!kwain (Bleek 1932: 58–9))

The lion turns itself into another thing; it becomes a hartebeest, becomes like a hartebeest, which it desires that we may head [*sic*] it. (It also becomes like a person (in a kaross)). And when we lie in waiting for it [the hartebeest], when we lie in front of it, when it comes up to up to us, it becomes a lion. Therefore terror kills us, when we see that a lion is that which walks up to us, and we have nothing with which we can do anything; while we feel that, it does not do nicely if it perceives us; for it slays us. (/Han‡kasso (LVIII. 23: 8075–8))

The lion goes about in the heaven, therefore he does firm stand above in the sky, he is the lion who talks, he eats people, he talks, he is a lion, he is a man, he has hair, he is a lion, his hands are a man's, they have hair. (//Kabbo (II.1: 277–8))

/Xue is not dead; he is alive and is a lion, has hair, much hair. (Tamme (XII: 9377 rev))

Lions are as much of a presence in the San's world of myth as they are in men's and women's real lives, on the hunting ground and out in the veld. In both temporal settings they are transformation-prone and, in encountering a lion, a First or Second Order person could never be certain about its ontological status (all the more so as shamans, too, were liable to lion-transform, adding to this animal's ontological ambiguity and inconstancy).

In both these times—the First and the Second Order—these creatures are a menacing danger and the principal factor in the underlying "inherent enmity" that Silberbauer reported among the /Gwi, with respect to the relationship of humans and animals to one another (1981: 63). It's an enmity /Xam myths trace back to the Early Race of humans, who hunted lions, killing them with clubs made of elephant and giraffe bones, and eating their flesh (Lewis-Williams 2000: 213) and whose mythology features a story—"The First /Xam Man Brings Home a Young Lion" (ibid.: 174–205)—which Wessels, following Hollmann (2005: 34), deems a "charter myth on the enmity between humans and lions" (Wessels 2013: 4). As explained by a /Gwi informant to Silberbauer, "we hate the lions, leopard and spotted hyenas because they will hunt us. The antelopes [listed by name] hate us because they see our fire at night and N!adimo has told them that these fires are to cook them" (ibid.). Sugawara, in an article on the same San people's experience of "being hunted", puts the spotlight on the lion, the /Gwi's prototypal *paa-xó* (lit. "bite thing" or "harmful animal"), dreaded by the /Gwi, whose vulnerability to this creature, and its kind, Sugawara deems a "fundamental condition", of danger and menace, of people's existence. Through it the predator-prey relationship between human and animal is reversed. This reversed relationship instils a primal dread in humans.[10] It is elicited by an animal that, as the "ancient dark enemy of human beings" (Biesele 1993: 149, writing about the Ju/'hoansi) preys on humans as humans prey on animals. It is a role reversal that places the human-lion-human relationship on a footing of equality, all the more so, suggests Sugawara, because the lion "not only hunts them but also sometimes is hunted by [them]" (2015: 3).

As a result of this "footing of equality", the human-lion relationship may among some San groups and in certain ecological and historical conditions, in fact, be one not of enmity but amity, manifested in mutual respect, interest in one another's affairs and even sharing. The item shared is of meat, which San obtain by resourceful scavenging from lions, after the latter have made a kill—"by choosing the psychological moment after the kill when the lions have had enough meat to satisfy their hunger, but not sufficient to make them full and lazy" (Silberbauer 1965: 49)—and rationalizing this act of meat acquisition as a form of "tolerated theft",

[10] A dread which, along with the primal and universal nightmare motif and genre—"chase dreams"—that derives from it, Thomas Wynn and Frederick Coolidge trace back to our Neanderthal origins (2012: 151–3).

indeed, of meat sharing by the lions, whom, according to "anthropologist-*cum*-game warden" Peter Johannes Schoeman, the Hai//om regarded as "our hunting dogs" (Schoeman 1961: 80; see also Rakitianskaia forthcoming). Indeed, among the /Xam there was even an element of reciprocity in this meat-sharing arrangement, who, when scavenging a lion's kill, were expected to leave certain parts—the carcass's upper backbone and heart—for the lions, as their share (Woodhouse 1984: 68). Reminiscing about the old days when Hai//om and lions were "best friends", Hai//om Stefans //Khamuxab reports that people

> even ate the meat caught by the lion, but he never killed us or became angry with us. They knew exactly who took that meat, and vice versa for us. (Le Roux and White 2004: 148)

As shown by Elizabeth Marshall Thomas, during the times of the Ju/'hoansi's "Old Ways", a "Bushman/lion truce" resulted from such equitably relational arrangements—resulting, as Thomas documents, in a low incidence of lion attacks on humans (2006; see also Thomas 1990, 1994: 131–4, 2003). Willemien Le Roux and Alison White received much the same lion lore from the San informants interviewed in their oral history project (2004: 148, 156), about amicable human-lion-human relations and of hunters obtaining "lion shares" from the animals' hunts, "at a time when lions and the San were best friends because they talked to each other in certain ways" (ibid.: 156). All three authors also note that this truce and amicableness exist no longer—the lions have now turned against people, preferring killing and eating them to "knowing about people" and interacting with them as fellow, non-lion, persons. As noted by Thomas, "in Nyae Nyae in the 1950s, the lions knew about people. The people knew about lions. Their relationship was stable … the Bushman/lion truce … exists no longer" (Thomas 2006: 127).

The human-lion relationship is thus one of ontological and moral ambiguity and emotional ambivalence: lions, who kill—and eat—people, *are* "also people". We see this in //Kabbo's asides to one of the narratives in "The Lion and Blue Crane" cycle, in which he elaborates on the nature of lion (referring specifically to the sinister Early Race lion !gu):

> He has feet with which he is a lion; he has hands with which he is a man, he talks. However, he eats people, when he is a man, because he feels he is a lion's man. (II.2: 306)

Lions are such—ontologically ambiguous and morally ambivalent—not only in the First Order of myth, when they were members of the Early Race, but also in the Second Order, as reflected from explanatory comments /Xam story tellers made about lions, which are a prominent topic in /Xam myth and lore (Hollmann 2005: 31–62). The most salient "lion's man" of the Second Order is the lion-transformed shaman, whose frequently sinister, dread-invoking nature and actions will be dealt with in Chap. 5. Ontologically, the phenomenon of a shaman's lion transformation underscores experientially a lion's potential human'ness.

The prominence of lions in the myth and real world of the /Xam derives in large measure from the fact that "they were comparable to humans in many ways" (Wessels 2013: 17). Human-lion similarities were perceived by San at both a general ontological level and in specific human anatomical and behavioural terms. The former is evident in Sian Sullivan discussion of "animal agencies" in the cosmology of Damara of western Namibia she interviewed, whose commentaries were focused primarily on the lion:

> Lions figure in peoples' realities as animals imbued with agency and intentionality …, lions are conceived as being able to see, recognize and represent the people they encounter and interact with. (Hannis and Sullivan 2018: 289)

And, as noted above and to be noted below, it is not only the sphere of humans that lions enter but also that of myth beings and spirits, especially the trickster who may be a transformed or transforming lion—"feet like a lion, hands like a man". It is yet another human-lion commonality that underscores the closeness of lions with humans and humans with lions.

Jeremy Hollmann discusses the various parallels that the /Xam saw between themselves and lions. Some are general ones, such as hunting and "sorcery" and like family structure and composition. The last is also found among the /Gwi who, notes Silberbauer, explain the "doing of lion families in terms of behavior typical in a /Gwi household", including attributing the role of hunting to the male (rather than lionesses, as do Western lion ethologists) (1981: 67). Other similarities pointed out by the /Xam pertain to more specific behavioural and ontogenetic traits, such as removing a hunted animal's stomach contents before eating the prey and identifying individuals' age and family status by the amount and colour on their bodies (Hollmann 2005: 33). To this the Ju/'hoansi might add a further detail: the similarity in the shape of the lion's footprints to those of humans, with "visible toe marks ... like footprints of the small Bushmen feet". This is one reason, Lorna Marshall suggests, why, contra the /Xam, Ju/'hoansi extend "the horror of eating human flesh ... to these animals" (Marshall 1999: 101), underscoring the human-lion physical similarity and ontological convergence.

Further to footprints, an anatomical and topographical feature of significance to the San as to hunting people generally, another similarity between lions and humans is that, according to the /Xam, both lions and sorcerers (shamans) "walk on hair", a reference to the lion's—and a lion-transformed shaman's—hairy footpads (Hollmann 2005: 36). It through this experience—"treading upon hair, the hair of the lion's own feet"—when walking that, according to Diä!kwain, the /Xam shaman !Nuin/kúiten "became a lion, ... that he thus felt, felt his body" (V.15: 5097; see also Lewis-Williams 2015: 19–21).

Another significant parallel, one mental rather than anatomical, that becomes evident from a number of real-life lion tales by Diä!kwain is that lions are capable of understanding human speech (e.g. Bleek and Lloyd 1911: 185, 191, 269; Lewis-Williams 2000: 176). This might be a pan-San notion, one found also among Kalahari San, such as the Ju/'hoansi, who communicate with lions, using a special respect mode of discoursing, when they come across them in the veld or when lions skulk around the night fires (Thomas 2006: 89, 168; Le Roux and White 2004: 156; Lee 15 November 2016, pers. comm.). Moreover, /Xam natural history lore has it that lions speak to one another, like humans (Hollmann 2005: 33) and the Ju/'hoansi reportedly believe they also can communicate with humans (Thomas 1990). Diä!wain told Lucy Lloyd how lions go about doing so, "sounding like a man":

> a lion will put his tail into his mouth and call seeking us with his tail in his mouth, when he wants to sound like a man, so that we do not hear that he is a lion. That is why he puts his tail in his mouth. (Hollmann 2005: 40)

This bit of /Xam lion lore is appended to a lengthy tale—one with the sound and feel of an urban legend, /Xam-cast and Karoo-set—Diä!kwain told Lucy Lloyd about a dangerous lion encounter his older sister had had as a young woman when far out in the veld, in which a lion stalked her and followed her footprints and her scent to a hill top onto which she had climbed for safety. When close by, "calling like a man", he importunes her to pull him up the rock face. He also addressed her by name—"O */a:kum* where are you?"—which he may have found out from an owl that had spied on her and her people. Lions, who, as observed closely during her field work among the Ju/'hoansi by Elizabeth Marshall Thomas, take a keen interest in human affairs (Thomas 1994: 134–7). We find this also in /Xam lion lore (Hollmann 2005: 34–5), such as the notion that lions employ owls—as well as crows and flies (that live in lions' armpits)—for such services (ibid.: 56–7). Through them lions will monitor, for possible violations of respectful deportment towards their kind, especially on the part of children whom consequently they will punish and lure into danger. The girl in Diä!kwain's tale, having as a little girl heard from her mother about the lion's tail-in-mouth human speech mimicry routine, was wary—"it cannot be a man who is calling, but it seems to be a lion"—and kept her silence. And lived to tell the tale, adding to the store of her people's lion lore.

Lion encounters of this sort, by men and women and children, out in the veld, or at night around the fire when the beasts can be heard roaring beyond the fire in the dark, eyes shining—at times, "like stars", a couple of which, according to /Xam myth, were at one time Early Race lions—are an inexhaustible font for stories. Lion's encounters and attacks are grist for "thrilling stories" among the /Gwi (Sugawara 2007: 88). Two such encounters were related in detail by George Silberbauer, both of them about lions at night, acting in uncanny ways, contrary to their normal behaviour that "neither the Bushmen nor any of us could explain" (1981: 24). In one encounter the ethnographer's nocturnal camp in the Central Kalahari at one time was surrounded by a dozen-odd skulking and roaring lions for 17 consecutive nights, psychologically rendering each of the humans huddled close to the fire "the mouse that the cat had been playing with" (1981: 24). In the second incident that occurred to a small party of San hunters out in the veld, a lion, silently, without any warning growls or grunts, attacked one of the men sitting by the fire biting him in the leg and seriously wounding him. Compounding the uncanniness and dread of this lion encounter—as lions rarely attack silently—was the fact that the same lion attacked the same man on two subsequent occasions, even after the men had moved camp and their wounded comrade several miles to safer ground. The men discussed this unsettling lion encounter at some length; its mythopoeic cast was that the lion had been sent by either the good creator-god N!adima or the, at times malevolent, trickster-divinity G//amama (ibid.: 55–6).

"Lions' doings", after all, "make things happen that we do not understand", providing the aura for stories with a mystical and mythic overlay. Lorna Marshall relates a similar nocturnal lion encounter—a prowling lion silently circling a moon-lit Ju/'hoan camp all night long—which terrified the people. This particular lion crisis, and its eventual resolution in the early morning when the lion departed, was explained in animistic terms, of the band's shaman's spirit having chased the lion away (Marshall 1969: 374; Thomas 2006: 154, 320–1, N 3).

Such stories abound in San oral literature, of close encounters with and narrow escapes from lions, all the more news- and story-worthy for their not infrequent "attention demanding" uncanniness (Guenther 2015: 296), such as Silberbauer's and Marshall's experiences. One such tale is about a /Xam hunter out in the veld in the evening lying down in a cave to spend the night and finding something soft and furry beside him—a sleeping lion! Like so many /Xam and San tales, this one, too, does not have a happy end: the lion chased down and killed the man, after he had tip-toed out of the cave away from the stirring, awakening lion. Another—a further staple in the close-encounter-narrow escape (or not) lion tale genre with roots perhaps back to Palaeolithic, cave-man and cave-lion times—is about a hunter being treed by a lion and falling asleep on the branch and down onto the lion who keeps vigil under the tree all through the night (or conversely, having a lion drop down on you—"fly"—from atop a tree during the day, while walking through the veld, something doubly unsettling as lions neither hunt during the day nor climb trees). Tall-tale tabloid epics, "inherent enmity"-charged, tell of lions abducting and killing hunters, or of hunters adopting lion cubs to train as hunting dogs only, again, to be killed by the latter, or of a lioness adopting and raising a baby girl after killing her human mother, to be killed later, by the girl and her newly-wed human husband.

The store of such experience-based lion lore is replenished by each band in each generation, through its members' encounters with these uncanny, cunning fellow-predators that are both like humans and unlike them and their "doings" are both within and beyond human comprehension. Lion tales become taller with each retelling, more and more the stuff of legend. Or of dreams, as in the case of //Kabbo, who starts his dream-narrative with such a narrative's classic-conventional opening as "I dreamed of a lion that talked"—conversing with his fellow-lion-hunters about a springbok's spoor that they were following—and whose appearance—black, "shod with hair", many-legged, long-tailed—so frightened the dreamer that he woke up "startled awake" (Lewis-Williams 2000: 131). Such dreams, and ever more legendized tales of lion encounters when awake, eventually merge with myth, and a time when lions *were* people, members of the Early Race. All this adds detail, texture and profile to the lion and lion-people of myth, lore and dreams.

As such lions displayed the full range of ontological and moral ambiguity of their Early Race fellows, blending their real-life and legendary fear-inspiring menace and power, which they hold and evoke in the First Order as they do in the Second, with risible foolishness and dupability, traits not attributed to real-life and time lions. Regarding the latter traits, the redoubtable lion duo *!Haue ta =hou* ("Belt") and *!Gu* ("Mat") stand out, a "pair of malevolent but foolish, wandering lions … outwitted almost everywhere they go" (who eventually became transformed into stars, the pointers of the Southern Cross) (Hewitt 1986: 106; see also Guenther 1989: 111–4). First Order lions, notwithstanding their anthrophagy, formidableness and irascibility, are bested by such other First Order beings as Jackal, Hyena, Muishond (pole cat), Field Mouse and Tortoise.[11] Another was an Early Race teenage girl who, through her good sense and quick thinking, was able to get the better of the lion duo in their attempt of catching her (Guenther ibid.: 111–4). This Early Race *kumm* resonates with the above-noted legendary memorate about Diä!kwain's sister, in her real-life lion encounter, attesting to the stamina and cleverness of women of Myth Time.

Let us narrow our focus on this arresting lion lore[12] to consider those stories that feature the lion in his transforming aspects (noting that, apart

[11] It is tempting to interpret this sort off presentation of the dangerous and irascible "king of the beasts" as yet another manifestation of levelling, the San's celebrated means for sanctioning an uppity band member and bringing him down a peg or two (occasionally practised to good effect also in the west, as by "Saturday Night Live" *vis-à-vis* presidential candidate, president-elect and president Donald Trump). Its culmination, in San oral tradition, is the *baas*-lion merger in the trickster tales of the Ghanzi farm Bushmen, in which Jackal-Trickster forever gets the better of his quick-tempered but slow-witted Boer employer (Guenther 1989: 129–34, 1999: 103; see also Wittenberg 2014).

[12] These highly readable—and tellable—narratives are all readily accessible, through the *Digital Bleek and Lloyd*, which has an expansive index entry on "lion" that itemizes and cross-references the many and varied tales, as does Sigrid Schmidt's *Catalogue* (2013a: 148–9). Lion lore, a substantial component of the oral traditions of today's Afrikaans-speaking /Xam descendants, can be found in folklorist José Manuel de Prada-Samper's recently collected anthology of these materials (2016: 99–136).

from transformation, these plot-rich, metaphor-laden stories reveal much else about San cosmology, culture and society).

What is notable here is that the transformation lions undergo in these tales is for the most part ambiguous in some way or other. It is often qualified by the story teller, with caveats and asides, and, in the narrative's unfolding, the lion transformation is sometimes seen to be incomplete and partial, without any clear indication as to whether what is going on is actual or simulated. If the former, the transformation may nevertheless lack in discreteness and distinctiveness, so that the transformed lion's ontological status is not clearly fixed or ascertainable. Moreover, the transformation may not be initiated by the lion itself but by another being of the First Order, who may transform himself or herself into a lion, fully or partially. This obscures the leonine nature of the transformed were-being, whose lion'ness is not its primary, original ontological state but secondary, acquired through transformation from another being. All this is consistent with the general inchoateness of the First Order and the Early Race, populated by therianthropes and galvanized by transformations, beings and processes without clear patterns and predictability. Lion transformations, of myth and legend, differ in this way from those that transpire in ritual which, as we will see in Chap. 5, are more distinct and discrete, less mystical and more "real"—a palpable experience to people experiencing or witnessing a leonine transformation, one visible, veritable and verifiable. It gives contours and confirmation to the phantasmic transformations of lions of Myth Time. Not only are the humans and animals of the Second Order more discrete and fixed ontologically than their slippery First Order counterparts, but the very process of ontological mutability, transformation, is more smooth and predictable in its unfolding. Inchoateness pervades the First Order, both structurally and processually.

A couple of examples are drawn from /Han‡kasso's exceptionally rich narrative corpus. In the one, titled "The Lion Has the Power of Turning Himself into Other Things" (for an excerpt see the second section epitaph), the "thing" featured is a hartebeest. The lion transforms himself into this much-hunted antelope species, in order to lure hunters out on the hunting ground to himself to terrorize or himself hunt the hunters, turning the tables on them. The story is similar to /Kaggen's ruse that was dealt with above, when the trickster attracted nubile girls to his seemingly dead hartebeest avatar, in order to frighten and chase them, not for meat but sex (which in San cosmology are cognates). It is not altogether clear from the story teller's account whether the lion "*becomes*" a hartebeest or "becomes *like* a hartebeest". /Han‡kasso uses both phrases, one suggesting metamorphosis, the other mimesis as the transformative or quasi-transformative process.

The same ambiguity is found in another account by the same story teller, as an aside to a lengthy tale about a !Khau (lizard) Early Race person being carried off by a lion (L VIII. 18: 7630 rev.);

> It (the lion) is wont to make itself (both male and female) into a man, it puts its tail over its head, when it goes to the vultures, when the vultures are eating a springbok; it trots (along) like a man, when it trots to the vultures, while it feels that [it] resembles a real man.

The qualifications—"like", "feels like", "resembles"—suggest that the man the lion has become, or is aspiring to become, is not as "real" as is the lion, whose lion'ness is retained, more or less. The qualifications suggest that the lion's humanness may be something mimicked rather than constitutive. Putting his tail over his head is a routine reminiscent of Diä!kwain's description of the lion, when stalking his sister in the veld, imitating human speech by putting his tail into his mouth. Both are, perhaps, mimetic actions to obscure the lion's most animalian trait, the better to realize the mimetic or metamorphic change into the human that he is in the process of undergoing. And even when this is achieved, as perhaps it was in /Han‡kasso's lion-to-man transformation account—actually rather than virtually as his account is somewhat unclear on this point—and the lion-man trotting along towards the vultures is in fact a man, his tail, evidently mutation-resistant, has stayed put after his transformation. The latter is thus only partial, having failed to transform away his immutable tell-tale tail, which he hides by stretching it up along his back and folding it over his head where its tuft blends in with his mane'y hair (a comb-over technique similar in design and complexity, as well as colour, to the iconic coiffure sported by the current American president). This sort of not fully complete or quite successful anthropo-morphing resonates with //Kabbo's statements about lions, on the one hand "also being people" (see chapter epigraphs), on the other, "people who are different", "lion's people" who "eat their fellows" and who "have hair". I note that the last—hair—is a distinguishing, idiosyncratic leonine trait that is so salient that "hair" becomes one of the respectful circumlocutions for lion which /Xam children were enjoined to use when talking about this animal (Hollmann 2005: 35).

We see the process of mimesis and metamorphosis, and the progression of the one into the other, in another one of /Han‡kasso's lion stories, in which, as in other such stories, a veil of vagueness and indeterminacy surrounds the animal's transformation into the other being. It is the story about a youth of the Early Race who became a lion (Lewis-Williams 2000: 209–10), through that same two-stage transformative process: his playmates, apparently on a whim, shoved the youth into the skin of a large field mouse they had hunted, in which they "held him very firmly", then telling him to "trot away". And away he trots, into the reeds, where he becomes—a lion! Not a field mouse, as might be expected in the context of a foraging culture in which people, as hunters or ritual or recreational dancers, may don animal skins or body parts, to stalk or mime game, and experience thereby degrees of ontological transformation into the animal when skin envelops them (as we will see in the next two chapters). Instead, the field mouse-karossed youth turns into a lion, who, having trotted off, then returns to chases his brothers back to their camp, where, in true lion fashion, he bites his sister-in-law in the head and drags her off, back into the reeds. The lion's transformation moves from what seems to be a form of field mouse mimesis,[13] of a human inside and through such a mouse's skin, to lion metamorphosis, injecting into the ontological make-up of the mighty lion a lowly field mouse aspect.

The plot element of the transformative (at times restorative) effect of an animal skin is found in other /Xam and Khoisan tales (Schmidt 2013b: 173); indeed, Lewis-Williams holds the motif of the transformative effect of a "skin envelope" on the being enclosed therein an instance *par excellence* of hunter-gatherers' cosmological "concept of enveloping, transforming animality" (Lewis-Williams 2010: 226). One is the /Xam myth, featuring /Kaggen, on another of his many misadventures. Hidden in the antelope-skin hunting bag of his son-in-law //Kwammang-a to be able to accompany the latter unseen on a visit to his Lion kin to obtain a share of quagga meat, /Kaggen, at the moment of danger when he is espied by one of the Lion children, transforms into a hare, as well as grows wings, affording his escape (Bleek 1929: 15–8; see also Lewis-Williams 2015: 118–9, 128).

[13]The element of lion-mouse hybridity resonates with a whimsical /Xam aetiological myth, in which the lion cajoles and bullies a field mouse to trade chests and voices. The lion being puny and thin, he desires the field mouse's chest "because it sounds huge, [being] broad inside, whereas the lion's chest was small". Pleased with the trade, "the lion roared with his new chest"—and, in the bullying fashion that is his wont, he "stands roaring at the Bushmen, the hills resounding" (Guenther 1989: 105).

We also find a reference to this motif in one of /Han‡kasso's lion transformation narratives (second section epigraph). Seemingly set in the Second Order, the story features a lion who transforms himself into a hartebeest. As hunters stalk the antelope, crawling on their bellies, it turns back into a lion and charges the hunters, bent on "slaying" the men. Inserted into the tale, as an elaborating aside to Lucy Lloyd that was possibly elicited by her, is the enigmatic comment "It [the lion] also becomes like a person (in a kaross)". While nothing further is said about this particular transformation, of a lion into a person (or, more ambiguously phrased, "becoming like a person"), it is tempting to conjecture that the lion's being enveloped by an animal skin, might, in the process of his transformation, have injected another ontological component to that process and compounded its ambiguity. The comment, perhaps, can be seen as an attempt by the /Xam storyteller to convey to his Western interlocutor one of the physio-mystical ways transformation can come about—"like a person in a kaross", the skin-to-skin enveloping effect of which on him or her has transforming effects.

A tale in which such an ontological identity transference, through the transformative effect of a skin envelope, is explicit is the KhoeKhoe story "The Lion Who Took a Woman's Shape" (Bleek 1864: 50–6). Having killed a woman out a-gathering in the veld, the lion eats her, "taking care, however, to leave her skin whole, which he put on, together with her dress and ornaments, so that she looked quite like a woman and then went to the kraal" (1864: 52). Evidently transformed into a human through this mimicry, presumably by the enveloping woman's skin, along with her social identity-defining garment and ornaments, and taken as such by her fellows initially, her "lion's nature" is eventually revealed to her people through certain behavioural quirks. Their anxieties are confirmed to her increasingly uneasy family and kin one night, when, watching her asleep, they saw that "some of the lion's hair was hanging out" (ibid.: 54). They kill her by burning her to death in the hut, in the process the lion-woman's heart jumps out. Her mother places the same inside a calabash filled with milk, thereby restoring her human form.

This sort of ontological ambiguity that attaches to lion transformation narratives, of the transformed lion being not quite that but containing within its being also a feature of another being (or two), is consistent with the therianthropic, at times chimerical nature of the Early Race, as discussed in the previous chapter. This fundamental ambiguity in San ontology and cosmology—being simultaneously "same as" and "other than"—is examined further in Vol. II (Chap. 3).

/Han‡kasso's tale about an Early Race youth's transformative experience inside the mouse skin is a lion transformation story in which that transformation was not of a lion into a non-lion creature but the other way round. It is a variation on the theme of leonine transformation that is found also in a number of other /Xam and San stories. The human-lion transformation of /Han‡kasso's youth of the Early Race resonates with one of the most frequently told /Xam stories, about the Dawn's Heart Star's beautiful young wife's emotionally agonizing and unsettling transformation into a lioness (or lynx) who, like the Early Race youth, hides in the reeds, where she undergoes her gradual metamorphosis. I will return to this marvellous story of transformation in Chap. 3 of Vol. II, as it is so superbly effective, thanks to the story teller's consummate narrative skill, in conveying the somato-psychic experience of ontological identity and being change.

Such stories, of Early Race persons' transformation into lions, are reiterations, through myth and lore, of what happens in the trance ritual, in which a human may become a lion. Diä!kwain long and haunting tale about a young man of the Ancient Race, a tale that appears to blend fact and legend with myth, is woven around this shamanic theme. The man, out alone in the hunting ground, is stalked, caught, stalked again after his almost successful escape and finally killed by a lion, who, in the tale's melodramatic conclusion, is himself killed by the man's band members. Just prior to the protagonist's and antagonist's Wagnerian demise, the lion, in talking to the man and his people, reveals himself to be a "sorcerer-lion" (Bleek and Lloyd 1911: 175–91).

In two /Xam stories, again by /Han‡kasso and set, it would seem, not in the First Order but the Second (the two orders, as noted previously, not infrequently dissolve one into another), it is not ancestral humans that morph into lions but animals. In the one (Hollmann 2005: 60, 61), the transforming animal is the wildcat, which the Bushmen hunt and which, in retaliation, will sometimes turn into lions that will come to people's nocturnal camp and drag sleepers away. Being Wild Cat-Lions, they leave no "lion's spoor", making people unaware of and inattentive to any lion danger when they check the camp ground prior to going to sleep, nor do they attend to the nocturnal fire kept alight when they sense a potential lion threat. When they go to sleep they thus "do not lie listening" and become all the more ready targets for mystical, small- and light-footed lion stalkers. In the other story (Lewis-Williams 2000: 167–8) it is a gemsbok who undergoes leonine transformation, right before the bewildered eyes of a hunter who has approached the grazing gemsbok out in the plain. Startled by the hunter, "the gemsbok … turned from him and became a lion. The gemsbok behaved in this manner: the gemsbok went"—and a lion took its shape, its long, sharp horns, the species's defining anatomical trait, vanishing before the watching hunter's startled eyes. Unsettled by what he saw, he walked around gawking at "the thing which was a gemsbok become a lion", and inattentive, in the process, to where he was treading and stepping, with his bare foot, on the poison arrow he had dropped (with dire consequences the story hints at but does not spell out).

Such non-lion to lion transformations further obscure the ontological status and identity of the transformed being, who is as much—or more? or less?—lion as he is gemsbok, or hartebeest, or human.

Or preternatural, myth-spirit being: "/Xue is not dead; he is alive and is a lion", asserts Tamme, one who "has hair, much hair". Hiʃe, counterpart, in Naro supernaturalism to /Xue, too, can appear in lion form, having been seen and described as such—"looking like a lion … walking upright"—by a shaman to Dorothea Bleek (1928: 24; also see Lebzelter 1934: 74).

As we will see in Chap. 6, shamans not only have visions of spirit lions but may become lions themselves, such that, within one crowded space—the dance circle—we may find transformed lions from two or three different ontological domains—of myth, of spirits and of the here-and-now. And added to the trickster-, spirit- and human-lions may be the real thing: lions, roaring in the dark beyond the fire-lit dance circle, eyes shining like stars. More fodder for lion stories.

Let us turn next from narrative to image, San rock paintings and contemporary studio art and prints. Some of these works of art strikingly depict transformation, with either graphic animation and vividness, or stylistic suppleness and subtlety, underscoring and complementing the story tellers' expressions of the cosmological theme of ontological mutability.

References

Bank, Andrew. 2006a. *Bushmen in a Victorian World: The Remarkable Story of the Bleek-Lloyd Collection of Bushman Folklore.* Cape Town: Double Storey Books.

———. 2006b. Anthropology and Fieldwork Photography: Dorothea Bleek's Expedition to the Northern Cape and the Kalahari; July to December 1911. *Kronos* 32: 77–113.

Barnard, Alan. 1992. *Hunters and Herders of Southern Africa*. Cambridge: Cambridge University Press.

Bennum, Neil. 2004. *The Broken String: The Last Words of an Extinct People*. London: Viking.

Biesele, Megan. 1993. *Women Like Meat: The Folklore and Foraging Ideology of the Kalahari Ju/'hoansi*. Johannesburg/Bloomington: Witwatersrand University Press/Indiana University Press.

Bleek, Wilhelm H.I. 1864. *Reynard the Fox in South Africa; or Hottentot Fables and Tales*. London: Trübener and Co.

Bleek, Dorothea. 1923. *The Mantis and His Friends: Bushman Folklore*. Cape Town: T. Maskew Miller.

———. 1927. Bushmen of Central Angola. *Bantu Studies* 3: 105–125.

———. 1928. *The Naron*. Cambridge: Cambridge University Press.

———. 1929. Bushman Folklore. *Africa* 2: 302–313.

———. 1932. Customs and Beliefs of the /Xam Bushmen. Part II. The Lion. *Bantu Studies* 6: 47–63.

———. 1933. Beliefs and Customs of the /Xam Bushmen. Part V. The Rain. *Bantu Studies* 7: 297–312. *Bantu Studies* 10: 163–199.

Bleek, Dorothea. 1934/53. !kun Mythology. *Zeitschrift für Eingeborenen-Sprachen* XXV (4): 261–283.

Bleek, Wilhelm H.I., and Lucy Lloyd. 1911. *Specimens of Bushman Folklore*. London: George Allen & Co.

Challis, Sam, Jeremy Hollmann, and Mark McGranaghan. 2013. 'Rain Snakes' from the Sequ River: New Light on Qing's Commentary on Rock Art from Sehonghong, Lesotho. *Azania: Archaeological Research in Africa* 48: 331–354.

Guenther, Mathias. 1989. *Bushman Folktales: Oral Traditions of the Nharo of Botswana and the /Xam of the Cape*, Studien Zur Kulturkunde No. 93. Stuttgart: Franz Steiner Verlag Wiesbaden.

———. 1996. Old Stories/Life Stories: Memory and Dissolution in Contemporary Bushman Folklore. In *Who Says? Essays on Pivotal Issues in Contemporary Storytelling*, ed. Carol L. Birch and Melissa A. Heckler, 177–197. Little Rock: August House Publishers.

———. 1999. *Tricksters and Trancers Bushman Religion and Society*. Bloomington: Indiana University Press.

———. 2002. The Bushman Trickster: Protagonist, Divinity, and Agent of Creativity. *Marvels and Tales Journal of Fairy-Tale Studies* 16 (1): 13–28.

———. 2014. Dreams and Stories. In *Courage of //Kabbo: Celebrating the 100th Anniversary of the Publication of* Specimens of Bushman Folklore, ed. Janette Deacon and Pippa Skotnes, 195–210. Cape Town: UCT Press.

———. 2015. 'Therefore Their Parts Resemble Humans, for They Feel That They Are People': Ontological Flux in San Myth, Cosmology and Belief. *Hunter-Gatherer Research* 1 (3): 277–315.

Hannis, Michael, and Sian Sullivan. 2018. Relationality, Reciprocity, and Flourishing in an African Landscape. In *That All May Flourish: Comparative Religious Environmental Ethica*, ed. Laura Hartman, 279–296. Oxford: Oxford University Press.

Hewitt, Roger. 1986. *Structure, Meaning and Ritual in the Narratives of the Southern San*. Hamburg: Helmut Buske Verlag.

Hoff, Ansie. 2011a. Guardians of Nature Among the /Xam San: An Exploratory Study. *South African Archaeological Bulletin* 66: 41–50.

———. 2011b. *The /Xam and the Rain*. Cologne: Rüdiger Köppe Verlag.

Hollmann, Jeremy C., ed. 2005. *Customs and Beliefs of the /Xam Bushmen*. Johannesburg/Philadelphia: Wits University Press/Ringing Rock Press.

Jolly, Pieter. 2015. *Sonqua: Southern San History and Art After Contact*. Cape Town: Southern Cross Ventures.

Lange, Mary E. 2011. Rock Art Research in South Africa. *Rozenberg Quarterly The Magazine*. Open-access Online Journal http://rozenbergquarterly.com/rock-art-research-in-south-africa, 10 pp. Excerpt from Mary Lange *Water Stories and Rock Engravings: Eiland Women at the Kalahari Edge*. Amsterdam: Rozenberg Publishers, SAVUSA Series.

———. 2015. *Water Stories: Original !Garib Narrations About the Water Snake*. Pretoria: Unisa Press.

Lange, Mary E., and Lauren Dyll-Myklebust. 2015. Spirituality, Shifting Identities and Social Change: Cases from the Kalahari Landscape. *HTS Teologiese Studies/Theological Studies* 71 (1), Art. #2985, 11 pp. http://dx.doi.org/10.4102/hts.v71i1.2985

Le Roux, Willemien, and Alison White, eds. 2004. *Voices of the San*. Cape Town: Kwela Books.

Lebzelter, Viktor. 1934. *Eingeborenenkulturen von Südwestafrika: Die Buschmänner*. Leipzig: Verlag Karl W. Hiersemann.

Lewis-Williams, David J. 1980. Remarks on Southern San Religion and Art. *Religion in Southern Africa* 1: 19–32.

———. 1981. *Believing and Seeing: Symbolic Meanings in Southern San Rock Paintings*. New York: Academic Press.

———. 2000. *Stories that Float from Afar: Ancestral Folklore of the San of Southern Africa*. Cape Town: David Philip.

———. 2010. *Conceiving God: The Cognitive Origin and Evolution of Religion*. London: Thames & Hudson.

———. 2015. *Myth and Meaning: San-Bushman Folklore in Global Context*. Walnut Creek: Left Coast Press.

Lewis-Williams, David J., and Sam Challis. 2011. *Deciphering Ancient Minds: The Mystery of San Bushman Rock Art*. London: Thames & Hudson.

Lewis-Williams, David J., and David G. Pearce. 2004a. *San Spirituality: Roots, Expression, and Social Consequences*. Walnut Creek: Altamira Press.

———. 2004b. Southern African Rock Paintings as Social Intervention: A Study of Rain-Control Images. *African Archaeological Review* 21: 199–228.

Marshall, Lorna. 1969. The Medicine Dance of the !Kung Bushmen. *Africa* 39 (4): 347–381.

Marshall, Lorna J. 1999. *Nyae Nyae !Kung Beliefs and Rites*, Peabody Museum Monographs, No. 8. Cambridge, MA: Harvard University.

Orpen, Joseph M. 1874. A Glimpse into the Mythology of the Maluti Bushmen. *Cape Monthly Magazine* [N.S.] 9: 1–13.

Pager, Harald. 1989. *The Rock Paintings of the Upper Brandberg. Part I: Amis Gorge*, ed. R. Kuper. Cologne: Heinrich-Barth Institut.

Rudner, J. 1956/57. The Brandberg and Its Archaeological Remains. *Journal of the South West African Scientific Society* 12: 7–44.

Rusch, Neil. 2016. The Root and Tip of the *//Kwanna*: Introducing *Chiasmus* in Three /Xam Narratives. *Critical Arts* 30 (6): 877–897.

Schmidt, Sigrid. 2001. *Tricksters, Monsters and Clever Girls: African Folktales – Texts and Discussions*, Afrika Erzählt. Vol. 8. Cologne: Rüdiger Köppe Verlag.

———. 2013a. *A Catalogue of Khoisan Folktales of Southern Africa*, 2 vols. Research in Khoisan Studies 28.1&2. Cologne: Rüdiger Köppe Verlag.

———. 2013b. *South African /Xam Bushman Traditions and Their Relationship to Further Khoisan Folklore*, Research in Khoisan Studies, 31. Cologne: Rüdiger Köppe Verlag.

Schoeman, Peter J. 1961. *Hunters of the Desert Land*. Cape Town: Howard Timmins.

Silberbauer, George B. 1965. *Report to the Bechuanaland on the Bushman Survey*. Gaberones: Bechuanaland Government.

———. 1981. *Hunter and Habitat in the Central Kalahari Desert*. Cambridge: Cambridge University Press.

Simay, Philippe. 2012. *What Images Show. An Interview with Philippe Descola*. Trans. Michael C. Behrent. Online Article, April 5, 2012. http://www.booksandideas.net/What-Images-Show.html

Solomon, Anne. 2007. Images, Words and Worlds: The /Xam Testimonies and the Rock Arts of the Southern San. In *Claims to the Country: The Archives of Lucy Lloyd and Wilhelm Bleek*, ed. Pippa Skotnes, 149–159. Johannesburg: Jacana Media.

———. 2009. Broken Strings: Interdisciplinarity and /Xam Oral Literature. *Critical Arts* 23 (1): 25–41.

Sugawara, Kazuyoshi. 2007. The Lion as the Symbol of the Other from the Perspective of the Gwi, a Hunting Tribe of the Kalahari Desert. *Biohistory Journal* (Spring): 1–2. Online Journal. https://www.brh.co.jp/en/simsishi/journal/052/research2-2.html. Accessed 19 Oct 2016.

———. 2015. On the G/ui Experiences of Being Hunted: An Analysis of Oral Discourse on the Man-Killings by Lions. Paper Presented at the 12th

International Conference of Hunting-Gathering Societies (CHAGS 12), University of Vienna, Vienna, September 7–12, 2015.

Thomas, Elizabeth M. 1990. The Old Way. *New Yorker*, October 15: 78–110.

———. 1994. *The Tribe of Tiger: Cats and Their Culture*. New York: Pocket Books.

———. 2003. The Lion/Bushman Relationship in Nyae Nyae in the 1950s: A Relationship Crafted in the Old Way. *Anthropologica* 45: 73–78.

———. 2006. *The Old Way: A Story of the First People*. New York: Farrar Strauss Giroux.

Thorp, Carolyn. 2013. 'Frog People' of the Drakensberg. *Southern African Humanities* 25: 245–262.

———. 2015. Rain's Things and Girl's Rain: Marriage, Potency and Frog Symbolism. *Southern African Humanities* 27: 165–190.

Thurner, Ingrid. 1983. *Die transzendenten und mythischen Wesen der San (Buschmänner) Eine religionsethnologische Analyse historischer Quellen*. Wien: Föhrenau.

Viereck, A. 1962. *Südwestafrikanische Felsmalereien*. Windhoek: South West African Scientific Society.

Wessels, Michael. 2013. Story of a /Xam Bushman Narrative. *Journal of Literary Studies* 29 (3): 1–22.

Willerslev, Rane. 2011. Frazer Strikes Back from the Armchair: A New Search for the Animist Soul. *Journal of the Royal Anthropological Institute* (N.S.) 17: 504–526.

Wittenberg, Hermann. 2014. The Story of //Kabbo and *Reynard the Fox*: Notes on Cultural and Textual History. In *Courage of //Kabbo: Celebrating the 100th Anniversary of the Publication of* Specimens of Bushman Folklore, ed. Janette Deacon and Pippa Skotnes, 93–98. Cape Town: UCT Press.

Woodhouse, Bert. 1984. *When Animals Were People: A-Z of Animals of Southern Africa as the Bushmen Saw and Thought Them and as the Camera Sees Them Today*. Melville: Chris van Rensburg Publications (Pty) Limited.

Wynn, Thomas, and Frederick Coolidge. 2012. *How to Think Like a Neandertal*. New York: Oxford University Press.

CHAPTER 4

Therianthropes and Transformation in San Art

Some very curious ideas, possessed by the Bushmen, which would probably otherwise not have come to light at all, have become known to us in the course of their [San rock art researchers] endeavors.
Lloyd (*1889*: 28)

We turn now to the other major domain of San expressive culture, art—both historic and prehistoric rock art and contemporary easel art—to explore how it depicts the two cosmological themes just examined in the context of San myth and stories, therianthropes and transformation. Also considered is what these depictions of ontological ambiguity might have meant to the artists and to what extent and in what way they might have drawn on myth's Early Race when the painters or engravers created their brand of therianthropes and their depictions of transformation.

A summative statement by David Lewis-Williams, the dean of San rock art studies, on the nature of European Palaeolithic cave art, specifically its elements of transformation and interaction, ontologically and phenomenologically, gets at the heart also of San rock art:

> The therianthropy of Upper Palaeolithic images suggest an intense kind of transformation, an interaction of both spiritual and material animality with humanity. (2010: 226)

© The Author(s) 2020

M. Guenther, *Human-Animal Relationships in San and Hunter-Gatherer Cosmology, Volume I*,
https://doi.org/10.1007/978-3-030-21182-0_4

Such resonance, through time and over remote and removed regions of the world, attests to both the antiquity and ubiquity of animistic cosmology (a point to which I will return in the next volume, Chap. 5).

Ancestral San Rock Art: Therianthropes

As presented in the myths, the Early Race of the First Order of Existence, for all their weird hybridity, do also display normalcy, the sort found in regular animals and humans in the real world. This is evident in their ontological (and social) make-up in which such Second Order normalcy is an inherent component of the First Order were-beings. Chimerical Mantis, despite of his ontological ambiguity and antics, is his band's elder or even its headman and in this at times harried and worrisome social capacity this "Dream Bushman" (Bleek 1923: vi) is also a real Bushman; indeed, as noted by Heinrich Vedder, "were you to remove from him his magical powers, what would come to the fore is a veritable Bushman" ("*ein richtiggehender Buschmann*" (1937: 423, quoted by Thurner 1983: 126, my translation)). In fact, it can be argued—and I'll return to this point again later—that it is their human'ness that prevails in the end and defines their being more than their animal'ness, after digressing from this default state for varying periods and in varying ways through their ontological antics and acrobatics. An example is Mantis suddenly—after a brief transformation into a hare—sprouting bird's wings and flying off after a scrape that backfired on him, or, on a whim, becoming a dead antelope, to be found by a bevy of nubile girls he lusts after butchered by them, only to reassemble himself again into his human form and chasing after them as a faun chases nymphs (Guenther 2015: 285–6). Or, as in the case of those members of the Early Race whose animality is implanted in their being, such as the Tick and Elephant People, Aardvark and Puffadder Woman and of course Mantis himself, this is really only a vague and diffuse ontological condition, manifested in certain idiosyncratic traits such as sucking blood, sporting a trunk, living and digging underground, brooding on eggs in a nest dug in Kalahari sand, or being—sometimes and sporadically—a "little green thing". While it is constitutive of their being, animal hybridity in most of the Early Race of myth is vague and diffuse, intimated rather than explicit, a matter, using Philppe Descola's terms, more of ontological "interiority" than of "exteriority". It becomes exterior in sudden flashes of transformation—the topic also of the next two chapters on other situations such flashes occur—when an animalian trait comes to the fore, more or less flashily and fleetingly in the Second Order.

All this is different in San art. Artists generally, as regards their expressive and aesthetic intent, are concerned first and foremost with exteriority—"an eager dwelling upon appearances, an engrossment in surfaces, and absorption in things … 'in themselves'", as noted by Clifford Geertz in a discussion in his classic essay on the symbolic nature of religion of the "aesthetic perspective" (1972: 111). In the context of San painters' and engravers' depictions of therianthropic figures, the focus of their artistic endeavours is on the "external body", an artist's "fundamental expressive instrument" according to Vivieros de Castro, in the context of Amazonian shamanism (1998b: 480), especially if that body belongs to a fantasy-beguiling hybrid being as it frequently does in Amazonian myth. San rock artists depictions of such beings dwell almost obsessively on that body's hybridity, comingling and conflating non-human traits and beings with riotous abandon, without subtlety or intimation, ambiguity and ephemerality, in a free-flow and free-fall of aesthetic expression of therianthropism.

The were-beings of San art have no match in San mythology, whose ontological hybridity, as noted above, is a good deal less explicit and less extravagant (with some exceptions, most particularly the ontologically trigger-happy, uber-protean /Xue). The stories about mythic therianthropes deal more with their at times outrageous antics rather than their outrageous bodies. As narrative rather than imagery, these bodies' interiority—its moods and motivations translated into actions and interactions—is what constitutes story's subject matter. Its subjects' external form and its wanton anatomical mutability is one narrative theme of many in San myth and lore. Instead of being openly flaunted as in painted therianthropes, in their mythic counterparts, the Early Race, ontological hybridity and fluidity reveals itself as either a potentiality of their being that manifests itself at certain dramatic moment in a story or it is a condition carried in their being diffusely and in specific and idiosyncratic anatomical traits which, in a number of stories—with a creative-aetiological twist—become more marked and fixed in the First Order being's Second Order recurrence.

The images, by San artists, intensify the element of hybridity of the beings of myth, as related by the story tellers, broadening and deepening our appreciation of the ontological ambiguity that pervades San cosmology, as well as, perhaps, adding to their store of "very curious ideas", as suggested by Lucy Lloyd when commenting on the rock art images she was presented with—and presented to a couple of her /Xam story tellers for comment—by such contemporaneous researchers as Stow,

Orpen, Schunke and Bain (Lloyd 1889: 28). While therianthropic representations make up only between 3 and 5% of all of the tens of thousands of images San artists have created (Jolly 2002: 85), the statement these images make about San cosmology's theme of ontological mutability is not commensurate with that very low incidence within San art.

In what follows I will describe the multiple ways this mutability is conveyed, through painted or engraved representations. By their nature, such representations are stills that can only suggest but not directly display transformation.[1] This is conveyed more effectively, as well as directly, in and of itself, by narratives, which are intrinsically flowing, in a story's unfolding, driven by the processual dynamics of narrative and narration (especially if performed by a skilled storyteller). Yet, as we will see below, some rock art images do attempt to do so, depict transformation through imagery, by means of certain visual effects employed by the artist when producing his or her image of theri-transformation. Both of these expressive forms reveal San art, through its therianthropic and transformation motifs, as a potent playing out of ontological mutability.

Like the myriad breeds of dogs are all variations on a basic, virtual, ideal-typical *ur*dog, a generic therianthrope-prototype underlies the many and varied ways this being is depicted by artists. Archaeologist Pieter Jolly, in a magisterial article on the San therianthropy, depicts this "archetypal therianthrope" (Fig. 4.1) and defines it as "a being on two legs with a human torso and limbs and antelope head" (2002: 86). The antelope species, while sometimes identifiable, as eland, rhebuck or other small antelope species, is usually generic and, in line with the archetype's antelopean ambiguity, the human part may also be indeterminate with respect to gender.

The multiple variations of this basic therianthrope attest both to the power of the imagination of the artists and to the salience of the theme of ontological mutability in the world view that guides their artistic endeavours. The many ways in which the archetype's human and antelope components are combined, Jolly notes, "sometimes almost defy classification". His article nevertheless attempts to differentiate between distinctive ways (nine of which are described) in which "antelope heads, torsos, legs,

[1] As Paul Shepard puts the point, in the context of the metamorphosis motif in Greek myth-stories and their representations on Renaissance paintings and engravings, "the artist has provided stills—which is the way we remember scenes, even of action—that represent transformation" (1978: 95).

Fig. 4.1 Prototypal San therianthrope. (Jolly 2002: 86, with permission)

hooves, tails and postures may all be combined in different ways with human heads, torso, limbs, hands, feet and postures" (ibid.). Another animal that is brought into the therianthropic mix is birds and Jolly again differentiates some eight ways in which winged therianthropes combine avian with human and antelope features, creating thereby chimerical beings consisting of three species (if not more, as traits from yet other beings can be added to the entangled mix).

The antelope component itself may undergo ontological variation, as it is replaced with, or has grafted unto it, components from other animal species. Such rare non-antelope therianthropes combine their human parts with such animals as baboon, elephant, feline, wildebeest, hippopotamus, snake, rhinoceros, ostrich and cattle[2] (Figs. 4.2 and 4.3). Instead of blending animal with human components, the depicted hybrids may con-

[2] The last was a not uncommon human-animal conflation among San rock art painters in the South-eastern Cape, who had a long period of contact with Nguni and Sotho pastoralists. Their impact on the autochthonous San and their economy and society is reflected in the "contact art" of the region, which includes depictions of cattle and horses, as well as their conflation with humans and animals (Jolly 2015: 254–61). Horses were a prominent motif also for /Xam engravers in the Northern Cape (Skinner 2017).

Fig. 4.2 Non-antelope therianthropes. (Jolly 2002: 87, with permission)
Top row: baboon therianthrope, KwaZulu-Natal; elephant therianthrope, Western Cape; lion (?) therianthrope, Lesotho; wildebeest therianthrope, Western Cape; hippopotamus (?) therianthrope
Middle row: snake therianthropes, Eastern Cape; rhinoceros (?) therianthrope, Lesotho; ostrich therianthropes, Western Cape
Bottom row: baboon therianthrope with human hunting equipment, Lesotho; cow or bull therianthrope, KwaZulu-Natal

flate different animal species, some real, such eland and horse, others imaginary, such as snakes with the head of a tusked canine creature or the torso and beaked head of some avian beast (Fig. 4.4).

The last two animalian hybrids point to another ontological realm artists drew on when depicting therianthropes: the other-world—of spirits or, perhaps myths, as well as dreams—or the reality-transcending liminal realm of ritual. They produced "imaginary creatures", another modality for the San artists for playing out the theme of ontological mutability. Jolly describes the bizarre features of such imaginary therianthropes:

> Their bodies may be greatly distorted – very elongated or in strange postures – or their features may combine clearly antelope features with human-

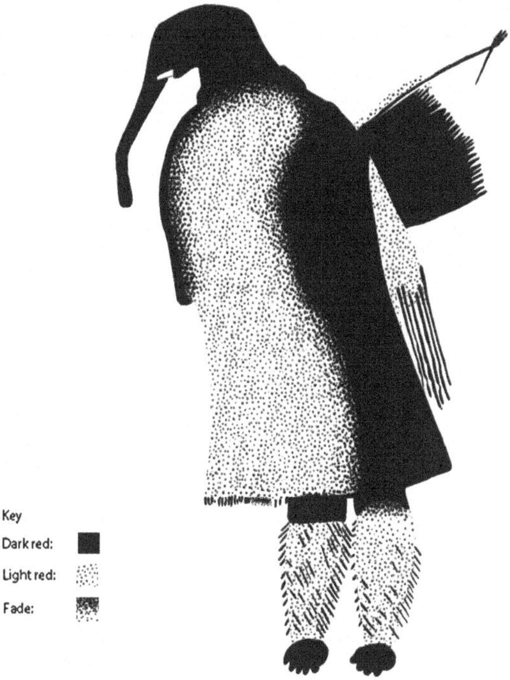

Fig. 4.3 Elephant therianthrope, Drakensberg. (RARI, with permission)

> like, but not fully human, torsos and limbs. Similarly ... human torsos and limbs may be combined with antelope-like, but not fully antelope, heads. Sometimes fantastic, bizarre and unidentifiable elements are combined with easily recognizable features of both humans and antelope. (Jolly 2002: 86)

Other than the antelope-headed humans, the artists' basic therianthropic body plan that conflates human and animal much the same way as what we come across in the Myth World, San artists also were fond of depicting chimeras (hybrid forms that are rare among the Early Race). Such beings—"fusion[s] of human with two species of antelope or with antelope and other animal species" (Jolly 2002: 86)—are an expression *par excellence* of "ontological entanglement", using Vivieros de Castro's term (1998a: 6). A striking example is seen in Fig. 4.5: a feline quadruped with human hind legs, one of which has a human foot and the other an equine hoof bearing porcine tusks on its snouted visage protruding from a large, leonine head. Another, presented by Lewis-Williams and Dowson,

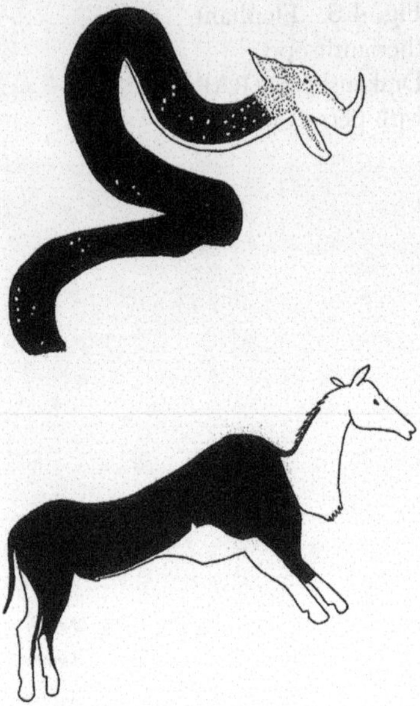

Fig. 4.4 Fantasy therianthropes. (Jolly 2002: 89, with permission)
Top: snake and imaginary tusked creature, Eastern Cape
Bottom: eland-horse, Eastern Cape

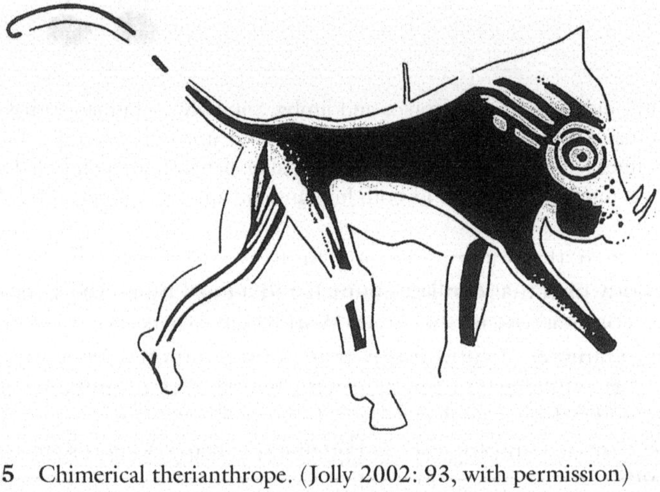

Fig. 4.5 Chimerical therianthrope. (Jolly 2002: 93, with permission)

is a striding humanoid figure with a bird perched atop its head, which the two authors describe as follows: "This figure's head has antelope ears, a curious 'tusk', an open mouth with teeth, and two facial lines" (1989: 85, fig. 38b, see also Jolly ibid.: 93).

Another is the horned serpent figure, a "hybrid conflation" which conflates not only antelope and snake but which may also be, in some of its cognates, the embodiment of the /Xam rain divinity !Khwa, in the form of a rain bull, his favourite incarnation (Solomon 2007: 152). "Blur[ring] therianthropic boundaries" (Rusch 2016: 13) even further, there are depictions that comingle this divine therianthrope with lightning (ibid.: 12, fig. 6) bringing to the ontological amalgam a meteorological component (not uncommon in San, especially /Xam mythology, in which wind and rain, as well as sun, moon and stars may all feature as anthropomorphized, meteorological and sidereal members of the Early Race). Each such vibrant representation of hybridity attests to ontological mutability and visually underscores, with striking imagery, the playing out of this theme of San cosmology in their myths and stories.

A noteworthy fact about San rock art and cosmology is that the oldest image—a stone slab fragment from the Apollo 11 shelter in the Huns Mountains of south-western Namibia dated at about 27,500 BP—depicts a chimera: a quadruped with feline features and human hind legs, along with bovid sexual organs and gemsbok horns (Vogelsang et al. 2010: 194; Fig. 4.6, Rifkin et al. 2015: 120). It appears that San cosmology's theme of ontological conflation has deep roots that go back not only to myth time's First Order of Existence but also to (pre)historic times of the Second Order.

Despite these dazzling re-combinations and permutations of the therianthropes of San ancestral art, there appears to be one constant ontological-anatomical feature in San artists' iconography: an *animal's head* and a *human's body*. In discussing this almost invariable feature amidst a sea of ontological variable ones with Pieter Jolly, the researcher notes that human-headed therianthropes are "extremely rare" in San art (pers. com. 19 February, 2016). He directed me to three images in Lewis-Williams' and Dowson's *Images of Power*, which depict, more or less clearly, human-headed and bodied male figures with antelope feet (Fig. 4.7). Another instance (Fig. 4.8), perhaps, as the figures are depicted headless, is what Edward and Cathelijne Eastwood refer to as "antelope women" (2006: 130–38), which are found in the central Limpopo Basin in north-eastern South Africa (Fig. 4.8), sometimes in panels together

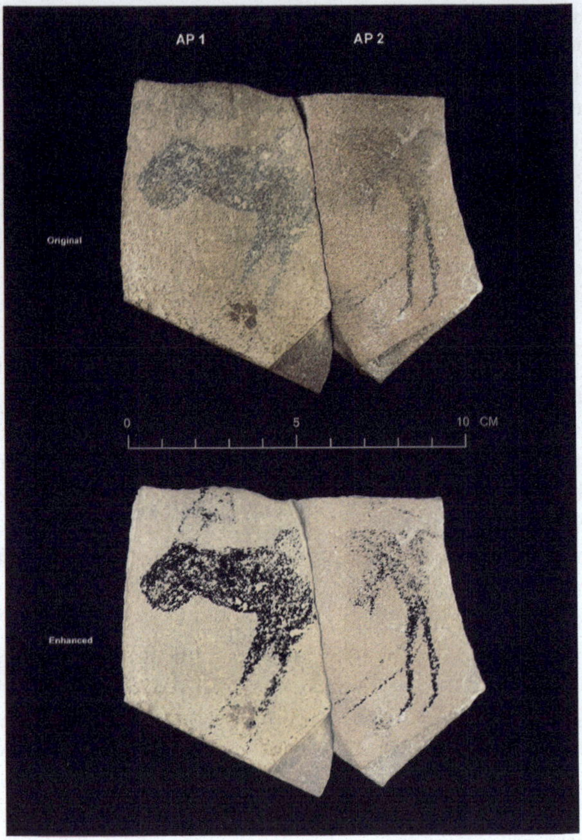

Fig. 4.6 Chimerical therianthrope from Apollo 11 shelter. (Rifkin et al. 2015: 120, with permission)

with Kudus that depict what the Eastwoods take to be a female initiation scene (ibid.: 138, fig. 8; see also Eastwood 2006). The antelope component of these clearly female humans (depicted with large pendulous breasts and buttocks) is evident in their legs and feet which, are "reverse-articulated", with the knees pointing backwards as in an antelope rather than forward as in a human. Moreover, the figures have wedge-like feet that resemble antelope hooves. Reversing the prevalent pattern of San—

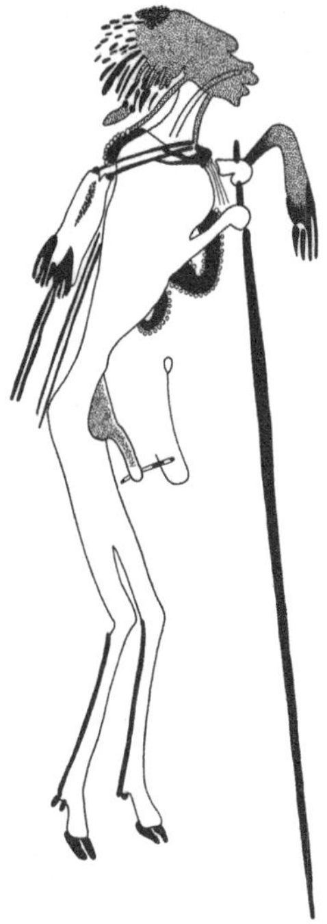

Fig. 4.7 Human-headed therianthrope, Drakensberg. (RARI, with permission)

and, possibly hunter-gatherer[3]—therianthropic art, of animal head and human body and extremities, such images display yet another variation on the theme of ontological mutability.

[3] In an examination of this aspect of therianthropic beings, Paul Shepard suggests that those with animal heads and human torsos and legs tend to be found among hunting peoples, with roots deeply back into human prehistory (its exemplar the Trois-Frères Sorcerer in southern France). Animal-keeping people, on the other hand, tend to depict such hybrid beings as human-headed and animal-bodied (exemplified by the Egyptian sphinx and the Graeco-Roman Pan and centaurs). He explains this difference in terms of different valorizations in people's cosmologies of animality and humanity. In the case of hunting peoples' animals, powerful and dangerous creatures, humans identify with animals through the mind (=head) and connect with the animal's spirit, gaining access to its power through the spirit of the animal rather than its body. Stock-keeping people, who have tamed animals and no longer draw on their power nor connect with their spirit, convey animality somatically and viscerally, as well as, in the case of Pan, libidinously (Shepard 1978: 96–7). The former, hunter's animal depictions retain the animal's autonomy of being while in the latter, the herders' human-headed animals have reduced the same, through domestication.

Fig. 4.8 Antelope women. (Eastwood and Eastwood 2006: 138, with permission from RARI)

An alternate and simpler explanation may be that such figures—human-headed and antelope-limbed—may not be depictions of therianthropes, but of transformation, of a human into antelope.

Ancestral San Rock Art: Transformation

We have already come across some images in the examination of the transformation theme in San myth and lore in the previous chapter, via the drawings by the story tellers !Nanni and Tamme, some of them annotated cartoons explicitly linked to their stories, specifically those about the protean trickster being. We saw /Xue depicted not only as transformed—into an elephant (Fig. 3.6) or !xane plant (Fig. 3.2)—but also as transforming, into trees. As such we find /Xue (or his wife) depicted in various mutational phases, the body being either a tree trunk, with feet growing roots and hands and head sprouting twigs, leaves and branches (Fig. 3.6), or intertwined with the trees' roots (Figs. 3.2 and 3.5) or branches (Fig. 3.4), sending fruit-bearing shoots up from the ground or radiating branches, replete with fruits, from the tree's canopy.

Here I will consider not pictures that are explicitly linked to stories, but, as in the discussion of therianthropes above, with images that, in some instances seem to stand on their own, tell their own stories, most, if not all of them unknown and unknowable. What these elusive "stories" have in common with those of myth is the theme of transformation, spun out in multiple and at times complicated ways, that, in their Dali'esque whimsicality and complexity, surpass the more static and staid therianthrope imagery. Indeed, as we will see below, it is one such, an especially "complex depiction" (Fig. 4.14), that elicited the somewhat exasperated rhetorical question—"How much of the art is now within our understanding?"—by two eminent San rock art researchers Lewis-Williams and Dowson (1989: 155), at the conclusion of one of their books on the meaning of San rock art, including therianthropic representations. I turn to the meaning question at the conclusion of this chapter.

Methodologically, there is no certain way to determine definitively whether an image of a therianthrope depicts the same statically, as either constitutionally hybrid or a transformed being, or dynamically, as one undergoing transformation or, as yet another possibility, as arrested somewhere in that process (as the previously encountered witch Ježibaba's were-raven, cat and rat assistants). As noted earlier, rock art images and engravings are stills; they arrest movement—unlike narrative, that can readily convey the same and does so most effectively in San myth and story, as seen in the previous two chapters. There are two ways I go about deciding whether a therianthropic image depicts transformation or merely constitutional hybridity. Both are impressionistic ways of determining this aspect of transformation, ranging from highly impressionistic, to moderately so.

The first is to look at a representation of such a figure in isolation, to gauge, from its overall appearance, its posture and gait, the tentativeness and nascence of its form, the extent to which certain body parts have mutated, whether or not the figure is forming or formed. This might strike some researchers as egregiously impressionistic, on the eyebrow- and hair-raising extreme, over-the-top, left of the "'aesthetic and expressive' to the 'verifiable and verbalisable'" spectrum of interpretations for Khoesan rock art (Solomon 2014: 712).

Fig. 4.9 Transforming therianthrope, human to antelope (?). (Woodhouse Collection, Special Collections, University of Pretoria, with permission)

In this vein, as a circumspectly couched "proximate" explanation of transformation of certain rock art images, situated left-of-centre on Solomon's interpretive spectrum, I present Fig. 4.9 as an exemplar: a human-antelope figure from the Drakensberg, a being not fully formed one way or the other, in a transitional ontological state. Antelope-eared and shaped head, somewhat human-shaped neck, torso and rump and human-kneed hind legs with feet that seem neither toed nor hooved, halfway between orthograde and pronograde and of wobbly gait and faltering steps and holding on to some branch or ledge, the figure seems not as yet quite to have found his or her feet, as the mantelope it seems to be morphing into.

Or vice versa: antelopes in myth and ritual are capable as well of mutating into humans, or back into the human they had been, before their antelope mutation. A case in point may be the therianthrope depicted in Fig. 4.10, from the Storm Shelter site in the southern Drakensberg, a transformation-informed rock art site, at which, contrary to other sites, "there seems to be an emphasis on the animal, rather than the human, components of therianthropes" (Blundell and Lewis-Williams 2001: 3). The image appears to depict this sort of human to animal transformation, of a pronograde antelope, whose hind legs and rump are in the process of assuming human shape and angle, all of it watched by the were-antelope with a backward glance.

Fig. 4.10 Transforming therianthrope, antelope to human (?), Storm Shelter, southern Drakensberg. (RARI, with permission)

Yet another transformation scene at the same zoomorphism-charged site is as striking as it is unusual. It depicts a polychrome figure, antelopean "in every aspect", who is seated on the ground, in human fashion, on his behind with his antelopean legs bent at the knee. The figure's head and forelegs are antelopean; like the head the figure's left leg is raised stiffly and is bent at the elbow like a human arm, its articulation confounding the foreleg/arm's antelopean aspect.

A being's transformation process is most evident in images that depict him—more infrequently her—wearing animal skins or karosses, as props during the shamanic tranceformation ritual and, as suggested by Jolly, fusing ontologically with their animal envelope and transforming into the animal from which it was obtained. Jolly illustrates this process with two striking paintings from the Eastern Cape, one depicting an individual figure, the other an active group scene replete with transforming beings from a variety of ontological taxomes. Because this transformative process that blends mimesis with metamorphosis will receive further, in-depth examination in subsequent chapters (and in Vol. II), I present both of Jolly's images, along with his annotations, as illustrations for that upcoming discussion.

The first figure, depicted by Jolly as a solitary individual (Fig. 4.11) but actually part of a busy group scene of what appear to be fully human figures (Fig. 4.12) in which the figure appears out of place, shows a striding human of indeterminate gender, walking with some effort and in opposite

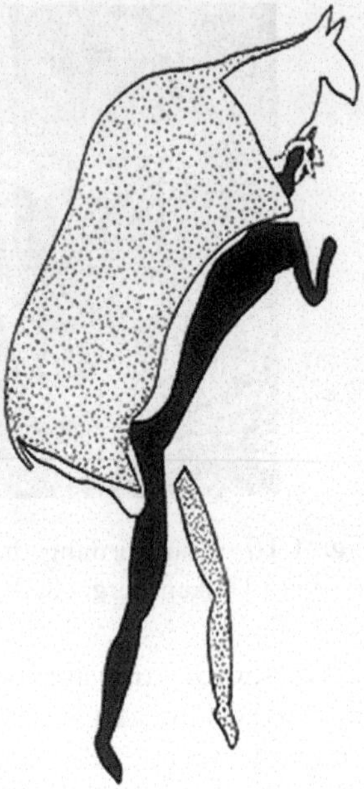

Fig. 4.11 Transformation scene of a man into the eland (skin?) he is carrying on his back. (Jolly 2002: 91, with permission)

direction to the other tall and dark-skinned, unambiguously human figures who appear to be dancing. The straining figure is seen shouldering the carcass or skin of a young eland on his back. The animal's throat and head cover that of the human figure as though absorbing the obscured human's head into the eland's. The animal's right leg consists only of its thigh, atop the man's thigh and in line with its bend and mass (something more evident in the photograph than the drawing). The eland's vanished (transformed?) left leg appears to have merged with the man's left leg, which, while still human in shape and knee's angle, has changed to the tawny colour of the eland's body, contrasting sharply with the reddish—human—colour of the right leg (Jolly 2015: 217). All this suggests transformation-in-progress, of different body parts, head and legs, none of it quite in synch. It is a pictorial version of what happened to /Xue in some of the transformation narratives encountered above, when feet, arms, fingers, head, sprout tree parts in different directions and at seemingly different pace, or to //Kabbo's story[4] about the arboreal transfor-

[4] See Chap. 3, Vol. II.

Fig. 4.12 Photograph of Fig. 4.11, in the context of the group scene it is part of. (Jolly 2015: 217, with permission)

mation of the hunter and *goura*-player in whose transforming body human and arboreal traits vie with each other. Jolly's apt caption for this image (Fig. 4.11) is in line with my reading of it: "man with eland skin, or carcass of a young eland, on his back, and in the process of transforming into this antelope" (ibid.: 91).

We see much the same scene in an image from the Drakensberg (Fig. 4.13) except that in this depiction of eland-kaross carrying shamans the head is fully transformed into its antelopean form while the human-shaped legs, projecting from the lower, straight edges of what still looks very much like an eland-kaross covering of the figure rather than an integral component of his form. This is the case especially with respect to the inner figure, as opposed to the outer one in which the eland back and torso seem less discrete. The two figures, especially the outer one, suggest a transformation-in-progress, in which eland-kaross has fused ontologically with the wearer's head and upper body (which also appears to sprout a second hoof-tipped forelimb), as opposed to the lower body where the kaross is still "cultural"—an artefact—and the legs still human. I am in

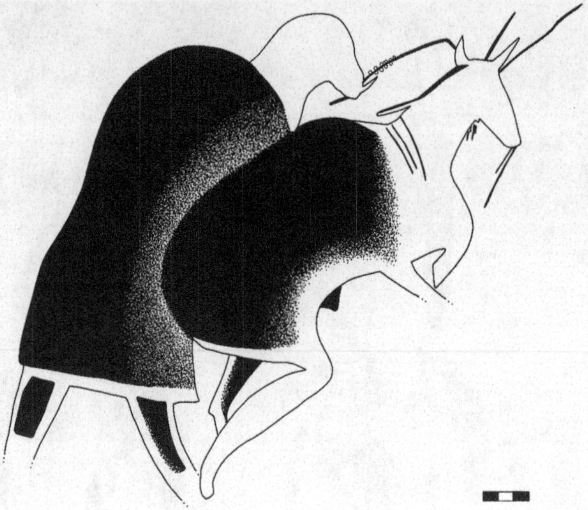

Fig. 4.13 Kaross-clad human-antelope transformation. (RARI, with permission)

accord with Lewis-Williams's general interpretation of this image's iconography: "shamans partially transformed into antelope" (1996: 130; see also 2015: 123). However, in place of the image's interpretation, as per the "trance hypothesis", that the transformation scene is a depiction of the shamanistic technique for "taking on the potency of an antelope", I suggest that the scene is yet another expression of the theme at the core of the artist's mythology and cosmology: ontological mutability (which, as I will argue in Chap. 5, may or may not be linked to trance).

The other image (Fig. 4.14) which Jolly drew from Lewis-Williams's and Dowson's *Images of Power* (1989: 154, 159) is an excerpt of a panel depicting a busy and surreal group scene of several dozen beings and shapes. The two authors present this "complex depiction" as the final image of the hundred-odd they discuss in their book and decide to "let it stand without detailed comment" (ibid.: 155), as an image with which to invite readers to ponder the question—"how much of the art is now within our understanding?"—the authors and the image raise. Here is Jolly's proposed answer, one that vividly conveys transformation as an unfolding process:

Fig. 4.14 Transformations scene, some of the beings wearing animal skins. (Jolly 2002: 91, with permission)

> people are shown merging with their karosses and transforming into mammals or reptiles, some of which have imaginary features. The figure at far right is shown with his front legs changed into antelope legs and hooves, which form part of his skin cloak and the figure at far left has also begun to fuse with his kaross. Both have adopted the four-legged posture of antelope, similar to the posture of dancers who have collapsed onto their hands and knees during trance. One of the other figures is in the process of becoming a cow or ox, another to be transforming into a snake with tusks and yet another into an unidentified animal with tusks. (Jolly 2002: 91)

Transformation-in-progress is suggested in similar fashion by the chimerical "human-antelope-feline therianthrope", as Edward and Cathelijne Eastwood refer to the "very strange creature" they found depicted at Shaman's Rock in the Mapungubwe area near the confluence of the Limpopo and Shashe rivers of north-eastern South Africa (Fig. 4.15). There is a certain tentative amorphousness to this semi-upright figure, as though it were still in the process of sorting out its multiple-species parts—"four antelope horns on its head, an animal's muzzle and the tail and general demeanor of a feline"—all of it in semi-upright stance, "akin to humans", with arms hanging down, half-ways between animalian quadrupedalism and human bipedalism (Eastwood and Eastwood 2006: 125). A similar indeterminacy hangs over the so-called Lion Man engraving at Namibia's /Ui-//aes (Twyfelfontein) rock art site, whose emerging (or submerging) humanity in what is a decidedly leonine being is evident in the humanoid five—as opposed to leonoid four—toes on one of his paws, as well as the end of his elongated tail (a sign, John Kinahan suggests, of tranceformation, when the subject experiences polymelia, a sense of possessing extra limbs [2010: 48]).

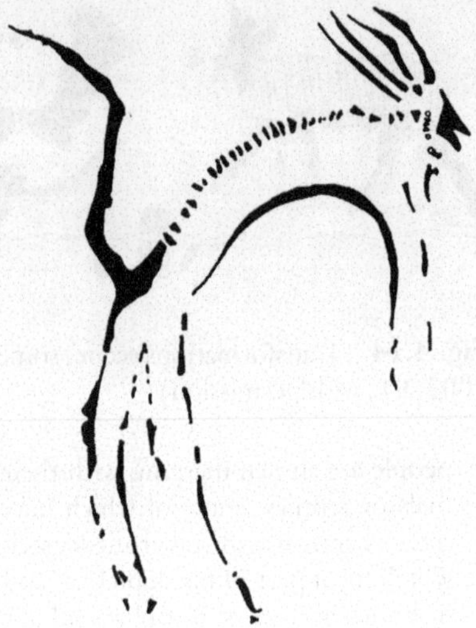

Fig. 4.15 Transforming chimera. (Eastwood and Eastwood 2006: 125, with permission from RARI)

The surmise that an image depicts transformation is somewhat less impressionistic when it shows multiple figures. An example, again from the Limpopo region, is a panel of wildebeest[5] that are depicted in various stages of transformation, conveying fluidity and process (Fig. 4.16). This comes across in the Eastwood's caption of a painting that, they suggest, depicts the transformation of shamans during a Wildebeest Medicine Song (which they found was still practised by some Kalahari Naro): "four of the therianthropes have the bodies and legs of human beings and the heads and front legs of wildebeests. Hooves replace the hands at the end of what should be 'arms'" (2006: 124). These ontologically mixed up features are unevenly distributed amongst the four figures, rendering some of them more, others less transformed: human feet and hands are found in the figures to the right which, apart from these features and a semi-erect stance, is fully wildebeest (including a tail). The same is found in another, smaller figure, except that it lacks horns. One fully erect figure in the centre front of the panel, on human legs and feet, tail-less buttocks and semi-

[5] Given its inherent, anatomically conveyed ontological heterogeneity and ambiguity—"head of buffalo, body and tail of a horse, with legs of an antelope" (Lewis-Willimas 2015a: 134)—the Black Wildebeest offers itself up to the imagination of the therianthropically inclined artist (or story teller).

Fig. 4.16 Transforming wildebeest-men. (Eastwood and Eastwood 2006: 124, with permission from RARI)

erect infibulated penis, stiffly holds out hooved front legs. His head is that of a gnu, including a full set of horns. The second such figure (on the left) is much the same, except it is squatting, in human fashion, rather than erect and resting its slightly forward-bent torso on a pair of stout, hooved antelope legs. His head sports gnu horns; however, the face as yet seems not quite or fully antelopean, having, what appears to be a long, drooping nose (in the process, perhaps, of morphing into a muzzle).

The surmise that an artist is actually depicting transformation is on firmer ground methodologically in multiple figure scenes that depict what seems to be the same being in different postures and transformational phases or stages. This can be seen, albeit only faintly as the image is itself blurred and somewhat faded, on what Jolly dubs "an antelope-human transformation scene" which he photographed in the Drakensberg (Fig. 4.17) and which the tracing in Fig. 4.18 helps to make more explicit. A row of ten figures is shown in "animation" style that depict both transformation and movement through the depiction of spatial depth. The tableau either depicts a row of creatures, more or less fully formed zoomorphs, walking in a row, off into the distance. Alternatively—and this may be unwarranted conjecture—it may depict just one figure, at various phases in its transformation from therianthropic antelope, becoming more and more erect and, in the middle figure, with human torso and antelope ears and hind legs, and then morphing back into its antelopean form.

Fig. 4.17 Antelope-human transformation in process. (Photo by Jolly (2015: 212), with permission)

Fig. 4.18 Tracing superimposed on Fig. 4.17

Fig. 4.19 Transformation into a bird. (Jolly 2002: 94, with permission)

Jolly offers the same sort of interpretation for the striking image he traced and photographed at Aliwal North in the Eastern Cape (Figs. 4.19 and 4.20), which "shows the transformation of a person with a quiver of arrows on his back into an antelope therianthrope with feathered legs, shown in a third stage of transformation, in a flying position, and finally transformed into a bird, the arrows having changed into wings" (2002: 94).

In the photograph Jolly published of this painting (Fig. 4.20), Jolly describes the scene as "a shaman grows feathers and transforms into an antelope-headed bird" (2015: 216).

Turning from the Cape and the Drakensberg in the south to the Brandberg in the north, we find a striking example of this sort of depiction of transformation (Fig. 4.21), at Amis Gorge in the Upper Brandberg (Pager 1989: 322, 482, plate 60). Depicted, in a panel of over 50 representations, are eight therianthropes (figs. 53, 55, 56, 58–61 on plate 60), arranged in a semi-circular row, all pointing, in their "coordinated action" of progressing metamorphosis, in the same direction, to the left. This is opposite to that of the eight purposefully striding humans, largely male

Fig. 4.20 Photograph of Fig. 4.19. (Jolly 2015: 216, with permission)

figures, whose clearly determined human status presents a foil to the indeterminate ontological status of the figures below. Pager describes this therianthropic scene with the following annotation:

> In a semi-circular order one sees, it would seem, several states of metamorphosis of human and buck into a therianthrope with features of both these species. First in line is a human with head décor (not indicating ears), carrying a bag and club under his arm. To his right is a buck, the only uncommon features of which are the front legs with human feet. Behind this buck one sees a standing human of whom only the body decoration and the "wig" are preserved. Below this, there are some figures in bent-over position displaying different kinds of human and animal features. Some of these could be masked men dancing but others are for the most part animals. Only one figure does not join in this apparently coordinated action … a therianthrope with human legs and the body of a buck from the waist upwards, running in upright position. (ibid.: 322)

Especially striking in this animation-style transformation sequence is the depiction of the legs of the bottom three figures, at the lower curve of the semi-circle, the transformation of which, from hooved to footed hindlimbs and straight antelope forelimbs to arm bent at the elbow. The

Fig. 4.21 Phases of transformation, Amis gorge, Upper Brandberg. (Redrawn from Pager 1989, plate 60)

double-bend of the left arm of the third figure from the top, human-legged and antelope-headed, invites comparison with the forelimbs of a *Mantis religiosa* and, taking an even further and bolder leap into conjecture, with the /Xam trickster (a leap, I grant, that is as unwarranted as it is beguiling given Mantis-/Kaggen's unique mythological status and limited range, along with more than a few other objections).

Further to trickster and Brandberg—and wild surmises—I point to a figure I traced on the Brandberg during a pilot study I undertook of the rock art of the Hungorob and Nuwuarib gorges of the Upper Brandberg in 1983. The figure (Fig. 4.22), heuristically, payfully—and quite gratuitously—labelled "trickster" by myself and my three field trip companions (Helmut and Marianne Reuning and Jo Walter) who had been up the Brandberg before and seen the figure, the figure seems to undergo transformation. This reading suggests itself by the being's fluid, multi-component, seemingly morphing form—yet another chimera it would seem: human legs, torso and arms, vaguely antelopean feet, and long spindly vaguely avian feet or human fingers, a heavily prognathous face with long, forward-pointing ears and an over-sized, trumpet-shaped penis. The last is the being's most prominent feature—as is the bow and quiver he carries around his waist.

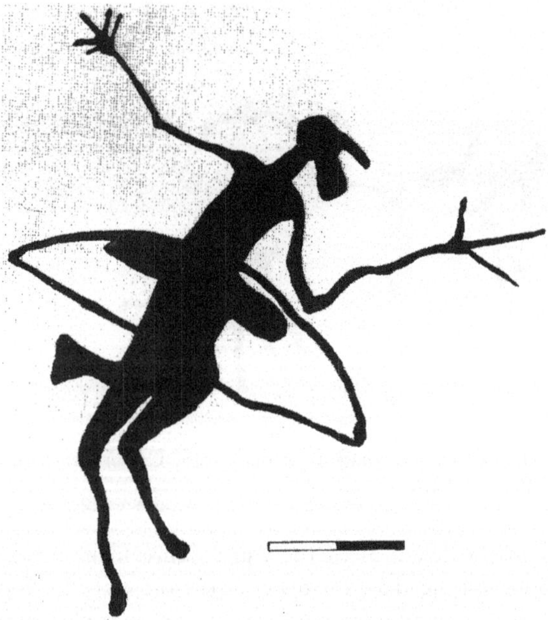

Fig. 4.22 Transforming "trickster"-hunter, Upper Nuwarib, Brandberg. (Source: author's tracing)

Shamans or Hunters, Trance or Transformation?

Whether the figure above is a depiction of a mythological trickster being—of which there seem to be very few notwithstanding the prominence of this being in myth[6]—or simply a shaman—or a hunter—undergoing transformation, what makes the image interesting is that it links the zoomorphically transforming being, who carries a hunter's toolkit, not to trancing but to hunting. A similar link is suggested by the other Brandberg procession of therianthropes we examined (Fig. 4.21), which has a hunter, carrying his hunting bag and club, and an antelope, both walking, leading the parade of transforming antelope-humans (or the transformation process of one such creature). It is a tableau that, once again, evokes a hunting rather than trancing scene, transformation by a hunter instead of tranceformation by a shaman. Notwithstanding Jolly's tranceformation-cast explanation for the complex therianthropic scene depicted in Figs. 4.19 and 4.20, here again the animal-transforming human, with "a quiver of arrows on his back", may not be a shaman at all but a hunter, undergoing the experience of deep sympathy with his quarry, which, as I will argue in a later chapter (Chap. 7), may approximate or culminate in, species-transforming metamorphosis and self-transforming altered states of consciousness.

Turning to the southern San once again, we come across paintings that suggest the same interpretation for therianthropic imagery, that is, in terms not of the shaman in trance but the hunter in transformation. What is depicted are scenes or tableaux of immobilized, dying or dead elands, surrounded by both bow-equipped hunters and, in the same scene, therianthropes, placing the latter into a hunting rather than trance healing context. These images can be seen in Patricia Vinnicombe's recently republished and now accessible *People of the Eland* (2001: 157, fig. 90; p. 175, fig. 102; p. 179, fig. 106; p. 182, fig. 109). The first is a complex panel depicting half-a-dozen eland, along with about a dozen humans. One of them is striding, one running, two are standing and two dancing. Another two—perhaps three—are hunting, two hunters with bows

[6]The absence of actual depictions of mythic trickster beings does not mean that this being—a cognate in so many ways to the oft-depicted trance dancer (Guenther 1999: 4)—was not an implicit presence in San rock art. Getting his clue from an intriguing note he found in one of Dorothea Bleek's notebooks on her 1911 journey to the Northern Cape to find descendants of her father's and aunt's /Xam informants stating that "/Kaggen makes the pictures", Lewis-Williams explores this presence, on the basis of the proposition that "the shaman-painter was, as it were, 'standing in' for the Mantis" (2015a, b: 170).

stretched and pointing towards a massive eland in the centre of the scene. One of the hunters—standing with a quiver-full of arrows on his back and eight arrows held in his hair—has antelope hooves and a buck head. Another, more fully transformed figure, has antelope legs and head, the latter sprouting a dozen arrows.

Another such scene, again from the Drakensberg, one exceptionally striking and aesthetically pleasing (Fig. 4.23) appears in Lewis-Williams' *Believing and Seeing*, the book in which he launched his "trance hypothesis" (1981). While the scene to Lewis-Williams depicts shamans in trance, drawing healing potency from the animal or animals he transforms—has transformed—into, an alternative reading suggests itself: a depiction, by a hunter-artist, of the mimetic-metamorphic sympathy mode hunters may experience at certain moments in certain types of hunts. What we see is two bow-carrying therianthropic hunters striding amongst five elands, whose hooves leave tracks that intermingle with those left by the hunters' antelope-hooved feet. The picture enigmatically and ambiguously—as is San myth's and art's wont—underscores and grounds mystical transfor-

Fig. 4.23 Therianthrope and eland tracks, Drakensberg. (RARI, with permission)

mation sensed by the hunter within what is for him physical evidence for a human or non-human being's presence: tracks.

Animal tracks appear on many other rock art paintings and engravings. Such images link San art and the artist who produces them to hunting directly, immediately and intimately (doing so, as we will see in Chap. 6, in terms of the partially transformative hunter-prey sympathy bond, which tracks and tracking may elicit). The metonymic connection between San art and hunting that is evident in painted or engraved animal tracks complements the more removed metaphorical connection of this motif to potency and trance (Ijäs 2017: 180–91).

It suggests as well that the "hunting magic" model for interpreting San rock art which the shamanism-based "trance hypothesis" rejects needs to be "reconfigured" (McGranaghan and Challis 2016; see also Thackeray 2005: 15–6), in terms of an "animism"-based model focused on the hunter-animal prey relationship rather than one based on shamanism and focused on the healer-human patient relationship. The Finnish visual artist and art historian Mikko Ijäs in his recent (2017) study of San rock art (along with European Upper Palaeolithic cave art) has done just that, suggesting that shamanic trance healing ritual developed from the experience of altered states of consciousness brought on by persistence hunting—"the hallucinatory hunting experience of transformation into an animal"—which he deems the primal hunting technique of Palaeolithic and Holocene hunters (and researched and participated in among Namibian Ju/'hoansi). This idea will be examined more fully in Chap. 6.

Contemporary San Easel Art: Therianthropes

Unlike the rock art of the past, by long vanished artists who cannot be consulted on their creative intentions, many of the modern-day artists are around and anthropologists and art historians and critics have talked to them about the whats and the whys of their pictures. Indeed, in tackling the meaning question of pre-colonial San rock art whose artists cannot be consulted on this elusive matter, it is not only instructive to enlist contemporary San peoples themselves but it is, to some seriously minded, culturally sensitive scholar (such as the South African Khoisan art critic Nyasha Mboti (2014), imperative to do so.

This has been done before, in Bleek's and Lloyd's time, by the same two researchers themselves, when they showed Diä!kwain and /Han‡kasso copies of rock art that had recently produced by some researchers (Orpen,

Stow and Schunke) and sent to Bleek for comment. Wilhelm Bleek showed great interest in some of them—"on account of their mythological bearings"—and their assumed complementarity to the myths and the likelihood, thus, to "throw light upon many things hitherto unintelligible" (Bleek 1875: 20; see also Bleek and Lloyd 1911: xiv; Lewis-Williams 1981: 7; Bank 2006: 302–39). Two of the sets of copies, obtained by Stow near Queenstown (Komani) in the Eastern Caoe in 1868 and 1870, depict therianthropes on one of which showing gemsbok-headed humans, /Han‡kasso ventured the opinion—one which Andrew Bank deems somewhat gratuitous as it might have been coached, subtly and inadvertently, by Lloyd through certain /Xam stories she was recording and mulling over at the time (ibid.: 316–7)—that the image (Fig. 4.24) depicted men wearing caps made from gemsbok heads (young ones, "the long horns being a little too long, they take horns from the young gemsboks, which are short"). The story tellers' engagement, through such second-

Fig. 4.24 Stow's copies of therianthropes. (Digital Bleek and Lloyd, image files Stow_015 & Stow_017)

hand images of San rock art shown to them by their interlocutors, not only elicited further stories, or commentaries on the same, but actually led some of them to producing images themselves, with pencils on paper, which Bleek and Lloyd provided to them. A number of these drawings were illustrations for stories (see previous chapter). Most of these are reproduced in Bleek and Lloyd's *Specimens of Bushman Folklore*.[7]

Ethnographers (myself included) have ever since shown San people they work with pictures of San rock art, much the same way they might read to them stories from books on San folklore—as did Dorothea Bleek, to her Naro informants when she did her field work at Sandfontein in 1921 and 1922, translating and reading to them stories from her father's and aunt's *Specimens of Bushman Folklore*, which "reminded some old people of this or that tale they had heard in their youth" (1928: 44).

Contemporary San artists, too, consult such books, as well as visit rock art sites, for ideas, motifs and inspiration in their own work. A striking example is the late ≠Khomani artist, Vetkat Regopstaan Boesman Kruiper (2014; see also Kruiper and Bregnin 2004; Lange et al. 2014: 376–7), whom art critic Nyasha Mboti sees as engaged "in conversation with rock art", expressed by the artist in what she refers to a "Kalahari style" ("in the sense of Kalahari visuality") (Mboti 2014: 484). A frequent motif in Kruiper's art, which reflects his engagement with ancestral San rock art, is animal-headed therianthropes. Fig. 4.25, one of many examples, depicts jackal-headed and tailed dancers. Mary Lange, who included this specific image in one of her publications (2011: n.p.), sees it as "emphasizing the liminal or transitional space between human and animal" (pers. com. 2019, 7 March, 2019). Vetkat's therianthropes have the head of jackals, indicative of Vetkat's ≠Khomani's people's "identification with the jackal and its outwitting of the lion (the Kgalagadi Park management and all potential oppressors, past and present)" something Lange found to be a prevalent theme in ≠Khomani story telling performance when she visited the people in the early and late 2010s (Lange et al. 2014: 376–7) (as I did among Ghanzi Naro, 50 years earlier). The artist reportedly drew on dreams in some of his art and, in that state, could himself become transformed into an animal, such as "a large grey cat moving in the twilight with a small warm light glowing from inside his temple" (Scheepers 2014:

[7] They were the subject of an exhibition, curated by Miklós Szalay (2002), at the University of Zurich's Museum for Ethnology in 2002 and, in 2003, at the South African National Gallery, Iziko Museum, Cape Town.

Fig. 4.25 Vetkat Kruiper's therianthropes. (Lange 2011: n.p., with permission)

22). He also connected the therianthropic motif to the trance dance, a recurrent theme in the work of this artist that infuses some of his pictures with sacred portent (Northover 2017). A number of his therianthropic dancers are depicted in an earthy—scatological, erotic—manner, evocative of the Rabelaisian tenor of some of these figures in San mythology (specifically the trickster beings).

The Naro and ≠Au//eisi artists at D'Kar village in western Botswana's Ghanzi District, who, under the auspices of an NGO (Kuru Development Trust) in the early 1990s, embarked on what has become a highly successful art project consisting of paintings and prints, too, engage to some extent "in conversation" with rock art, albeit to only a limited extent and without any marked or direct effect on the images they create. They do so not only through books, which are readily available in the library section of their museum-*cum*-gallery at D'Kar, but also through "inspirational trips" Kuru organizes, to such rock art sites as Tsodilo and Oliphantsfontein in Botswana and the Brandberg and Twyfelfontein in neighbouring Namibia.

Regarding the last of these "cultural animation" initiatives, which occurred during the period in the mid-1990s when I carried out a research project on the Kuru art project at D'Kar, I met with the group during their debriefing session the day after the trip to and up the Brandberg, when the trip members picnicked on the beach at Swakopmund and relaxed from the ordeal of the climb, of what is Namibia's highest mountain top. It was evident that the trip had indeed animated some of the participants. They related to and engaged with the paintings and engravings and agreed that the artists must have been San and wondered how they might have survived on the mountain, as the Kuru people were unable to find any familiar edible plants anywhere in the region. Different explanations were offered for the pictures by the Kuru travellers: that the animal and human depictions recounted certain events, such as hunts or the arduous experience of climbing the mountain; that the paintings were used to demarcate a family's or band's living shelter; that they kept a tally of the animals hunted; that they depicted scenes of the trance dance or the trance experience. As if to underscore the last reading of the paintings, the group performed a trance dance on the mountain that same evening as lyrically described in the report on the trip that was written by Kuru staff for the Cultural Centre:

> That night, the mountain echoed once again with the long lost rhythms of ancient dances. Next to a small fire two women were clapping and singing their yodelling songs, accompanied by the stamping rhythms of the men's feet as they danced their familiar dances to the mountain which keeps the memories in the heart of its age old rocks. (KDT 1995: 15)

During the debriefing session I asked two of the artists if they had gotten any ideas for their paintings from the pictures at these two sites, or if they had perhaps made sketches while they were up on the mountain or at the Twyfelfontein cliffs. They said they had not and that in their next paintings they would not paint any of the things they had seen at these two places, that is, not in any deliberate way, as qualified by two of the artists who explained to me that, in as much as the pictures were "in their heads", they probably would in some way affect what they drew or painted. Thus, while the sites did not serve as a source for specific motifs or ideas in any outright way, in an indirect, desultory fashion they might have left an imprint, both stylistically and substantively, on the work of some of the Kuru artists (one of whose favoured motifs, animal hoofs, the then-director of the Kuru art project, Pieter Brown, attributed to the artist's visit to the rock art site at Twyfelfontein some years previously). Instead of providing concrete subject matter right then and there, the trip to the rock art site valorized and confirmed to the artists that what they produced was indeed Bushman art, pictures that are in tune, in a general way, with an established San art tradition.

The therianthrope motif was, at the time I conducted my study, a part of that tradition in the Kuru oeuvre, at an incidence of about 3.5% (which coincides roughly with its frequency in ancestral rock art). It was used with different artistic intentions by different artists, either narratively, to depict a mythological theme or a dream, or aesthetically, as a whimsical, beguiling image or design.

An example for the former was the artist Qoma who tended to refer to such were-beings as "the Old People", the title he gave to a painting depicting, among other hybrid beings, a winged, animal-eared humanoid, a heavy-snouted or beaked, bird-tailed creature with human arms and hands and what appears to be—the artist had died a few years before my study and I was thus unable to interview him—a short tree with leaf-handed arms and a jackal-like head (Fig. 4.26). Another of this artist's pictures juxtaposes a gaily-patterned gemsbok and two kori bustards with an avian mammal that recapitulates, in its form and markings, the latter two birds and the buck (Fig. 4.27). My reading of Qoma's art, as informed with myth, specifically its aspect of ontological mutability, resonates with Jessica Stephenson, an art historian who studied the Kuru art project in the late 1990s to early 2000s and recently described this artist's work in the following terms:

Fig. 4.26 "Old People", Qoma. (Kuru Art Project Archives, with permission)

> In his works, Qoma populates the First Order, the primal world of San cosmology and of ambivalence, with hybrid human-animal beings and fantastical plant forms that float on an unidentified ground, moving in and out of time and space ... By setting narratives within the cosmological landscape of the First Order, Qoma draws on the transformative and subversive power of myth and religion. (2016: 202)

When I studied the Kuru art project in the mid-1990s I found the motif of therianthropes to be more frequent in the work of the male artists than the female ones (where it was all but absent) and unequally distributed among the former (Guenther 2003: 138–83). Some of the men never depicted this motif, others did so with the imagination and flamboyance of the anonymous /Xam painters, in the form of such wildly hybrid beings as a buck-headed bird, bird-footed springhare, bird-fish, bird-mammal with flower tail, bird-footed humanoid, caterpillar dog, bird-footed horse, crocodile-headed warthog. An example of an artist given to such were-beings is the late Thamae Setshogo, as in the top row of one of

Fig. 4.27 "Gemsbok and Birds", Qoma. (Kuru Art Project Archives, with permission)

his pictures that depicts a row of two-legged, antelope-footed, crocodile-tailed, giraffe-necked and headed chimerical beasts (Fig. 4.28).

I close this gallery of Kuru therianthropes with one more example, from the work of one of the women artists, in which, as noted above, this motif was rarely used. The artist is Coex'ae Qgam ("Dada"), whom we will meet again in the next section as she was much drawn to the related motif of transformation which *sui generis* would lead to depictions of therianthropic figures. Titled "Birds and a Tree" (Fig. 4.29), the depicted birds, with their widely spaced legs and their horizontal stance, appear as much mammalian as avian.

The difficult-to-determine question as to whether or not these mammavian figure are therianthropes, or "stills" of a transformation event of either ontological taxome into the other brings us to the next section, transformation.

Fig. 4.28 "Untitled", Thamae Setshogo. (Author's photograph with permission)

Contemporary San Easel Art: Transformation

Some of the contemporary San pictures are vibrant with transformation, a theme more prominent and elaborated than that of therianthropes in the work of two of the artists, which, as just seen, too can be found in the modern San oeuvre. These two artists are fascinated by transformation and dwell on this theme and explore it in intriguing ways. They do so either formally or substantively, as an element of style and aesthetic expression, or of myth and ritual—as well as, in one instance (Fig. 4.41), obliquely to rock art, further linking the two bodies of San art, past and present.

When interviewing the 15 Kuru artists in the mid-1990s, while they were working on their paintings (Fig. 4.30 depicting Qwaa and Dada[8] at work), or afterwards, looking at recently completed work or pictures on

[8] The artist at right is first-Dada's cousin Coe'xae Bob, a fellow-artist.

Fig. 4.29 "Birds and a Tree", Dada. (Author's photograph, with permission)

Fig. 4.30 Qwaa and Dada (L) at work. (Author's photograph)

exhibit on the walls of the Kuru NGO's museum at Dekar, I routinely asked about the motif of therianthropes and the element of transformation. The latter, faint or absent from the work of most of them, was striking in that of two of them, one a woman, Coex'ae Qgam, the other a man, Xg'oa Mangana, whose *noms de plume* (*de brosse*?), respectively, were Dada and Qwaa, applied as signatures to their paintings and prints. These two artists were in their senior years—older by one or two generations than most of their colleagues—at the time of my study, and both have since passed away.[9]

Both were also interested in *hua*, the ancestral myths and stories of their people, which they told to their grandchildren. Qwaa quite explicitly brought *hua* into his art, along with shamanic dreams and visions: "men with bird heads" and "beings with upper-bodies only and very large, round eyes", as he related to the South African artist Catharina Scheepers in a conversation she had with Qwaa. At the time she also talked to Dada who in remarking on a baboon painting Qwaa was currently working on that "Qwaa is drawing these baboons today because all of last night he was walking with them" (Scheepers 2014: 21). Two-thirds of his paintings, in my mid-1990s' sample, consisted of paintings with either explicit or oblique mythological content, which he identified and, being as interested in the ancestral stories as he was fond of telling them (and of talking in general), at times dwelled on the mythological aspects of a certain picture when talking to me about it. One of these, one rich in transformation, was in the process of painting when I was there in 1995 allowing for probing discussion of the images as the painting developed, over the course of close to four weeks (Figs. 4.30 and 4.31).

Both artists were also, or had until recently been, trance dancers. When he was a grownup man Qwaa learned trance dancing and curing, which he effected by "pulling [i.e. sucking] sickness out of a person's chest". He became skilled and sought after as a dancer and curer until stiff legs and sore knees forced him to give up the dancing and curing trade. One of Qwaa's early self-portraits depicts him in a trance state, arms outstretched and concentric circles—"buttons"—whirling about (Guenther 1998: 125,

[9] After their deaths biographical accounts were written on both of these artists, attesting to the significance of their work and its impact on both contemporary San art production and on current social, political and community issues, including San identity politics to which, in some instances, contemporary art is linked (Guenther 2006; Thomas 2016). On Dada see Golifer and Egner (2011), on Qwaa Guenther (1998).

Fig. 4.31 Qwaa's lion-shaman climbing *ghudabe* tree. (Author's photograph, with permission)

fig. 4). He continued to attend dances, though, and he maintained his interest in trancing, even though he had switched in recent years from this strenuous mode of curing to one more suitable to his age, herbal medicine. I was present once at a session of "shop-talk" with another trance dancer, at which he sang the dance song and gave forth the *kow-hi-dile*—the "death shriek" (to show that he could still do it). He was also an accomplished dengo (thumb piano) player, occasionally accompanying his playing with a hummed or sung tune, from one or another of the trance curing songs that were used by shamans of the Ghanzi region, a creative activity that may have been instrumental in eliciting altered states and evoking myth times.[10] An interest he acquired from his career as trance dancer was a preoccupation with the spirit world. He has had more than one encounter with the spirit and veld god //Gãuwa (Dxãwa), either at night, when walking through the veld or village, or at a dance, where he is always nearby ... as always when you dance you are together. He knows the proper way to approach this quixotic and oftentimes dangerous spirit being, by whispering your addresses to him, rather than talking aloud. To do such is to arouse his anger and to glare at you balefully, his head swelling in size (like the hood of an agitated cobra). Being under his angered glance will constrict your throat and make you choke and bring you sickness. One of his paintings depicts //Gãuwa, in just that state of dangerous vexation (Guenther 2006: 168, fig. 5).

Dada learned to trance dance from her father and younger brother Cg'ase as a young woman. Her shamanic activities began with the songs—including the Ju/'hoansi-derived *nqabe* (giraffe) dance—which she sang for other trance dancers until she turned to dancing and trancing herself. She described to me the stabbing pain of the spirit arrows that enter into her sides when she dances, and the boiling sensation in her stomach. She then gets up and dances and touches the sick person. She can see through the dark, Dada informed me, "the opposite of sun glasses, as though you are wearing spectacles that penetrate the dark"; moreover, when in trance you can look through people, "like x-ray vision". She dances until she collapses, when the spirit goes out of her body and her eyes look inside, not outside. Her spirit will now go away, to god and to the spirits of the dead. These Dada refers to as "*dxãwa*" (*//Gãuwa*), and regards as secondary

[10]As reported in a fascinating paper on Ju/'hoansi dengo playing by Megan Biesele (1975) for some dengo players playing their instrument may be a means for going into trance. Her article features one such player, the Ju/'hoan shaman "Jack". Biesele's article invites comparison—and further field research—with players of the same instrument among the Shona of Zimbabwe, as per the account of Shona *mbira* (thumb piano) music making by ethnomusicologist Paul Berliner, who himself learned to play the instrument during his field work. Shona players reportedly use the instrument not merely as a means for producing musical sound but also, "animistically", as a non-human musical partner with whom the player engages, reflexively, in a musical dialogue. So animated, with soul and creative agency, as partner in the player's duet, the instrument is capable of evoking mythic time, of ancestors or recently departed relatives, all accompanied at times with intense emotions ranging from sadness to joy. Included on the played mbira's emotional scale is a "state of consciousness transformed"—"like that of smoking marijuana", a Shona informant explained to Berliner (1978: 133).

spirits, or, collectively, as a secondary healing spirit.[11] To Dada "everything that is happening [at the trance dance] is from God; *//Gãuwa* is only in the background" (a view that may reflect the Christian beliefs she acquired from the Dutch Reformed Missionary and mission church active at D'Kar at the time). Dada is still actively involved in trance dancing; indeed, a week before my interview with her about these matters in July 1997, she and her brother had both treated a sick woman at D'Kar. She has done so on many occasions, ever since she learned trance dancing as a young woman. She danced and cured at various farm villages throughout the Ghanzi. And beyond: at one time, 20 years ago, she and Cg'ase went to Gaborone, on the occasion of then-prime minister Seretse Khama's illness (to which he succumbed in 1980), to dance so that he would get better.

I surmise, with more or less explicit confirmation from the two artists' commentary on certain pictures, that the prevalence of transformation in the work of these two artists may be linked to their advanced age, their cultural conservatism, their interest in stories and their shamanistic activities and proclivities. While in many ways novel and in some ways "invented", the pictures of these two contemporary artists, produced in a studio operated by an NGO for purpose of community development and as a source of income for the artists and revenue for the organization, also shows continuities, conceptually, substantively and stylistically, with traditional San cosmology and religion.

Qwaa's mythological pictures may be elaborate and evoke narratives from the artist that are either expansive and embellished—and not always consistent as there were times when Qwaa would give me more than one voluble account of the same picture—or that are subtle and understated. The former is illustrated by Fig. 4.31, a big canvas on which we see the artist at work (in June 1995) (Fig. 4.30). The rows of splendidly decorated tortoises in the upper third section of the picture "have no *huwa* in them", Qwaa told me. The bottom part of the picture does. It depicts baboons on a flowering *ghudabe* tree found in certain regions of Ghanzi, a tree baboons are drawn to and will frequently climb because they like to eat its nutritious and health-conveying fruit. Gemsboks and elands, too, like to eat it, Qwaa told me, but not humans, who do not like the taste. Trance dancers, too, are drawn to the tree and its medicinal fruit, which they use in curing. This fact is the key to one of the figures shown on the picture, the one on the right side of the tree, beside the one encircled by

[11] Attesting to the wide inter-personal variability of San belief, Dada's idea here seems somewhat at variance with what other Naro believe (Guenther 1999: 95–125).

the bottom tree branch. Even though the painting was titled "Wild Cat and Baboons" after its completion and does indeed depict the latter animals, this one figure is not, Qwaa points out, actually a baboon, but a trance dancer who has transformed himself into a baboon-like lion—a "dancer lion"—and is climbing the *ghudabe* tree to eat its fruit, as it will give him strength as a healer. Given his penchant for "walking with baboons" at night, it is tempting to view this image as autobiographical, of Qwaa as dream- or trance-transformed trance dancer, on an out-of-body journey on a shaman's errand.

Another picture "with a story in it" is seen in Fig. 4.32. At the face of it, it seems simply a depiction of veld animals, an impression confirmed by the picture's title, "Jackal, Sprinhare and Ostrich". However, in the course of a discussion of these pictures, Qwaa told me that he "just drew this picture the way he did because there is a story on top of the picture" and that story is "what the picture really means" and how he would talk about it to his grandchildren should they ask him the story about it. It is the story about a leopard, a spot-less one, depicted by the animal figure on top, the one beside the dark-backed mammal to the right, which is the

Fig. 4.32 "Jackal, Springhare and Ostrich", Qwaa. (Kuru Art Project Archives, with permission)

Fig. 4.33 "Naro men and animals from long ago", Qwaa. (Author's photograph, with permission)

jackal. Even though it has no spots, the animal beside the jackal—who already has his markings—is a nevertheless leopard, as he looked at one point in the "First Time" in the mythological past, when animals looked different from what they look now, and spoke and acted like humans. The myth-leopard he shows on his picture, Qwaa said, is actually in the process of transformation, into an animal of the "present time". It has already shed its human traits and become a general animal; as it transforms itself further it will become more and more like a leopard, down to its spots. The other animals, too, have recently transformed themselves from the mythological beings they had been before; they are further along in their transformation than the leopard. The ostrich is lagging behind as its feet have partially retained their human shape.

The mystical and mythological transformation theme appears again in two thematically similar pictures, titled "Naro men and animals from long ago" (Fig. 4.33) and "Naro men and Kalahari creatures" (Fig. 4.34). As

Fig. 4.34 "Naro men and Kalahari creatures", Qwaa. (Author's photograph, with permission)

explained to me by the artist, the two large, fuzzily outlined and vaguely spectral mammals in the former picture, once again, are jackals from the "First Time" and the human figures in the picture are all //Gãuwani (or "Satani"[12])—who, in two cases, are in the process of changing themselves into animals, the one into a tortoise, the other into a bird. "And what about the rectangular shape?" I asked, pointing to it, on the bottom left of the picture. "That's a wallet", he said, with a chuckle, but without further elaboration (adding, even to this picture of "long ago", an element of modernity, in his own, wry fashion[13]).

[12] Most of San residents at D'Kar tended to conflate their traditional trickster figure with this figure (or, in some instances, even with "Jessu Kriste") from Christian religion to whom they had been, or were being, introduced by the resident evangelist (Guenther 1997).

[13] This sort of image, a combination or conflation, or, most frequently juxtapositioning, of modern with traditional elements was one of this artist's hallmark traits, giving his pictures both a post-modernist and, as I argue elsewhere, post-colonialist twist (Guenther 1998).

His picture of Naro men and Kalahari creatures (Fig. 4.34) is also teeming with ontologically inchoate //Gãuwani creatures, one of which, on the left, is changing into a bird. The bird in the bottom right corner is a fully transformed bird, Qwaa points out (the extended, somewhat undulating neck of which, that makes the bird tilt forward and downward, complements the image of an approaching snake, suggestive of yet further ontological entanglement). As for the large square shape on the top right, it, like the wallet in the other picture, is from the real, modern world: it is a table cloth.

Turning to Dada, we meet here an artist who, while elaborately employing the transformation element in some of her paintings, does this in ways that are quite different from Qwaa. Her artistic purpose is altogether different: transformation is for her a matter not of substance—as it is for Qwaa—but of style, governed more by aesthetic choice, rather than narratively or conceptually, by myth and cosmology. This is underscored by a practice when employing this stylistic technique the artist is fond of: combining abstract shapes and patterns—one of her favourite motifs—with representational ones, such as birds or plants, pots or musical instruments.

One such combination is exemplified by her painting titled "Disco Dancing" (Fig. 4.35), a busy, colourful scene painted with bold strokes

Fig. 4.35 "Disco dancing", Dada. (Author's photograph, with permission)

onto a linen sheet. The painting depicts three rows of about 30 people, men, women—a couple of them with babies on their steatopygous buttocks and a couple of them pregnant—and children. All are engaged in lively dancing. In amongst this busy human scene are two sets of shapes, block letter Ms[14] and wiggly ovoid shapes. The former appear below the first row of dancers, some of whom, the male ones, standing with arms stretched out or akimbo and legs apart. The M-shapes recapitulate, in reverse, the shape of the men's legs, which, in the paired dancers on the top left, form the same letter-shape. The ovoid shapes, if looked at from right to left, become more and more abstract, starting as the heads, shown in profile, and necks of two singing figures. The singing head profiles become more abstract, as—discrete and for the most part severed from their bodies—they are transformed into wiggly, irregular ovoids, shaped more or more discernibly, like the number 8.

Dada was fond of such transitions and transformations, of beings and objects into other beings or objects or into irregular shapes. This "morphing technique" allows for much creative scope and playfulness and brings much originality and diversity of forms and colour to the artist's work. We see it at play in, for example, her guitar picture (Fig. 4.36), the prominent subject of her painted cloth, at the centre of the composition. It fills the central third of the canvas and is presented in realistic form and in black outline and colours. To the left and right are two and three rows of smaller shapes. The two rows on the extreme right are 15 small guitars, all of them quite stylized, and painted, in various ways, in the same colours as the central guitar. The third, inside row on the right, to the left of the stylized guitars, consists of five shapes which depict the body of the guitar, in even more stylized fashion, as the eye moves from the bottom to the top of the row. At the top the shapes curve to the left and continue in the same sweep over to the outermost left row, in which the body of the guitar morphs into that of a pot. The rows of a dozen pots, each with a circle in the centre, as in the guitar, is also surrounded by flower petals or sunlight rays, which is the decorative pattern on most of the morphed guitar-pots. They culminate in one large pot, the size of the central guitar's body and recapitulating its black lattice markings. In it stands a tiny human figure and from it grows another recognizable shape, a squat bottle, the final shape in this morphing progression from guitar, through pot, to bottle.

[14] Letters, as well as numbers, are occasionally used by this Kuru artist (as well as others) in this formal way, for their shape, as aesthetic embellishment.

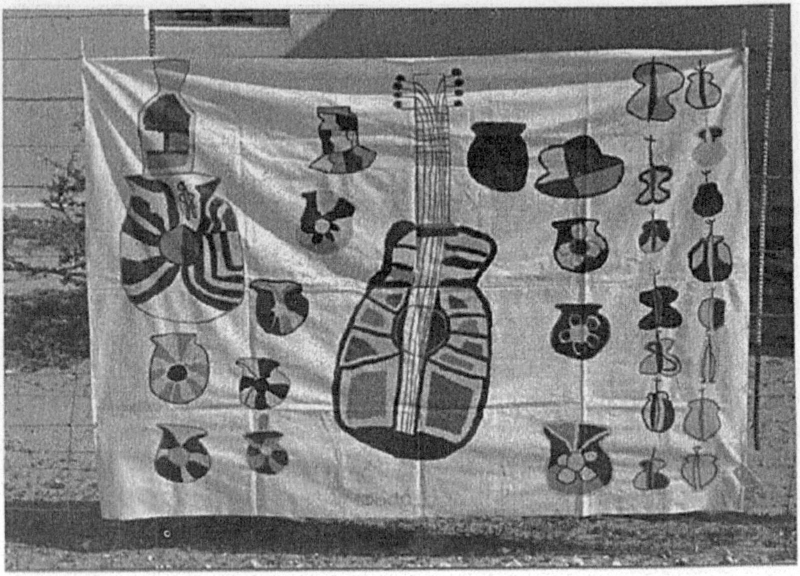

Fig. 4.36 "Guitars, pots and bottle", Dada. (Author's photo, with permission)

Let us look at another example of this playfully ingenious stylistic technique, in which two symbolically opposite objects, stemming from the domain of men—a bow and arrow—and women—a hut—are integrated through a series of intervening shape-transformations. Fig. 4.38, a Dada painting on exhibit in Windhoek in 1997, depicts, at the right of the canvas, a fairly literal representation of an oversized bow and arrow, the latter held to the bow string by a man with his right hand, waving his other hand in a "here-look-at-me" stance, which guides the viewer's gaze and makes the depicted object explicit. The bow motif gets more abstract and removed from its literal referent as the viewer's eyes move across the picture, to the left of the bow, across a number of differently coloured fields and into the direction of the pointed arrow. At the left side of the composition the bow, more distorted and stylized than on the right, gets turned around and the string is thick and overlaps the ends of the bow, the midsection of which is thin and string-like. At the extreme left, a row of red shapes picks up the bow theme, but with an altogether different referent: the bow shape is now presented horizontally, in three vertically arranged

Fig. 4.37 "Man, bows and arrows and huts", Dada. (Author's photograph, with permission)

shapes. In the first and third, on the top and the bottom, it is a roof hut in which stand, sheltered, one and two human figures. Inside the second, the hut roof shape, we see an arrow, in vertical position, directly opposite the horizontal arrow poised on the bow string at the extreme right. Its vertical shape is recapitulated by a vine, on the bottom left, its twigs drooping, in line with the roof slope (Fig. 4.37).

Dada uses the morphing technique in a number of other pictures, sometimes in ways that are both more subtle and more explicit, as the depicted subject looks like two or three things at once, such as string of pods that, brightly coloured, look like beads—as well as transformed or transforming birds as a bird foot each spreads from the stalks of the bottom

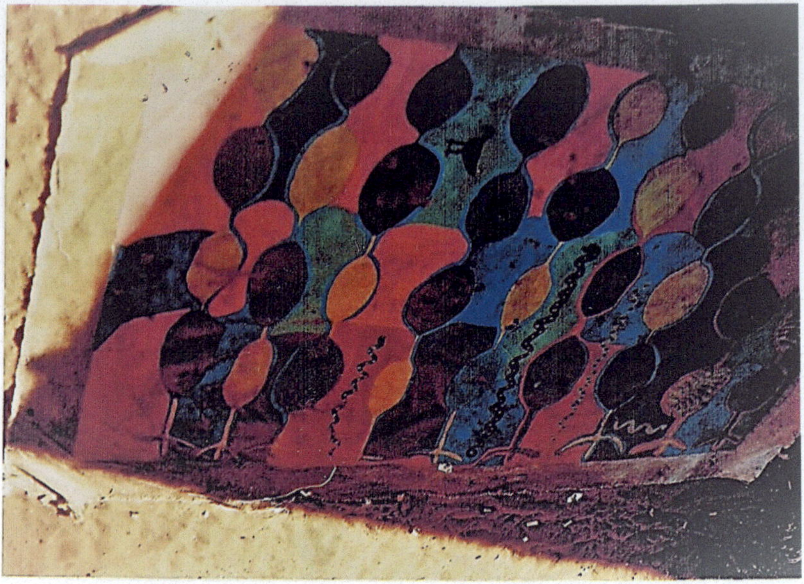

Fig. 4.38 Pods, beads, birds, Dada. (Author's photograph)

six pods (Fig. 4.39).[15] The bird motif is recapitulated by the bird depicted in amongst the pods; moreover, three wiggly, worm-like shapes running parallel to the pod strings add yet another ontological register to the scene (Fig. 4.38).

As noted in the preceding section, transformation, of the subtle, incipient kind, is seen in the three birds in Dada's print titled "Birds and a Tree" (Fig. 4.29). With their two legs wide apart and at opposite ends of their bodies and their short, blunted beaks these pronograde avian creatures have the look as much of mammals as of birds (or a hybrid conflation of the two). In another of Dada's bird paintings (Fig. 4.41) we see birds morphing into a turnip-like plant shape, which recapitulates some of the other plants on the scene, the pictures main elements. The birds inside the

[15] The painting is in rough condition because Dada used it as a floor mat in front of her house, to sit on in the sun, under the roof overhang's shade. I found this to be the fate of a number of paintings Kuru artists had failed to sell, using them for their own practical purposes: the cotton duck canvas is well suited to the purpose Dada used it for. As well as for weather-proofing of draughty hut walls (Fig. 4.40). Yet another artistic intention!

Fig. 4.39 San hut with painted canvas as weather-proofing, D'Kar, June 1996. (Author's photograph)

top panel are shown in profile, which is this, and all the other artists' principal way of depicting birds. Inside the bottom panels the depicted birds are shown in an unusual pose, from behind, looking back at the viewer, "over their shoulders". So positioned, the birds' bodies take on the shape of a pod or turnip, an impression confirmed in one rendition of the backwards looking birds, the one on the left. In giving its body shape a 90° downward twist, removing the feet and changing the head into a cluster of leaves, as well as painting it in the same colour and adding a similar cross-hatched pattern, the connection between the bird, and the "turnip" beside it, becomes explicit (Fig. 4.40).

Returning again to myth and its feature of transformation, which is explicit in some of the pictures by Qwaa and hinted at in some by Dada, the latter artist also, as story teller, is very much aware of this element. Indeed, during the time of my field work one of Dada's versions of a pan-San story was published by the NGO as a bound portfolio of artists' prints, all of them tied, more or less directly, to the story's plot (Kuru Art

Fig. 4.40 "Birds and plants", Dada. (Author's photograph, with permission)

Project 1994).[16] It is a transformation story, embedded in a gruesome tabloid melodrama about in-law animosity. In Dada's version of the tale, linked, according to San place lore, to one of the rock art sites the artists visited on one of their "inspirational" outings, the tale's protagonist is the beautiful, recently married woman Qauqaua. The site is near Mamuno in western Botswana and features rock engravings of what are folklorically regarded as Qauqaua's footprints, along with a large shiny boulder that is held to be the transformed body of the story's protagonist. After stabbing

[16] In 2018 a number of the Kuru artists, when asked by Maude Brown, the project coordinator, to produce a number of paintings on the theme of creation and Primal Times, painted pictures replete with transformation, on such primal matters as "all elephants derived from the elephant" and "when elephants were people", as well as representations of therianthropic bucks, jackals, hares and birds (Chris Low, 3 May, 2018, pers. comm.).

her husband with a red-hot awl in retaliation for cutting off her mother's head, Qauqaua and her small daughter are chased through the veld by her husband's brothers. They finally catch up with the woman and the girl and kill them both. As Qauqaua's blood "pours out soaking into the ground", a portion of it turns into the morama plant—the *veldkos* staple of Ghanzi San people—and another into the wild potato as her body morphs "into a smooth, shiny stone as beautiful as Qauqaua's skin" (ibid.). Qauqaua's daughter turned into a guinea fowl, "still calling *tsa tas tsa tsa tsaaaaaa*". It was the same sound the girl had uttered during her and her mother's flight, while being carried on her mother's back looking backwards, as a warning whenever she caught sight of their pursuers, prefiguring her imminent transformation into a guinea fowl.

Of the drawings contributed by Kuru artists to this project, only one was explicitly concerned with transformation, depicting this theme with powerful graphics. That artist was Qwaa and the picture, early on in the story (Fig. 4.41), depicts a steatopygous ostrich-legged, human-bodied, hare-antelope-headed chimera (the figure seated on the right), in amongst other mytho-fantastical beings (one of which seems itself in the grips of transformation, into a Gregor Samsa'esque animal-legged, human-armed insectoid with a human-shaped head and face the human features of which are receding). It is interesting to note that this image, replete and aquiver with transformation and mythological goings-on, is not related in any evident way to Dada's story, which is about transformation only at the end of

Fig. 4.41 Chimeras and fantastical beings, Qwaa. (Author's photograph, with permission)

the tale, at the heroine's death and metamorphosis. Nevertheless, for reasons I was unable to ascertain as the portfolio was produced during my research period (off site, in Johannesburg), Qwaa chose to provide this picture to the story.

In their hybridity and fluidity—and their whimsicality—the beings depicted by Qwaa for Dada's story would be at home very much also in the myth world of /Kaggen or /Xue or on the rock surfaces of the Drakensberg to the south and the Brandberg to the north.

The Meaning Question

Given its long-standing and on-going prominence in San rock art studies, we need to deal with the elusive question of what San art might mean, specifically the therianthropic and transformation images. This matter was already touched on above; at the end of the discussion of the transformation motif in San rock art where it was suggested that in explaining what such images might mean the hunter and transformation should be considered as much as the shaman and trance. I will here place this suggestion into its wider theoretical and polemic context.

In this exercise I will bring to the discussion what I learned about contemporary San art from artists I interviewed, in hopes that their perspective might help elucidate the many elusive and much debated question of what these ontologically ambiguous and fluid figures of San rock art might mean. Are they depictions of mythological beings, as they were for Qwaa, as well as a couple of the other Kuru artists? Or of a trance or dream experience? Or of transformation, psycho-somatically, through either the imagination or the body, "the painterly solution to what has been experienced, or what is known but has not been experienced" (Skotnes 1996: 243), thereby making the experience of ontological change more graspable and less bewildering to the shaman, initiand, dancer or hunter? Or something less portentous—a playful display of hybridity and mutability of beings and things, aesthetically motivated, rather than by myth, trance or transformation, as for Thamae Setshogo and for Dada? Or—something I have dealt with elsewhere (Guenther 2006; see also Thomas 2016[17]) and

[17] Interestingly, this recent study of contemporary San art, in terms of post-colonial and resistance theories, features Dada on its cover, squatting on the ground, at work on a painting set in front of her. Chapter five of Thomas's book is on "resistance through self-representation", with its primary focus on the Kuru art project (2016: 172–212).

likely outside the interpretive framework of ancestral San rock art[18]—a statement on cultural identity, in the context of post-apartheid identity politics?

While we do not and cannot know anything definite about the artistic intentions of the San rock painters and engravers of the past, it is nevertheless reasonable to assume that the artists of the past had intentions not dissimilar to those of their present-day descendants, all the more so as the latter's mythology, cosmology and religion—via "persistent habits of thought" and "memorate knowledge habits and knowledge characteristics"—abidingly resonates, and reveals extensive continuities with the past (Low 2014: 350; see also Biesele 1993: 11–15). San, of the present and the past, share a certain mythological, ritual and cosmological complex—which elsewhere I have dubbed "the Khoisan Religious Tradition" (1989: 33–3)—which leaves its mark on the images produced by today's San artists, notwithstanding their respective stylistic and substantive idiosyncrasy and diversity, and divergence in multiple ways from ancestral rock art. And it may provide the student of the latter with soundings and directions on the sense and meaning of San rock art and suggests the outlines for a tentative interpretive framework.

However, in as much as the past artists are not around for questioning, all we can do on the meaning question is surmise, our surmises backed by commentaries of contemporary artists on their art (as well as, as noted above, eliciting more or less enlightening comments from them on ancestral art they themselves view, either *in situ* or in books). Yet, even this—asking artists to comment on what their own pictures mean—I oftentimes found problematic when I carried out my study of the Kuru art project. Indeed, some of my interpretative comments above on the contemporary Kuru art were surmises, based not on what the artists—such as Dada—told me but what I saw in her art—such as birds transforming into pots or dancer's legs into capital letters M. I found a few artists entirely disinclined to provide comments on their pictures, beyond shrugs or content-less, disengaged, and dismissive verbal equivalents ("don't know", "maybe, maybe not", "no story here", "no reason, really"—or, from one artist, "The reason I paint is because I don't tell stories; I don't know

[18] Or perhaps not: a recent study of the rock art of the Strandberg, */Xam-ka-!xoe*'s prominent landmark, links its imagery (specifically horse depictions) to regional processes of identity negotiation amongst eighteenth- and nineteenth-century Khoesan and non-Khoesan indigenous and settler groups of the Northern Cape (Skinner 2017).

how"[19]). Having artists available for exegetical insights into their pictures may thus not necessarily provide the key to the art's meaning, nor be the seemingly indispensable and unfailing tool for unravelling the same.

One or another of the ancestral artists, if asked about it, would probably have, like Qwaa and one or two other of the contemporary San artists,[20] linked a certain painting to its corresponding myths.[21] Indeed, as noted above, two of the /Xam story tellers did this in a fashion, when shown copies of rock art by Bleek and Lloyd. However, which myth plot or character is linked to which rock image cannot be known as the artists have all vanished. Thus, given the lack of evidence for any one-to-one correlation

[19] My attempts to engage some of the Kuru artists in conversation about the deeper levels of meaning their pictures might hold for them resonated with what I subsequently read about a north-west American First Nation carver named No Bull—an appropriate name as shown in the following vignette (around which hangs the whiff of an ethnographic tall tale): Introduced at an exhibition to a visiting audience by a Western commentator steeped in Jungian and Christian symbolism and showered with probing questions and comments about the symbolic polyvocality and mystical depth of one of the artist's fish carvings the carver is importuned to offer his own commentary on his carving. Standing up from his seat the No Bull steps forward, and pointing to the object, true to his name, offers the following commentary:

> "This here", said No Bull, pointing to the eye of the fish, "is the eye". "And this, this here", he said, indicating the fin, "is the fin". "And this right here", he concluded, tapping the tail, "is the tail." Whereupon No Bull returned to his seat and sat down. (Berman 1999: 169)

[20] A recent study on the Kuru art project dwells on the story telling aspect of much of the Kuru pictures, about both ancestral myth and lore and events from the historical and recent past (van der Camp 2012). The link of pictures to stories is one of the key themes also in an exhibition and publication of South Africa-based !Xun and Khwe artists (Rabbethge-Schiller 2006).

[21] The question on the relationship between San rock art and myth, their convergences and divergences, has been widely discussed, especially in the /Xam context (Lewis-Williams (2015b), where a rich body of myth exists side-by-side with an even richer body of paintings and engravings. The myth-art connection has been examined either in general terms (e.g. see Guenther (1990, 1994; Deacon 1994, 2002; Lewis-Williams 2015a: 149–72)), or more specifically with respect to its certain myth motifs (e.g. see Eastwood and Eastwood 2006: 91–111; Mguni 2006, 2015), in the context of a polemic critique of the interpretive monopoly of the shamanic/trance hypothesis, principally between, respectively, Anne Solomon (1997, 1999, 2000, 2007, 2008) and David Lewis-Williams (1998, 1999). For an attempt to reconcile the two camps, some 25 years back when the debate was at its highest pitch, see Jolly (2002: 100–1).

between specific myths and specific images, not even a clear depiction anywhere in the art of myth's ubiquitous, larger-than-life trickster, there is general agreement amongst San rock art scholars with Lewis-William's recent statement on the matter, that "the straightforward idea that the rock art 'illustrates' the myths is clearly not viable" (2015a: 149) and disagreement, thus, with Andrew Bank's optimistic suggestion that "a Bushman painting will frequently help us unearth a myth, legend or fable" (2006: 310). What connects myth and image, suggests Lewis-Williams, is not their respective narrative elements but shared conceptual ideas and beliefs that are inherent in San ritual and cosmology (and accessible through ethnography). These appear as "nuggets"—meaning-laden words that "invoke reticulations of fundamental beliefs and associations that may not be explicitly expressed in the text" (Lewis-Williams 2015a: 43, 2016)—that appear in either a narrative or imagistic garb. These, elaborates David Witleson, come up not so much as specific "scenes from mythical narratives" than they do, more broadly, as "mythical subject matter … wide in scope [that] pervades different social contexts and encompasses varied cultural beliefs" (2018: 3). Myth and image, each in its own way, spin out the fundamental cosmological themes and concepts of the story tellers' and artists' culture.

This conceptual framework shared by myth and art—what South African archaeologist Anne Solomon refers to as the "cultural logic that constitutes them" (Solomon 2007: 102)—is seen by Lewis-Williams to be informed with trance and its manifestations and metaphors and its human-animal hybrid "tranceformations" (Jolly 2002: 101) and by Solomon with myth and the spirit world (the ancestral therianthropic beings from Myth Time, including the transformation-prone rain divinity !Khwa). It is these two positions that define the "classic" debate on the meaning of southern San rock art in southern African archaeology.

In his article "Therianthropes in San Rock Art", Pieter Jolly critically evaluates these two interpretive and at times polemic paradigms with respect specifically to therianthropic and transformation representations, along with three other interpretive approaches (2002: 89–95). One is hunters' disguises, an interpretive paradigm that has fallen out of favour (and which, as noted above, I consider worthy or reconsideration). Another, which I will also consider (in Chap. 7), related to the animal disguise paradigm through the element of mimicry, is dancers donning animal masks and animal costumes in dances that simulate and kinaesthetically embody the animal of whose potency the dancer partakes through this ritual or ludic mimetic performance. Another interpretation is cast in terms of Solomon's Myth Time therianthropes' preternatural counterparts, spirits of the dead, especially of dead shamans, who have undergone animal transformation. Jolly also integrates and reconciles what are at times sharply polarized points of view, concluding that they all apply, either individually—one or another each to different specific images—or, more fruitful and commendable for analytical purposes, in combination, as complementary rather than antithetical interpretive approaches.

I am in accord with this position, as I am with Jolly's point in the concluding discussion of his article: that what integrates these interpretative approaches to these diverse therianthropic images is myth. As noted, the linkage to myth is not through any specific myth plots or mythological characters, however, but through "deep conceptual linkages" inherent in San myth's "major themes" (Jolly 2002: 101). These themes—perspicaciously identified as "the primal unity of people and animals", "access to a privileged realm free of the bounds of reason" and the "ability to 'bridge the categories' by changing from one physical state to another, even from life to death"—also inform this book, especially the first (the other two, dealt with more tangentially here, having been considered more fully in my *Tricksters and Trancers*). The primal human-animal unity theme of myth, as we will see in the next two chapters, informs also ritual and the hunt, such that the two worlds, of myth time and here-and-now reality, "inform each other in a process of constant mutual feedback" (ibid.: 101). They do so in multiple ways, one of them through the actions of, on the one hand, the shape- and reality-shifting Early Race's ultimate were-being and *ur*-shaman, the trickster and, on the other, his tranceformed counterpart in the curing ritual, the trance dancer.

The corollary of the conceptual linkage of trance to myth and art for proponents of the trance paradigm is that their two expressive processes, storytelling and painting, respectively, are each a "ritual practice" and a part of the /Xam shamans' status-defining and enhancing arcane knowledge and ritual toolkit they used for healing and rain making (Lewis-Williams and Pearce 2004b: 193–7). Indeed, as suggested by Richard Katz, on the basis of his interviews of Ju/'hoan trance dancers, supplemented by TAT-like drawings he presented to them, that these ritual practitioners tended to have richer "fantasy lives", which psychologist Katz sees linked to their trance experience "since fantasy life and kia are both altered states of consciousness [wherein] experiences with one state may affect the other" (1982: 236). In their responses to the TAT-like images, Katz found shaman subjects, in normal, non-trance state of mind, responding "in an emotional, intuitive way, willingly, at times eagerly, telling stories, which express elaborate fantasies"[22] (ibid.). Through the expressive performances, frequently, publically and dramatically, of storytelling and image-making, by shamans, the mythological and cosmological premise of ontological ambiguity and flux inherent in, and a variation of, the theme of "primal unity of people and animals" pervades people's lives and experience, as we will discover more fully in the next three chapters.

These chapters also touch on hunting as this non-ritual practice, too, is a variation on the "primal unity" theme, manifested in the hunter-prey sympathy bond. This may signal incipient transformation and result, as noted above, in a depiction around this experience, by a hunter-artist, rather than shaman, of a therianthropic image, especially one blending human hunter with animal prey. The phenomenological basis for the therianthropic images is seen here to be not a shaman's trance or transformation experience but a hunter's sympathy bond with the hunted animal. John Parkington favours an interpretation of the art along such lines over the trance paradigm as it allows for a more cogent interpretation of the rock paintings of the region he has investigated, the Western Cape, the region that was the /Xam people's home territory. He points to the low degree of correspondence in /Xam myth and what was presumably /Xam art between animal mythological characters and pictorial representations: "mantisses, birds, porcupines, small carnivores, dassies—are extremely rare in Western Cape rock paintings". Nor do they account for the animal aspects of therianthropic representations; indeed, Parkington ventures to assert, "they are *not* 'people of the early race'" (2002: 41–42, my emphasis).

Instead, depictions of such were-beings, as well as animals, were linked to the hunt, specifically the bond between hunter and prey (and its gender corollary, to which I will turn in Chap. 7, between prey and woman/wife). The painting of animals, including human-antelope conflations, via therianthropic or transformation depictions, when executed by a hunter rather than trance dancer or a story teller (or listener, or dreamer), as no doubt was often the case, the that and what and how of painting had little or nothing to do with trance, myths or spirits but much or everything with a hunter's sympathy bond with an animal. Yet, as already noted and to be explored more fully in Chaps. 6 and 7, this bond also contains myth- and

[22] My field work with the Kuru artists bears out Katz's impressions: of the dozen-plus artists I interviewed about their paintings and prints the most voluble were Qwa and Dada, both of whom were also, or had been, trance healers.

ritual-derived elements through which the instrumental practice of hunting, too, becomes a manifestation of the ontological "primal unity" theme of San cosmology expressed to its fullest extent in San mythology. Moreover, an important point to which I will return again in the next chapter, is that trance and transformation are synergetic experiences of the psyche, mind and imagination.

In sum, the various explanations put forward for the question on the meaning of San ancestral art, specifically its therianthropic and transformation motifs, are all applicable to what is a diverse canvas of motives and motifs—altered states, mimetic ritual, mythic characters, spirits of the dead, hunter-prey sympathy. Around each a key has been crafted, over generations of scholarship on the mystery of meaning of a body of art that is as enigmatic as it is beguiling.

There is, in addition, a master key that opens all the other locks: ontological mutability. It is manifested in species conflation and transformation, both inherently beset with ambiguity and flux. And it is here, paradoxically, where we find the "cultural logic" of San cosmology, notwithstanding the inherent ambiguity and paradoxicality of ontological mutability: therianthropes and transformation are its "logical"—that is, meaningfully consistent or warranted—expression, more or less extravagantly so in myth and art, both set in and consistent with the inchoate world of the First Order and reverberating—less fulsomely, yet also, on occasion, flashily—through the real world, in ritual and ludic dancing and hunting.

We will return to some of the images of human-animal conflation and transformation surveyed here, in the body of contemporary San artists and, more pointedly, that of the rock painters and engravers of the past. Some of the images by ancestral San artists relate to the phenomenological dimension of transformation, as a process that humans—shamans, dancers, initiands, hunters—experience in various situations in which they encounter an animal. Images such as depicted by Figs. 4.9, 4.10, 4.11, 4.12, 4.13, 4.14, 4.15, 4.16, 4.17, 4.18, 4.19, 4.20 and 4.21, and more abstractly and formally some of Dada's "morphing-style" pictures, suggest this process, transformation, to be a progression, from *virtually* being an animal, though ludic or ritual enactment, to *actually* being one—all through a progression through phases that may be experienced with both the vagueness of form and sharpness of bewilderment that accompany dreaming. Such images objectify and give contours to these phases, affording, thereby, a measure of control and understanding to bewildered novice shaman and initiand, dancer and hunter, as well as teller of and listener to stories—and as dreamer. Pippa Skotnes, one of the leading experts on the /Xam corpus and herself an artist, some of whose work is inspired by and resonates with ancestral San art, sees the depictions, on rock art, of transformation as "a particularly visual, painterly creation" and, as such, "the painterly solution to what has been experienced, or what is known but has not been experienced, it is a reification of belief"[23] (1996: 243). To the shamans and initiands, dancers and story tellers as well as hunters, who experience transformation at some level of awareness, this experience may, at some point, be emotionally unsettling as their being, in the course of such ritual, expressive, oneiric or practical experiences transforms and re-reforms. It may do so incipiently and faintly, through a menarcheal girl's or eland-dancer's sense human self-disassociation and animal other-identification signalled through physiological and mystical stirrings in parts of a hunter's body. Or, very much more unsettling, through a shaman's full-blown metamorphosis, into a snarling lion, with canine teeth growing in his mouth and hair growing from the back of his neck.

These sorts of transformations, which are experienced not conceptually, aesthetically or virtually—through the mind and the imagination—but affectively, somatically and actually—through the body—are the topic of the next three chapters. They present descriptions of the manifestations of transformation within the real-life arenas of ritual, dance and hunting, providing the ethnographic context for probing, in the three core chapters of Vol. II, the phenomenological dimensions of transformation for people undergoing the experience, both viscerally and vicariously, through participation, as performers or onlookers, in ritual and hunting or, via the imagination or dreams, in myth.

[23] Note here the Finnish visual artist and philosopher Mikko Ijäs, comments on art production, or "image making" (in his book on San mythology and rock art and its linkage to Altered States, especially as brought on by endurance hunting): "My experience as an artist has led me to believe that image-making might be seen as an action of the human body, a sort of complex co-action of the eye, hand and the whole body ... a functional and perceptual activity." Linking this somatic process of image-making to specifically to the San and hunting (and, by extension, trancing), Ijäs notes that "drawing employs the same physiological features that are used for throwing rocks, knives or javelins. Drawing is one of the simplest ways of understanding the functions and relations between our perception system and our own actions" (2017: 26, 23).

References

Bank, Andrew. 2006. *Bushmen in a Victorian World: The Remarkable Story of the Bleek-Lloyd Collection of Bushman Folklore*. Cape Town: Double Storey Books.

Berliner, Paul F. 1978. *The Soul of the Mbira: Music and Traditions of the Shona People of Zimbabwe*. Berkeley: University of California Press.

Berman, Morris. 1999. *Wandering God: A Study in Nomadic Spirituality*. Albany: State University of New York Press.

Biesele, Megan. 1975. Song Texts by the Master of Tricks: Kalahari Thumb Piano Music. *Botswana Notes and Records* 7: 171–188.

———. 1993. *Women Like Meat: The Folklore and Foraging Ideology of the Kalahari Ju/'hoansi*. Johannesburg/Bloomington: Witwatersrand University Press/ Indiana University Press.

Bleek, Wilhelm H.I. 1875. *A Brief Account of Bushman Folk-Lore and Other Texts*. Cape Town/London/Leipzig: J. C. Juta/Trübner & Co./F. A. Brockhaus.

Bleek, Dorothea. 1923. *The Mantis and His Friends: Bushman Folklore*. Cape Town: T. Maskew Miller.

———. 1928. *The Naron*. Cambridge: Cambridge University Press.

Bleek, Wilhelm H.I., and Lucy Lloyd. 1911. *Specimens of Bushman Folklore*. London: George Allen & Co.

Blundell, Geoff, and David Lewis-Williams. 2001. Storm Shelter: A Rock Art Discovery in South Africa. *South African Journal of Science* 97: 1–4.

Deacon, Janette. 1994. Rock Engravings and the Folklore of Bleek and Lloyd's / Xam San Informants. In *Contested Images: Diversity in Southern African Rock Art Research*, ed. Thomas A. Dowson and David Lewis-Williams, 238–256. Johannesburg: Witwatersrand University Press.

———. 2002. Relationship Between the San Drawings and Rock Art. In *Der Mond dals Schuh: Zeichnungen der San/The Moon as Shoe: Drawings of the San*, ed. Miklós Szalay, 67–89. Zurich: Scheidegger & Spiess.

Eastwood, Edward B. 2006. Animals Behaving Like People: San Rock Paintings of Kudu in the Central Limpopo Basin. *South African Archaeological Bulletin* 61 (183): 26–39.

Eastwood, Edward B., and Cathelijne Eastwood. 2006. *Capturing the Spoor: An Exploration of Southern African Rock Art*. Cape Town: David Philip.

Geertz, Clifford. 1972. Religion as a Cultural System. In *The Interpretation of Cultures*, ed. Clifford Geertz, 87–125. New York: Basic Books.

Golifer, Ann, and Jenny Egner. 2011. *I Don't Know Why I Was Created: A Biography of Dada, Born Coex'ae Qgam*. Gaborone: Eggsson Books.

Guenther, Mathias. 1990. Convergent and Divergent Themes in Bushman Myth and Art. In *Die Vielfalt der Kultu: Ethnologische Aspekte von Verwandschaft, Kunst, und Weltauffassung*, ed. Karl-Heinz Kohl, Heinzarnold Muszinski, and Ivo Strecker, 237–254. Berlin: Dietrich Reimer Verlag.

———. 1994. The Relationship of Bushman Art to Ritual and Folklore. In *Contested Images: Diversity in Southern African Rock Art Research*, ed. Thomas A. Dowson and David Lewis-Williams, 257–273. Johannesburg: Witwatersrand University Press.

———. 1997. Jesus Christ as Trickster in the Religion of Contemporary Bushmne. In *The Games of Gods and Men: Essays in Play and Performance*, ed. Klaus-Peter Koepping, 203–230. Hamburg: Lit Verlag.

———. 1998. Farm Labourer, Trance Dancer, Artist: The Life and Works of Qwaa Mangana. In *The Proceedings of the Khoisan Identities & Cultural Heritage Conference*, ed. Andrew Bank, 121–134, South African Museum, Cape Town, July 12–16, 1997. Cape Town: InfoSource.

———. 1999. *Tricksters and Trancers Bushman Religion and Society*. Bloomington: Indiana University Press.

———. 2002. The Bushman Trickster: Protagonist, Divinity, and Agent of Creativity. *Marvels and Tales Journal of Fairy-Tale Studies* 16 (1): 13–28.

———. 2003. *The Kuru Art Project at D'Kar, Botswana: Art and Identity Among Contemporary San*. M.s.

———. 2006. Contemporary Bushman Art, Identity Politics and the Primitivism Discourse. In *The Politics of Egalitarianism: Theory and Practice*, ed. Jacqueline Solway, 159–188. Oxford: Berghahn.

———. 2015. 'Therefore Their Parts Resemble Humans, for They Feel That They Are People': Ontological Flux in San Myth, Cosmology and Belief. *Hunter-Gatherer Research* 1 (3): 277–315.

Ijäs, Mikko. 2017. Fragments of the Hunt: Persistence Hunting, Tracking and Prehistoric Art. Doctoral Thesis, Aalto University: School of Arts, Design and Architecture, Department of Art.

Jolly, Pieter. 2002. Therianthropes in San Rock Art. *South African Archaeological Bulletin* 57 (176): 85–103.

———. 2015. *Sonqua: Southern San History and Art After Contact*. Cape Town: Southern Cross Ventures.

Katz, Richard. 1982. *Boiling Energy: Community Healing Among the Kalahari Kung*. Harvard: Harvard University Press.

KDT. 1995. *Kuru Conquers the Brandberg*. Dekar: Kuru Development Trust.

Kinahan, John. 2010. The Rock Art of /Ui-//aes (Twyfelfontein) Namibia's First World Heritage Site. *Adoranten* 8: 39–51.

Kruiper, Vetkat R.B. 2014. *Mooi Loop: The Sacred Art of Vetkat Regopstaan Kruiper*. Pretoria: Unisa Press.

Kruiper, Belinda, and Elana Bregnin. 2004. *Kalahari Rain Song*. Pietermaritzburg: University of Natal Press.

Kuru Art Project. 1994. *Qauqaaua, a San Folk Story*. Johannesburg: The Artists' Press.

Lange, Mary E. 2011. Rock Art Research in South Africa. *Rozenberg Quarterly The Magazine*. Open-access Online Journal http://rozenbergquarterly.com/rock-art-research-in-south-africa, 10 pp. Excerpt from Mary Lange *Water*

Stories and Rock Engravings: Eiland Women at the Kalahari Edge. Amsterdam: Rozenberg Publishers, SAVUSA Series.

Lange, Mary E., Miliswa Magongo, and Shanade Barnabas. 2014. Biesje Poort Rock Engravings, Northern Cape: Past and Present. In *Courage of //Kabbo: Celebrating the 100th Anniversary of the Publication of* Specimens of Bushman Folklore, ed. Janette Deacon and Pippa Skotnes, 363–382. Cape Town: UCT Press.

Lewis-Williams, David J. 1981. *Believing and Seeing: Symbolic Meanings in Southern San Rock Paintings*. New York: Academic Press.

———. 1996. 'A Visit to the Lion's House': The Structure, Metaphors and Socio-political Significance of a Nineteenth-Century Bushman Myth. In *Voices from the Past: /Xam Bushmen and the Bleek and Lloyd Collection*, ed. Janette Deacon and Thomas A. Dowson, 122–141. Johannesburg: Wits University Press.

———. 1998. *Quanto?*: The Issue of Many Meanings in Southern San Rock Art Research. *South African Archaeological Bulletin* 53: 86–97.

———. 1999. 'Meaning' in Southern African San Rock Art: Another Impasse? *South African Archaeological Bulletin* 54: 141–145.

———. 2010. *Conceiving God: The Cognitive Origin and Evolution of Religion*. London: Thames & Hudson.

———. 2015a. *Myth and Meaning: San-Bushman Folklore in Global Context*. Walnut Creek: Left Coast Press.

———. 2015b. Text and Evidence: Negotiating San Words and Images. *South African Archaeological Bulletin* 70 (201): 53–63.

———. 2016. The Jackal and the Lion: Aspects of Khoisan Folklore. *Folklore* 127: 51–70.

Lewis-Williams, David J., and Thomas Dowson. 1989. *Images of Power Understanding Bushman Rock Art*. Johannesburg: Southern Book Publishers.

Lewis-Williams, David J., and David G. Pearce. 2004a. *San Spirituality: Roots, Expression, and Social Consequences*. Walnut Creek: Altamira Press.

———. 2004b. Southern African Rock Paintings as Social Intervention: A Study of Rain-Control Images. *African Archaeological Review* 21: 199–228.

Lloyd, Lucy. 1889. *A Short Account of Further Bushman Material Collected. Third Report Concerning Bushman Researches, Presented to Both Houses of Parliament of the Cape of Good Hope by Command of His Excellency the Governor*. London: David Nutt.

Low, Chris. 2014. Locating /Xam Beliefs and Practices in a Contemporary KhoeSan Context. In *Courage of //Kabbo: Celebrating the 100th Anniversary of the Publication of* Specimens of Bushman Folklore, ed. Janette Deacon and Pippa Skotnes, 349–361. Cape Town: UCT Press.

Mboti, Nyasha. 2014. To Exhibit or Be Exhibited: The Visual Art of Vetkat Regopstaan Boesman Kruiper. *Critical Arts: South-North Cultural and Media Studies* 28: 472–492.

McGranaghan, Mark, and Sam Challis. 2016. Reconfiguring Hunting Magic: Southern Bushmen (San) Perspectives on Taming and Their Implications for Understanding Rock Art. *Cambridge Archaeological Journal* 26 (4): 579–599.

Mguni, Siyakha. 2006. Iconography of Termites' Nests and Termites: Symbolic Nuances of Foundlings in Southern African Rock Art. *Cambridge Archaeological Journal* 16 (1): 53–71.

———. 2015. *Termites of the Gods: San Cosmology in Southern African Rock Art.* Johannesburg: Wits University Press.

Northover, Richard A. 2017. Interrogating the Sacred Art of Vetkat Regopstaan Boesman Kruiper. In *Indigenous Creatures, Native Knowledge, and the Arts: Animal Studies in Modern Worlds*, ed. W. Woodward and S. McHugh, 59–84. London: Palgrave Macmillan.

Pager, Harald. 1989. *The Rock Paintings of the Upper Brandberg. Part I: Amis Gorge*, ed. R. Kuper. Cologne: Heinrich-Barth Institut.

Parkington, John. 2002. *The Mantis, the Eland and the Hunter Follow the San….* Cape Town: Creda Communications.

Rabbethge-Schiller, Helga, ed. 2006. *Memory and Magic: Contemporary Art of the !Xun and Khwe.* Johannesburg: Jacana.

Rifkin, Riann F., Christopher S. Henshilwood, and Magnus M. Haaland. 2015. Pleistocene Figurative *Art Mobilier* from Aollo 11 Cave, Karas Region, Southern Namibia. *South African Archaeological Bulletin* 770 (201): 113–123.

Rusch, Neil. 2016. Sounds and Sound Thinking in /Xam-ka !au: 'These Are Those to Which I Am Listening with All My Ears'. *Cogent Arts and Humanities*, Open Access Journal. http://dv.dororg/10:10.108012331983.2016.123615

Scheepers, Catharina. 2014. I Had a Dream. In *Rêves de Kalahari: L'art du Kuru Art Project/Kalahari Dreams: The Art of the Kuru Art Project*, ed. Leïla Bararcchini, 20–23. Neuchâtel: L'Usage du Temps.

Shepard, Paul. 1998 [1978]. *Thinking Animals: Animals and the Development of Human Intelligence.* Athens: The University of Georgia Press.

Skinner, Andrew. 2017. The Changer of Ways: Rock Art and Frontier Ideologies on the Strandberg, Northern Cape, South Africa. Unpublished Master's Thesis, University of the Witwatersrand.

Skotnes, Pippa. 1996. The Thin Black Line: Diversity and Transformation in the Bleek and Lloyd Collection and the Paintings of the Southern San. In *Voices from the Past: /Xam Bushmen and the Bleek and Lloyd Collection*, ed. Janette Deacon and Thomas A. Dowson, 234–244. Johannesburg: Wits University Press.

Solomon, Anne. 1997. The Myth of Ritual Origin? Ethnography, Mythology and the Interpretation of San Rock Art. *South African Archaeological Bulletin* 52: 3–13.

———. 1999. Meanings, Models and Minds: A Reply to Lewis-Williams. *South African Archaeological Bulletin* 54: 51–60.

———. 2000. On Different Approaches to San Rock Art. *South African Archaeological Bulletin* 55: 77–78.

———. 2007. Images, Words and Worlds: The /Xam Testimonies and the Rock Arts of the Southern San. In *Claims to the Country: The Archives of Lucy Lloyd and Wilhelm Bleek*, ed. Pippa Skotnes, 149–159. Johannesburg: Jacana Media.
———. 2008. Myths, Making and Consciousness. *Current Anthropology* 49: 59–86.
———. 2014. Truths, Representationalism and Disciplinarity in Khoesan Researches. *Critical Arts: South-North Cultural and Media Studies* 28: 710–721.
Stephenson, Jessica. 2016. Picture the Past, Creating the Future: Art of the !Xun and Khwe Cultural Project. In *Memory and Magic: Contemporary Art of the !Xun and Khwe*, ed. Helga Rabbethge-Schiller, 17–25. Johannesburg: Jacana Media.
Szalay, Miklós (ed.). 2002. Der Mond dals Schuh: Zeichnungen der San/The Moon as Shoe: Drawings of the San. Zurich: Scheidegger & Spiess.
Thackeray, J. Francis. 2005. The Wounded Roan: A Contribution to the Relation of Hunting and Trance in Southern African Rock Art. *Antiquity* 79: 5–18.
Thomas, Roie. 2016. *Bushmen in the Tourist Imaginary*. Newcastle upon Tyne: Cambridge Scholars Publishing.
Thurner, Ingrid. 1983. *Die transzendenten und mythischen Wesen der San (Buschmänner) Eine religionsethnologische Analyse historischer Quellen*. Wien: Föhrenau.
Van der Camp, Ankie. 2012. *Bushman in the Tourist Imagery*. Newcastle-upon-Tyne: Cambridge Scholars Publishing.
Vedder, Heinrich. 1937. Die Buschmänner Südwestafrikas und ihre Weltanschauung. *South African Journal of Science* 34: 416–436.
Vinnicombe, Patricia. 2001 [1976]. *People of the Eland: Rock Paintings of the Drakensberg Bushmen as a Reflection of Their Life and Thought*. Johannesburg: Wits Press.
Vivieros de Castro, Eduardo. 1998a. Cosmologies: Perspectivism. In *Cosmological Perspectivism in Amazonia and Elsewhere*, ed. Eduardo Vivieros de Castro. Four Lectures Given in the Department of Social Anthropology, Cambridge University, February–March, 1998. HAU: Masterclass Series, 1. *Journal of Ethnographic Theory*, 19 pp. Online Journal. http://www.haujournal.org/index.php/masterclass/article/view/107/135
———. 1998b. Cosmological Deixis and Amerindian Perspectivism. *The Journal of the Royal Anthropological Institute* (N.S.) 4: 469–488.
Vogelsang, Ralf, Jürgen Richter, Zenobia Jacobs, Barbara Eichorn, Veerle Linseele, and Richar G. Roberts. 2010. New Excavations of Middle Stone Age Deposits at Apollo II Rockshelter, Namibia: Stratigraphy, Archaeology, Chronology and Past Environment. *Journal of African Archaeology* 8 (2): 185–218.
Witleson, David W. 2018. Frogs or People: Dorothea Bleek and a Genealogy of Ideas in Rock Art Research. *Azania: Archaeological Research in Africa* 53: 1–25. Online publication.

CHAPTER 5

Transformation in Ritual

San religion contains two major rituals, the trance curing dance and the female and male rites of passage. Both are astir with transformation, experienced in various forms and to various degrees, from potential and incipient to fully realized, through mimesis and metamorphosis, and transitional states of ontological mutability in-between. The two rituals are also linked to myth and its First Order and Early Race, one of whose members, the trickster-animal being, whose mythical, mystical, mercurial presence underscores the aura of ontological ambiguity that pervades the trance curing ritual and the girl's and boys' initiation rites and sites. Mythic and mystical links connect these two rituals, along with metaphorical, metonymic and somatic ones. These, as noted in, and to be elaborated on in the preceding and following chapters, are reiterated by hunting.

The Healing Dance: Trance, Tranceformation and Transformation

> He (the sorceror [sic]) sniffs at a person with his nose, as the man lies; then he beats (? the air). He bites people with his teeth; when other people seize him with their hands, then he bites the other people. The others hold him down and rub his back with fat, as he beats [shakes, trembles].

© The Author(s) 2020

M. Guenther, *Human-Animal Relationships in San and Hunter-Gatherer Cosmology, Volume I*,
https://doi.org/10.1007/978-3-030-21182-0_5

> Lion's hair comes out on his back, people rub it off with fat, they rub pulling the hair out. Then the man leaves off beating, when the lion's hair has come off his back. They rub it off, so that the lion's hair falls off.
>
> Then the man (the patient) gets up, because he is well. He sits down because he is well, he sits talking because he is well. They sit drinking there, he lies down, for he is well. (//Kabbo (Bleek 1935: 1–2))
>
> The people must look out for his vertebral artery, for he would turn into a lion if they did not by singing make it lie down. This is why people must help him quickly, so that his vertebral artery may lie down, for if a sorceror's blood vessels do not lie down, he grows hair, he becomes a beast-of-prey, he wants to bite people. ... he would bite them were he to become a beast of prey. (Diä!kwain (Bleek 1935: 23–4))

Of the two mystical events that happen at the San curing dance, trance and transformation, the former has received more attention from anthropologists and archaeologists than the latter. The key event of the first, experiencing, through trance, an altered state of consciousness, is phenomenologically linked to the second, experience, through transformation, an alternate state of being. This is experienced most directly and intensively through trance-induced or transce-perceived transformation, which, following Jolly, may be referred to as "tranceforming" (2002: 91). Researchers subscribing to the "trance hypothesis" on San ritual have tended to see it as such; however, as we will see, transformation is something San may experience also outside the trance dance context.

Around trance, by means of that hypothesis, a general, all-encompassing paradigm became formulated for understanding not only this ritual of the San people but also, via metaphors or "nuggets", their art, mythology and belief. Its employment as the dominant interpretive tool of San ritual, expressive culture and cosmology, thanks to the prodigious quantity—and quality—of the writings by its architect and chief proponent David Lewis-Williams, has led to some misconception about the San healing dance.

One is an overemphasis of the trance component itself: the dance, even its healing component, can be, and frequently is, performed without any trancing; moreover, trancing can be a mimicked and ludic activity rather than a ritual one, sometimes played and acted by non-dancers, even children when they "play Healing Dance" (Marshall 1999: 61; see also Guenther 1999: 83, 2005). Moreover, real trancing—*!kia*—is not exclusive to the healing dance. Nor is the other feature linked to it by trance hypothesis proponents, *n/om*,[1] the trance dancer's healing potency generated by dancing and trancing, which may reside within and manifest itself outside this ritual and context (such as hunting, as we will see in the next section). The same applies to transformation itself, the third feature of the healing dance deemed integral to it; this, too, as mentioned above, is not exclusive—let alone reducible to—the healing dance nor necessarily linked to trance.

Turning to the first two, I note, with regard to *n/om*, the ambiguity of this physiologico-mystical San concept, as reflected from its multiple meanings, many of them outside the context of the healing dance. Richard Lee lists the following:

> medicine, energy, power, special skill, or anything out of the ordinary, menstrual blood, African sorcery, herbal remedies, a vapor trail of a jet plane, tape recorders, and travelling at high speeds. (2003: 130, footnote 5)

A propos menstrual blood, Lee elsewhere mentions another meaning for the term, "the ripening of young maidens" (1968: 33), thereby linking the healing ritual to the San's other ritual, initiation. Yet another link is to hunting (a point to which I will return below).

As for *!kia*, the ASC trance experience, in its "deep trance" state rather than the one playfully simulated by children or by story tellers when they elicit the "death shriek" of a trancer just before collapse as a histrionic element of a story they tell, this, too, can be found in other ritual or quasi-ritual contexts (such as initiation rites and *dengo* playing, as seen above and below). Moreover, neither of the bodily sensations that accompanies trancing, shaking and dizziness—"like a drunken person",

[1] *!kia* and *n/om* (or *n/um*) are Ju/'hoan terms, having first been described in the context of these San people (Lee 1968; Marshall 1969, 1999: 39–90; Katz 1976, 1982). They have since assumed a certain generic status, as analytical terms to designate a feature of San supernaturalism that is found in different forms and with different vernacular designations amongst most if not all San groups. Ubiquity is matched by antiquity: depictions of what appears to be trance dancing are found on ancient rock art scenes. As the matter is somewhat ancillary to this book, I will not deal with the symbolic, mythological, ritual and experiential aspects of these two key ingredients of San trance dancing, as well as the connection between them but refer the reader to the sources cited in the footnote above in which these elements are discussed in detail (most extensively in Katz 1976). More recent accounts, couched in New Age'ish phenomenological terms, of trance curing, again among the Ju/'hoansi, are offered by the anthropologist-*cum*-"creative therapist" Bradford Keeney (1999 and; with Hilary Keeney, 2015).

as described to me by the Naro trance dancer //Aicha !Khoma//ka (Xgaiga Qhomatcã) in an interview in 1997—is exclusive to the healing dance. While the former, called *thara* by the Ju/'hoansi, maybe what induces the trance state in the healing dance—a form of trance referred to by Bradford Keeney as "kinetic trance" (Keeney and Keeney 2015: 52)—in its "kinaesthetic" (Katz 1982: 129), mystical or *n/om*-associated form, shaking (or trembling) may be somatic sensations that story tellers or—as seen below—hunters may also experience, with different degrees of intensity. Indeed, it is a kinaesthetic sensation that is likely linked as much to transformation as it is to trance (Low 2013).

As for trance and transformation, these two experiential processes need to be kept apart, both conceptually and phenomenologically. On both scores we find convergences and divergences. Conceptually both are altered or alternate states and, in this sense, transformations; however, the one is of a person's consciousness, the other his or her being. The latter transformation is essentially ontological; however, it is infused also psychologically, with much the same emotional and affective intensity—including terror—brought on by the trancer's experience of self-disassociation and transcendence, along, as noted above, with the bodily reaction of shaking (Katz 1976: 287–8; 1982: 92–116, 229–49). Moreover, both operate within the symbolically, mystically and sensually charged field of ritual liminality (Guenther 1999: 182–92).

Yet, that said, transformation is more than "tranceformation"; it occurs in other contexts as well and to persons other than trance dancers and experiences other than "transformative *hallucinations*" (Ijäs 2017: 13, my emphasis), including to certain individuals held by their fellows to be afflicted with a peculiar propensity to morph into a certain animal, such as a hyena or a baboon (such as the ≠Au//ei man Oba described in the ethnographic vignette by Reneé Sylvain; see Appendix 2). Writing about the /Xam, who also practised the trance curing dance, an arguably pan-San cultural trait (Solomon 1997, 2011; Lewis-Williams 2015a: 62), and about their accounts of transformation in myths and tales, memorates and commentaries about their lives and the hunting ground, Neil Bennum notes that:

> transformation was not the extensive province of the trancing healer … Things share states … A perception of the liminal nature of things and a readiness to accept the possibility of transformation were the hallmarks of the culture of the /Xam-ka !ei. (2004: 358)

Making much the same point, Pippa Skotnes notes that "transformations were a part of the range of experiences described by the /Xam as a normal part of their existence" (1996: 243). Indeed, so much so that, while "sometimes imbued with symbolic significance" they also were "a banal part of everyday life" such that, "in their own terms, there was no need to explain these phenomena as only emanating from some supernatural power, or as an epiphenomenon of a social constructed ritual (and, indeed, they did not)" (ibid.: 243–4).

Let us get back again to the trance dance, the arena within which San experience, most frequently, affectively, somatically than in any other context both psychological and ontological transformation, into altered states of consciousness and alternate states of being, so much so that the American anthropologist of dance, Kimerer DeMothe, in an essay on the San trance dance, refers to it as "a nonverbal, visceral act of transformation" (2015: 128). Regarding the ontological variant of being-change, it is here that transformation is experienced or witnessed in its most full-fledged and quintessential form—into the most dreaded predator of the veld, dreams and myth.

The shaman's lion transformation is the template for all other transformations, of myth and lore, ritual and the hunt and when it is seen to happen—quite rarely as it is beset with danger and dread—it is one of the more memorable phenomena in a person's life. Either experienced or witnessed first-hand—or, as a more diluted experience, second-, third- or fourth-hand, through veld-set and broadcast urban legends or rumours—lion transformations, by all accounts an extraordinary phenomenon, becomes valorized and normalized. All the more so as the first-hand accounts may be based on what seem, and likely are, startlingly real experiences by those who have experienced and witnessed them and described to those who have not (including—most[2]—anthropologists), in grippingly

[2] Some anthropologists have tried their hands at the experience, most notably Bradford Keeney, who, in his own trance curing, reports that, in addition to experiencing altered states of consciousness he also experienced the lion transformation, as a "visionary occurrence" (Keeney and Keeney 2015: 151). At least two other anthropologists—from more conventional theoretical and methodological quarters including cultural-materialist ones—have experienced animal transformation (albeit not into lions), as well as altered states. An example for the latter is the consummate ecological empiricist Jiro Tanaka who participated in a //Gana gemsbok dance to such a degree of absorption that he reportedly found himself "melt [ing] away into the other world as he emulated the game animal by voicing deep grunts" and kinaesthetically miming the animal's actions "with rapid arm movements and

graphic prose. An example is the Ju/'hoan healer Tshao Matze's description to Richard Katz and Megan Biesele of his own experience, of deep immersion, ontologically, "inside that lion":

> When I turn into a lion, I can feel my lion hair growing and my lion-teeth forming. I'm inside that lion, no longer a person. Others to whom I appear see me just as another lion. (Katz et al. 1997: 24)

The lion's hair, Tshao Matze describes, growing on his body, one of the animal's defining traits in San mythology, as well as, among /Xam and this portentous animal's respect name, appears as the prominent feature in all of these leonine transformation accounts, especially when blood vessels swollen, it sprouts and spreads from his back and neck (the site on the male animal's body of its iconic, species' defining mane). Add fangs and clawed paws to the mix and the transforming man's lion'ness is complete.

Or of a *woman*: lion'ness sometimes manifested itself as lioness, as San women, too, were capable of undergoing leonine transformation (as well as lions into women, as seen earlier). Examples are the benign /Xam sorceress—and Diä!kwain's aunt—Tãnõ-!khauken who, when lion-transformed, watched over her people at night (Hollmann 2005: 267–72) or, further back in time, the nameless Saan huntress from myth time who hunted for her Koranna master by changing into a lioness. Detecting a wild horse (either a zebra or quagga, depending on the version[3]) not far off in the distance, "hair began to appear at the back of the woman's neck her nails became claws and her features altered. … [S]he threw off her skin petticoat … and a perfect Lion rushed into the plain." While her startled Hottentot master escaped up a tree, the lioness killed one of the wild horses and, having "lapped its blood", returned to her previous spot, behind the bush where she had left her petticoat, slipped it back on and

stomps" (1996: 27, note 2). In a similar vein Richard Lee makes mention of his "one attempt to enter *!kia*", noting his struggle, physically and psychologically, with its aspects of "physical exertion" and "acute fear of loss of control" (2003: 133). Lee's fascinating account is tantalizingly brief and it is not stated how far into trance he was able to get, given these obstacles, and whether transformation was an element of his experience, specifically its aspect of fear and loss of control.

[3] One of them is in Bleek's *Reynard the Fox in South Africa* (1864: 57–8). His source was /Akunta, a Khoe-speaking /Xam informant to whom Bleek had dictated the story which he had himself obtained from Alexander through another Khoe source (Bleek 1864: 52–8). The latter was Captain James Alexander, who had obtained the story from his Khoisan guide when travelling to the Orange River in 1837 (1838: 197–8; see also Deacon 1994: 249).

once again "the woman took her proper shape" (Alexander 1838, vol 2: 197–8; Bleek 1864: 57–8).

As seen from the epigraphs at the opening of this chapter section, for a /Xam *!giten* (shaman) this latter transformation, from lion back to human, was accompanied with no small measure of dread and uncertainty, requiring ministrations from people around him, such as rubbing his sprouting lion hair with healing potency-laden antelope fat. The same dread surrounds a contemporary Kalahari San shaman's lion transformation, for presumably the same reasons. One of these is that an inexperienced, unruly leonine shaman could be a danger and menace to people present at a dance performance (see below). The other is that if for some reason the transformed shaman were to be incapable of reversing back to his humanness and remained trapped in his lion'ness, he might be killed by either humans—Bushman or Black hunter or a Boer commando—or by real lions. A shaman may be drawn to lions, kindred beings in his lion state, and attempt to seek out their company and mingle with them. However, Ghanzi San lore has it that after sensing the ontological otherness of the were-lion the real lions will reject him from their midst and their kind, with vicious assaults and killing bites (much the same fate as that suffered by the lion-woman, in the KhoeKhoe lion-human transformation story noted in the section above, whose people sensed lion'ness through residual leonine traits in her behaviour and appearance). This sort of behaviour of lions in such a mystical setting has ethological backing as male lions, which live together in coalitions formed through kin links amongst them, will attack and sometimes kill outside males that try to enter their group (Grinnell et al. 1995: 96, 102).

As already discussed, descriptions of lion transformation in the anthropological literature all pretty well seem to fall not into the first-hand but the second- or third-hand category, based as they are on accounts—vivid and dramatic ones as exemplified by the Ju/'hoan snippets provided above—related to the ethnographer by San informants, either as an experience he or she (a trance dancer) had had themselves, or as something witnessed or described to them by others. In my own field work I have collected several such accounts from San informants, awed, starry-eyed were-lion believers. Other San informants I (and other ethnographers) have talked to about lion transformation were more guarded, acknowledging the possibility of such a thing happening, very rarely.

If ever: I talked to San individuals who were altogether dismissive of lion transformations; the matter was something they either didn't believe exists or deemed possible, or would attribute to outside groups, as something other people do. The Naro generally attribute the skill of leonine transformation to the ≠Au//eisi, as part of the general stereotype of fierceness and vindictiveness (Guenther 1999: 18). The linguistically related G/wi hold the same ≠Au//eisi stereotype, especially its aspect of turning themselves into lions and preying on people (Silberbauer 1972: 320), whereas the linguistically distinct !Kõ *vis-à-vis to* whom the Naro are the San out-group attribute the same stereotype to the Naro (Heinz 1975: 28). The San, in turn, collectively are wont to stereotype their Black neighbours, again collectively, in terms of the same stereotype (Schmidt 2013, vol. 2: 733).

Or, instead of ascribing lion transformation to people as something they *do*, this phenomenon may at a semantic, folkloric level be deemed as something people *are*. Thus, to the Naro the faunal category *n/ie /wa*, clawed and predatory animals of whom the lion is the ultimate exemplar, includes Black people and to the /Xam, as developed by Mark McGranaghan, lions, as "violent and angry 'beasts of prey' who went about at night to kill people" were, using //Kabbo's words, "people who are different". The moral, anti-social otherness of lions *qua* other-than-human people was generalized also to some humans by the /Xam, to "people who displayed such qualities of greed or violence [who] were "equated as 'different' (*/xarra*) and equated with lions (*/xa*)" (2012: 211, 204–5, quoted in Wittenberg 2014: 80). This, as noted by Hermann Wittenberg in a discussion of the conflation in Khoisan orature of the lion of traditional stories with the Boer of colonial folklore, is an instance of "lexical similarity reinforc[ing] an overlap in meaning" (ibid.).

A phenomenological question is raised at this point (to which I will return in Chap. 4 in Vol. II), about whether or not and in what way lion transformation is "real" as opposed to imagined. In San ethnology this question needs to be examined from both the emic and etic perspective. Is lion transformation, from the indigenous perspective, a nightmarish reality, its embodiment nurtured and sustained by myth and ritual that depict the lion as the ultimate "anti-social agent" (McGranaghan ibid.)? Or is it a negative, boundary-maintaining "exoteric" (Jansen 1959) ethnic stereotype by an individual or group about an out-group? As anthropologists, does an examination of this matter necessitate a replay of the classic anthropological "hocus-pocus vs. God's truth" debate generated by Carlos Castaneda four decades back (Guenther 2018)?

In my own field work at D'Kar village in the Ghanzi district of western Botswana, I heard half a dozen second- and third-hand accounts from people about lion transformations—including one first-hand transformation account, by an ≠Au//ei trance dancer, albeit not into a lion, as his wife "doesn't want him to do so dangerous a thing", but into a non-poisonous snake (called *n!am di tsoro* in Naro)! There was only one occasion at which I actually witnessed what people who were present at the event and with whom I talked about it afterwards considered a lion transformation. It occurred at a trance dance on 30 May 1974, in the evening around a dance fire, less than half an hour into the trance dance ritual. The dancer, a man with the (fictional) name of Sebetwane, was in his mid-40s, of dark skin colour attributable to his Kgalagardi father (his mother was ≠Au//ei). The episode was relatively brief—a little under an hour—and its climactic conclusion abruptly ended the trance dance which had been scheduled to last for several hours. For all of its intensity and histrionics and the danger and dread it evoked in the attendants at the dance, its leonine transformation aspect seemed—to me, a cultural outsider—rather subdued and somewhat contrived and, for all of its dread and drama, somewhat anti-climactic. (My reaction here is not dissimilar to that of the Arctic explorer Knud Rasmussen when he witnessed an Iglulik shaman's bear transformation; see Vol. II, Chap. 5.) The reason may be that the ritual performance did not contain full trance, thereby reducing its duration and intensity. Furthermore, it contained other histrionic and ritual elements—erotic and scatological play as well as Kgalagardi sorcery—which all diluted and distracted from dance's central feature, the dancer's transformation into a lion.[4]

The element of dread surrounding this event that I observed and to a somewhat lessened degree sensed myself reflects the general view San hold of lion-transformed people, especially shamans. Moreover, in San groups closely associated with Bantu-speaking people, such as the Ghanzi San,

[4] Because what I observed is the closest I ever came to a "real" rather than related lion transformation in my own field work, and because it appears, as far as I am aware, to be the only first-hand description in the ethnographic literature, I provide an account of this event, excerpted in unedited form from my field notes (see Appendix 1). Apart from its intrinsic ethnographic significance, this account is germane to the theme of this book and thus instructive in its full-length rendition. For the same reason I include (with the author's permission) another "empirical" animal transformation account, into a baboon, which was observed by my colleague Renée Sylvain among ≠Au//eisi in the Omaheke in neighbouring Namibia some 20 years later (see Appendix 2).

sorcery elements from the latter people[5] may become folded into San trance and transformation practices, thereby deepening and broadening the sensed dread. Indeed, San groups of the Ghanzi District's Central Kalahari Game Reserve, whose trance healing dances appear not to contain any lion transformation but who attribute such arcane skills to other San groups (Maruyama, pers. comm. 18 November 2016), explain lion attacks on, and killings of, people in sorcery terms, triggered by "persistent anger" of people towards one another, as well as curses directed against persons. As explained by Kazuyoshi Sugawara from the perspective of the central Kalahari G/ui, the negative effects of such harboured malice "spread over" from inter-personal relationships to "interactions among humans, animals, and things". Sugawara sees it as a negative manifestation of what he calls "corporeal syntony" (2015: 2), the latter an instance of the possibly pan-San cosmological principle of "invisible agency" which, like the related Ju/'hoan concept of *n!ow* (Marshall 1999: 168–73), people are capable of actively operating, towards both constructive and harmful ends (Marshall ibid.: 319, note 1).

A shaman-lion is generally considered by San people to be malevolent; reminiscent of Yeats's fearsome lion-man astir in and departing the poet's *Spiritus Mundi*—"gaze blank and pitiless as the sun … moving its slow thighs, while all about it reel shadows of the indignant desert birds"—such a being "stalks the desert in search of human prey" (Lee 1968: 46), either killing them outright or invading their dreams as one of the culture's ultimate nightmares. Also unsettling, as seen in Chap. 3, are lions one might encounter when awake, in the veld or near the camp or waterhole, lions that are "not proper lions of god" (Katz 1982: 227), even though they are "real lions", "trance-lions that are not 'norma'l" (ibid: 115). This ambiguity of being is matched by ontological fluidity, as after their lion prowl they will mutate back into their human form—in the normal course of events,

[5] While witchcraft and sorcery have become practices that affect San people and that some of them practice in their own fashion, the Ghanzi San told me that this matter is not "a Bushman thing" (Guenther 1992; see also Marshall 1999: 233). Like other regions of southern Africa, now and in the past centuries and millennia back, Ghanzi is ethnically pluralistic, and Bantu-speaking peoples are close neighbours with whom San interact extensively (including intermarriage). One of their accultarative acquisitions is witchcraft and sorcery, which San, especially their healers, have learned to accommodate themselves to and cope with, in part by integrating certain supernatural and mystical concepts—one of them animal transformation (Jolly 2002: 90)—into their own cosmology (Guenther 1986: 60–1, 1992; Lee 2003: 137–40).

provided they haven't been killed by "real normal" lions, or had a leg bitten off by a hyena, returning back into human form minus one leg, as happened to one of the ≠Au//ei shaman Reneé Sylvain interviewed in the Omaheke. In accord with their ontological ambiguity, such men-lions are wont to act in abnormal and uncanny ways. One of them is flying through the air—and, in the process, clearing the height of the supposedly lion-proof *djora* (pole-palisade) Bukakhwe San surround their huts, to "catch you", as related by the Bukakhwe man Tanaxu Khôâkx'oxo in a nightmarish lion tale (Le Roux and White 2004: 89). Apart from transformed shaman, the other mystical-ontological state a lion one encounters might assume is that of a mythic and spirit being, in the form of none other than the trickster-god, who might make his unsettling presence known at a trance or initiation ritual (Bleek 1928: 24)—after all, as we learned above, "/Xue is not dead; he is alive, he is a lion"!

Yet, once again in accord with ontological ambiguity and conceptual incongruity, such shaman-lion-shamans may also be harmless. Their actions here range from such innocuous antics as using their leonine transformation power to play jokes on one another—for instance, entering, as lion, another shaman's nocturnal hut and giving him or his wife a good fright (Katz et al. 1997: 24–5)—to such beneficent deeds, among the / Xam, as guarding their people's camp at night against danger (such as Diä!kwain's shaman-aunt), or, among the Ju/'hoansi, as a trance-lion, putting his as of now bred-in-the-bone leonine hunting prowess to good use by carrying out nocturnal hunts to provide meat for his people during periods of meat shortage through failed hunts (Le Roux and White 2004: 133). Other prosocial actions Ju/'hoan shaman-lions might take was chase off real lions that threaten a nocturnal camp or "check up on relatives in faraway places and bring healing when necessary" (Marshall 1969: 347; Katz et al. ibid.: 24). As noted above, the last beneficent action was also the theme of a painting by the trancer-painter Qwaa, depicting a lion-transformed shaman climbing a tree towards baboons sitting on its branches and evidently unperturbed by the were-lion's approach, in order to obtain from them healing medicine needed to stock up his store or to use in a current treatment (Fig. 4.31).

Further to a shaman's benign animal transformations, I return once more to rock art panels of southern San which depict an animal a San shaman might transforms himself that evokes not dread and danger but boon: the antelope. First among equals here is the eland, the antelope species most likely to be merged with a human in rock art depictions of antelopean therianthropes. The frequency of such depictions suggests that the eland was as much a "tranceformation animal" as was the lion among southern San (Lewis-Williams 1981), as opposed to the Kalahari where the only animal a trance dancer transforms into is the lion. A corollary to the /Xam shaman's becoming an eland, in spirit, is that he also could hunt eland, when in trance (Lewis-Williams 1988: 5), as does his mythological counterpart //Kaggen, who, in addition to creating the eland and being its (and other antelopes) Spirit Protector, could also be an eland as well as shoot eland (as well as allow it to be shot). Among the Ju/'hoansi shamans did not trance-transform into eland in the same direct way they became lions. Instead, in mystically sensing or somatically feeling the "spirit presence" of the animal that inspired the dance—eland, giraffe, gemsbok, wildebeest—and the song that is sung and miming the same in the dance more or less explicitly and consciously, they might "catch the eland" (or giraffe, gemsbok or wildebeest). "Catching the eland" in this context is not meant in any hunting sense as it is among the /Xam, unless it holds such an association at a subconscious level given that the animal so "caught" is a prime game species. Instead, it is "catching the feeling" of the animal, in a somatic-mystical sense, of "dancing out" the animal's spirit, animated by the "caught" sense of it. It is drawn to the dancer, along with ancestors and mythological beings, in the context of mind-disassociating and imagination-firing ritual liminality and altered states of consciousness that invited mystical presences (Marshall 1969: 364; Keeney and Keeney 2015: 85–95). As we will see below, the eland is the species also prominently featured—sensed and danced, as well as "shot"—in the transformation moments at Kalahari San initiation rites.

While transformation is the most culturally salient mode of relating to animals, it is not the only way. Others are more subtle, operating with varying degrees of affective intensity, through mimesis rather than metamorphosis. The link may be tenuous, as in Kalahari San curing dances that are named after animals and in the dance steps and gestures which, in a fashion that ranges from indistinct and imperceptible to flashily realistic, imitate the animal that the dance and song is named after (Barnard 1979: 74; Marshall 1969: 358–9, 1999: 73–9). Vocalizations may also be uttered that imitate the animal; sometimes these may be part of the song. Props may be used by a Ju/'hoan dancer that are derived from animals that may

Fig. 5.1 ≠Au//ei trance dancer wearing hornbill beak. (Source: Gentz 1904: 158)

or may not be linked to the dance's ritual healing purpose,[6] such as wildebeest, eland or giraffe tails, skin caps, ostrich feathers (Marshall ibid.: 69). Among the ≠Au//eisi to the south it is reported, in what is perhaps the first detailed first-hand account of a trance curing dance described in vivid detail by the colonial soldier Lieutenant Gentz near Rietfontain (1904: 156–8), that one of the frenzied trancers, the *Großdoktor* (lead healer), wore a hornbill head attached to a headband (ibid.: 158, Fig. 5.1). Gentz's explanation is that this "adornment" was an emblem of his special ritual status.

However, the fact that the use of this animal prop was evidently restricted by the dancer exclusively to this particular healing dance may suggest some sort of mystical connection between the dancer and the animal whose head he attached to his head. It is reminiscent of the springbok scalp caps some /Xam dancers donned, for instance, Diä!kwain's shaman-aunt Tãnõ-!khauken who was a "shaman of springbok" (Lewis-Williams 2015a: 175) who would make a "springbok head's cap, that

[6] Marshall notes that today such accoutrements are used by Ju/'hoan dancers "only for adornment", a surmise I would agree with on the basis of my own observations among Ghanzi San. Equally, I would agree with her caveat that in the past "there might have been some significance in the use of these objects", especially when considered in the light of /Xam ethnography and southern San rock art that suggests that the use of ritual accoutrements by dancers, in both ritual and ludic contexts, might have connected them ontologically to the animal. I deal with this matter in the next chapter.

she might put it by, that she might listen whether she would not hear the springbok's story" (Hollmann 2005: 269). There were other such "Game *!Giten*", linked to such specific animal species as ostrich or locust, over which (s)he was believed to have control (Hewitt 1986: 296–9), through some invisible attunement or agency. The animals so featured were drawn, among Kalahari San, from a wide range of species, from puffadder and black mamba, through ostrich and kite, aardvark and ratel (honey badger), eland and wildebeest and buffalo and elephant to the all-important giraffe, mystically linked to rain and the weather and, via their skin markings, to certain cloud formations in the sky (Eastwood and Eastwood 2006: 99–101).

The last animal is at the centre of the most widely practised trance curing dance of Kalahari San, the Giraffe Dance (Marshall 1999: 76–9; Keeney and Keeney 2015: 86–91, 196–7). It is practised by both men and women; indeed it was a Ju/'hoan woman named Be (Beh) who, together with her husband /Ti!kay, in the 1950s re-configured what is likely an ancient San curing song (Biesele 1993: 67–9; Lee 2003: 132; Guenther 1999: 82). It had reportedly come to be in a dream, of two giraffes walking to the site of a freshly killed eland butchered by its hunters. She "instantly caught the feelings" for the two giraffes, Be's sister /Kunta Boo reported to Bradford Keeney, "and started to run like they were running. She imitated their movements" (Keeney and Keeney 2015: 194). The sensed connection of a Ju/'hoan dancer expressed to Lorna Marshall—"when he (a man) dances the Giraffe Dance he becomes a giraffe"—is also experienced somatically, both in his external body—"we sing it for the tail, it begins at the tail, the way he waves it"—and internally, when, as a Ju/'hoan dancer told Keeney, "God threw him inside a giraffe's intestine [and] … he went through the anus of the giraffe from behind" (Keeney and Keeney 2015: 115).

This is where he danced, inside "the organ", note the Keeneys' perceptively, "that embodies the most transformation". Of all the non-lion transformations shamans undergo in trance, the one that is closest, in its kinaesthetic—and gastro-intestinal (!)—intensity and its metamorphic, rather than merely mimetic effects on the trance dancer's being is the giraffe dance. This may be the reason giraffes are such a prominent feature on the rock art panels of the Brandberg of Namibia[7] as well as the Tsodilo

[7] In Namibian archaeologist John Kinahan's sample of the rock art of the shelters and boulders of the Brandberg's Hungorob ravine giraffes were the most commonly depicted

Fig. 5.2 Giraffe and eland, Tsodilo Hills, Botswana. (Source: author's photograph)

Hills of Botswana, where, along with an eland and, on a nearby panel, a rhinoceros, it is one of the three iconic images of the Kalahari's most significant rock art site, in plain view high up on a flat hill surface (Fig. 5.2).

While trance dancers—especially those, among Kalahari San, who become lions—exhibit the ability to transform ontologically to the highest degree and in paradigmatic form, transformation in some lesser form and degree is something experienced by San other than trance dancers and in contexts other than the trance dance. Initiation rites, ludic dancing and hunting are three of these contexts, which, in addition to the trance curing ritual and the pervasiveness of oft-told myths and stories—and dreams—of transformation—implant ontological mutability within San belief and experience.

I note here an important caveat, to which I will return again, and expand on in Vol. II (Chap. 4): the prominence within San cosmology and world view of transformation (and, more generally, ontological mutability) does not mean that people cannot differentiate themselves from ani-

and most widely distributed animal species. Springbok and gemsbok ranked second and third, respectively, with eland, kudu and lion, as well as therianthropes, sharing third place. The most frequent and widely distributed subject at the Brandberg were humans (Kinahan 1991: 20).

mals or fail to recognize and retain their own human identity and acknowledge the other-than-humans' alterity, nor does justify any essentialist, as well as evolutionist, portrayal of the San as an ethnic collectivity in such terms, as a people with an especially marked penchant[8] for non-human transformation, as did some earlier writers on the San.[9] Whatever

[8] That this sort of penchant may be seen also in the odd person within one's own community—for which the above-noted cultural stereotype may serve some of its members as an explanatory device—became revealed to me in a fascinating account my colleague Renée Sylvain related to me of an event she witnessed during her field work among the ≠Au//eisi of the Omaheke in eastern Namibia in 1996. It was about just such a person, an ≠Au//ei man named Oba who displayed the peculiar behaviour of evidently undergoing physical and mental transformation into an animal, in the light of day surrounded by people who watch his frenzied, seemingly obsessive zoomorphic transformation with dismay and puzzlement. The animal this farm Bushman—who was not a shaman—transformed into was a baboon. His apparent metamorphosis is insightfully described by Sylvain in her field notes (see Appendix 2 for the relevant excerpt). The details are graphic and gripping: "fully a baboon now … loped across the veld, bounding up trees … knuckle-running". And when just then a real baboon appeared running across the scene a mytho-magical moment became added to the unfolding drama, giving it extra poignancy and realism, and intensifying the consternation of people as they watch and try to deal with what they see transpiring. To what extent the ≠Au//ei man's behaviour is in fact anything like an "obsession" in any clinical sense—some sort of psychosis? An epileptic seizure? And, if so, might the same be linked to the prevalence and pervasiveness of transformation in his culture's world view? All these are questions, from a discourse and paradigm, culture and personality no longer current, that I lack the data and expertise to consider. I like the "psychological" explanation for Oba's penchant for baboon-morphing suggested by Sylvain: that "communicating psychological distress engages the expressive resources the Bushman culture offers—and so Oba turned himself into a baboon. Who could blame him? Life as a human hasn't proven to be such hot shit so far." All this resonates with the life situation of Ghanzi farm Bushmen when I did my major field work in the late 1960s, especially the last bit: the term in general use when referring to life on the farms was "sheta"—a regional Naro and ≠Au//ei neologism derived from the English "shit" (Guenther 1986: 50).

[9] Starting with the Victorian Andrew Lang who, in his discussion of the Cape and Maluti /Xam (1901: 34–40), dwelled on Bushman mythology, specifically its "peculiarity … the almost absolute predominance of animals" (ibid: 38) and its extensive anthropomorphization of the same. This, to Lang, exemplified, "the eternal confusion of savage thought" (ibid.: 37), a key notion of the Early Animists. It was a notion about Bushman myth and world view that was re-echoed in some of the early Bushman ethnography. One such ethnographer was Viktor Lebzelter who, in his account of religious beliefs and practices of (unspecified) northern Kalahari San groups, reports them to "believe that humans and animals are able to change one into another … and [that] many animals are held to be different outward manifestations (*Erscheinungsformen*) of humans" (1934: 64). It is not clear whether this statement is based on Lebzelter's own field work or derived from Dornan, who reports much the same information, couched explicitly in terms of Old Animism-style "transmigration" in

the transformative experience may be for a shaman, initiand or hunter, wherever it may lie on the mimesis or metamorphosis spectrum, his or her human identity is never obliterated at such moments of ontological ambiguity.

Initiation Rites: Liminality and Transformation

Animal transformation is one of the number of manifestations of liminality, which, in San initiation rites—and such rites generally—(Van Gennep 1908; Turner 1970)—is a state of ambiguity that is especially pronounced symbolically during the rite's transition phase, expressed in multiple ways through inversion, reversal, suspension or transformation. What is affected by these classic "anti-structural" markers of ambiguity in San initiation rites are such social and cultural aspects of identity and status as gender, age and decorum, and the ontological aspect of species identity and integrity. I dealt with the first set elsewhere in the context of both San initiation and trance ritual (Guenther 1999: 173–5, 182–92), where I also argued that species transformation, the theme of my discussion of San initiation rites in this chapter, is the more radical of these two types of transitions undergone by the initiand as it is a change of ontological being, rather than one of social position. The latter type stays within the human/culture sphere, transformation moving along a transitional track that runs through that sphere's social domain, from one status and role to another. The former, ontological transformation, takes the initiand outside that social-cultural domain, along an outside track, into domains of other beings, who, in their non-human being and actions (and tracks!), are not human and not of the "this world", that is the cultural world.

Having, in my previous discussion of the menarcheal rite, dealt with the initiand's social transition from child to woman, as well as sexual agent and potential wife, I here deal with her transformation from human to antelope. The two, as we will see, are also interlinked, symbolically and phenomenologically: the latter, human-antelope transformation, is the ontological equivalent of the former, child-woman transition. In the context of San—as of many other hunting people's—relational ontology a key manifestation of the human-animal relationship is the metaphorical

his ethnography on the Masarwa of north-eastern Bechuanaland (1925: 152). In elaborating on his own information Lebzelter cites Dornan, all of it drawing Bushmen into the Old Animism discourse.

and metonymic merger of antelope and prey with wife (and its corollary, the equivalence of the hunter's chase with the sexual chase or hunting with wooing and sexual intercourse). The boy's initiation, into a man and hunter, too, is linked to this theme as the rite—of becoming a hunter, of a large game antelope/woman—readies him for marriage, of a wife/antelope.

The girls' and the boys' initiation rite is thus a crossing station, of tracks from the ontological and phenomenological domains of humans and culture and that of animals and nature. As in the similar case of the trance dance dealt with in the previous section, where a social event, synergistic healing by an entranced human, can become an ontological event of lion transformation, we will in this section follow the latter track, the initiands' transformation into another ontological being, rather than her or his transition to another social position. Moreover, in blending animalian traits into her or his being and thereby becoming in effect therianthropic, the initiand's condition resonates with myth beings of the First Order. This very subtle, subliminal[10] intersection of myth with reality becomes more explicit and real when one of these myth beings—the trickster—makes an appearance at the rite (which, as we will see below, he is wont to do, as also at the trance dance). This not only underscores the mythic portent of the initiation experience but also underscores the aura of liminality and ontological mutability which, as seen in the previous chapter, this myth character exudes. In so doing, the ritual's ambiguity is reiterated and reinforced, not only by suspending or blurring the ontological human-animal divide brought about by transformation but also the divide between myth time and place and the historical present, through the appearance on the scene of a mythic being. At initiation, one of the more dramatic moments for the intersection of myth with reality comes about, each investing the other with the experiential phenomenon of the "really real".

The animal that generates ontological mutability at a San youth's transition from childhood to adulthood, both for girls and boys is the eland. This is why David Lewis-Williams, in his classic *Believing and Seeing* (1981), refers to this antelope species as an "*animal de passage*" (also adding marriage to the transitional "triptych" of San rock art and myth, as

[10] *Contra* B. and H. Keeney see Ju/'hoan initiation rites, especially the menarcheal rite, as a key "contextual frame" for initiands' and participants' "re-entry into First Creation", for spiritual and physical revitalization and to recharge *n/om* (2013: 71–75, 2015: 135–39, 186, 203–4). The Keeneys' take on the matter is backed up with commentaries from San informants that suggest people's explicit awareness of this instance of myth-reality intersection.

this passage event, too, has metaphorical and metonymic links to this meat- and symbol-laden antelope species). The eland has a strong presence, both symbolically and sensually, at the girl's menarcheal rite—so much so that, among the Kalahari Nharo the rite is in fact called *dùù*,[11] eland, "because", a Naro informant told me, "it is such a beautiful and important animal which Bushmen like and cherish". The same is the case with respect to the boy's rite, referred to by Lewis-Williams as the "first buck" ceremony, a ritual centred on that buck's prototypal representation, the eland. The boy—usually a lad or young man in his upper teens or early twenties—hunts an eland in a protracted ritual that socially underscores his transition to manhood through his demonstrated ability to hunt this ideal prey animal (and in the process, as we will see below, becomes ontologically linked to it).

The Female Rite

The highlight at both of these rituals, especially at the girls', is eland transformation. It is enacted kinaesthetically through dancing, which, through its emotional intensity, mystical portent and metaphorical polysemy, may so intently focus the initiand and the people present and participating at her or his ritual on the eland that all may "catch the feeling" of the eland and momentarily and more or less subliminally merge identities with the animal. This process is not as embracing, both of body and mind, as is the trance dancer's lion transformation; however, it may nevertheless pervade the girl's—and, to a lesser extent, the boy's—being and identity. At the moments in her rite the girl is regarded to be an eland, and, having "caught the feeling", perhaps may also feel herself to be such. Ontological boundaries other than that between human and eland may be blurred as well by

[11] Or *dùù gxoo*, eland bull (Lawy 2016: 239), semantically emphasizing the maleness of the eland at Naro female initiation. As already noted, this is one of a number of rite's liminal aspects—gender inversion—which I have discussed elsewhere (Guenther 1999: 174–5; see also Marshall 1999: 197, for a discussion of why among the Ju/'hoansi it is *not* an eland bull dance, as "male participation in the dance is not absolutely essential to the ritual"). I note, as before, that in the Naro language *dùù* is the word also for rain, pointing to what is likely a pan-San association; for instance, among the /Xam, where, as discussed above, the eland is one of the rain divinity !Khwa's incarnations. Like the Kalahari eland, the Karoo !Khwa is a strong presence also at the menarcheal rite. The association of the eland with *n/om* that was discussed earlier is also direct in the Naro instance: there are two *dùù* dances, one at menarche and the other at a trance healing dance.

the initiand, a human "perpetually verging on the animal world", as noted by Roger Hewitt with respect to menarcheal girls who, in that state, "could move from human society to the natural world beyond social order" (2008: 106). We saw this earlier, in Chap. 2, specifically the myths about !Khwa which typically ended climactically with her and her people, along with their material things, turning into frogs and such other amphibious, ontologically ambiguous "rain's things" as tadpoles, crabs and turtles, while their things revert back to the shrubs, reeds or antelopes from which they had been manufactured.

Eland'ness is experienced by the principal and the attendants at the girl's rite in two ways, internally—"in spirit"—and externally—"in body". The former, in terms of the San view of things, applies to the girl's experience. There is a moment when she is inside the menstrual seclusion hut, which traditionally was located at a remote place, a far distance from the camp, which contributed towards rendering seclusion a "frightening experience for girls" (Lawy 2016: 244). Here a menarcheal girl "became eland". In that state she also attracted eland to herself—and close to her people enabling hunters to set out on an eland hunt, with increased expectation of success. Indeed, she herself became a hunter and she "shoots an eland" (Lewis-Williams 1981: 51; Keeney and Keeney 2015: 127–8)—a playing out, symbolically and ontologically, of her ambiguous species state and gender status, as human-animal and woman-man, respectively. This figure, depicted spread-legged, holding a bow and arrows in hands raised above her head and antelope-eared, is a motif of at least three paintings from different regions of the Drakensberg (Jolly 2015: 211; Vinnicombe 2001: 152[12]). As noted by Lewis-Williams, she was, among the /Xam, "spoken of as if she were a hunter *and* as if she were an eland" (ibid., author's emphasis). Her hunter status was expressed through such symbolic and ritual acts as handling or shooting men's hunting weapons and medicines, as well as hunting dogs (Guenther 1999: 174–5).

Her antelope state was acknowledged expressively and symbolically, by mythical beliefs and narratives, and through certain ritual acts. Among contemporary Naro, these consist of such symbolic and ritual practices as requiring a girl undergoing *dùù* to wear special clothing made from ante-

[12] One of the three such figures depicted by Vinnicombe (Fig. 87c) she deems "clearly male". Like Jolly, Carolyn Thorp (2013) regards them as female figures, in an initiation setting; however, the animal change undergone by the girl initiand is into a frog from its tadpole phase, symbolizing transition, metamorphosis and danger, as well as rain and hunting.

lope skin and daubing her face and parts of her body with antelope markings (Lawy 2016: 264–5[13]). As to be elaborated on later, when discussing hunting and ludic dancing, both of these practices can be seen as mythosomatic measures for the transfer of ontological and/or spiritual essence. Another, more sensuously physical way, whereby eland essence is transferred to the menarcheal girl is when she is rubbed with eland fat (Guenther 1999: 176). Boys, too, as will be seen presently, receive this symbolically and somatically potent ministration during their initiation.

That "the girls will feel very close to the eland", indeed, at moments in their seclusion-transition phase, "feel like and eland" and attracts spirit or real elands to themselves are notions that appear to be basic cultural knowledge among Ju/'hoansi elders, as per accounts given about this matter to Bradford Keeney by some of them (Keeney and Keeney 2015: 127–8). This theme is reiterated and reinforced by mythical and preternatural narratives, that have the girl, as spirit eland, re-enter the First Order of Existence to visit its Eland Beings (ibid.) or move out, as eland-in-spirit, into the veld to join "the herd" (Eastwood and Eastwood 2006: 136). As eland she is also seen to become "prey", as opposed to "hunter" and, instead of "shooting an eland", she, herself can become the target of hunters' arrows (Biesele 1993: 196; Keeney and Keeney 2015: 128, 139), all of it further entangling her ontological and social state and status (Parkington 2003: 144; Jolly 2015: 210–1).

As perceptively discussed by Megan Biesele in her monograph *Women Like Meat*, this metaphorical fusion of women with prey—specifically antelopes—is pervasive in Ju/'hoan symbolic culture. So much so that men would metaphorically conflate meat and fat with women, a notion reiterated by the Ju/'hoan woman Ti!ae ǂOma when she told Bradford Keeney that the girl, at initiation, is sometimes the eland or meat for the men to hunt, elaborating that "since the eland is one of the most desired meals, being an eland is also desirable to men. This awareness is also part of what she feels" (Keeney and Keeney 2015: 123). Men's heads may be imbued

[13] See also Keeney and Keeney (2015: 126), who report that eland designs, resembling the animal's red forehead tufts, are painted on the Ju/'hoan girl's forehead and cheek. Ju/'hoan girls were painted with animal patterns also outside the menarcheal ritual context, for aesthetic rather than ritual considerations, as described by a Ju/'hoan woman to her interlocutors, during an interview about the old days: "The San used to make marks on the skin to look like a zebra because of its beauty. We started from the face down to the legs. They also did the same thing with beads, especially when they made rings and belts and the band used for the head" (Le Roux and White 2004: 85).

with such notions especially after a successful hunt returning to camp to appreciative band members, including the hunters' wives. This scenario presents the most likely context for "fleshing out" this metaphor, as described by Biesele:

> Coming home after a successful hunt … a hunter would greet his wife with special fervour. He would "praise the meat", Dahm N!a'an said, lying next to his wife with his face between her breasts. He would see her buttocks and her legs and would be happy "because the meat had fat and was fat." (1993: 196–7)

So carried away with the "women like meat" metaphor can a man become in such a Ju/'hoan domestic post-hunt idyll, that, "joyfully immersed in the things which tie animals and women together, … it is hard to tell … which meat—animal or woman—is being discussed. The metaphors tying women to the enchanted, hunted prey are so intricate as utterly to defy untangling" (ibid.).

A situation that underscores the metaphorical identification of women as game and meat in the context of hunting is when a hunter returns from an unfinished hunt, in which he has wounded a large prey animal and has returned home for the night, to resume the hunt next morning and, he hopes and expects, finish it by dispatching the poison-weakened animal. As a corollary of the proscription on meat eating that is placed on him, as engaging in this predatory, carnivorous might jeopardize his bond of sympathy with the herbivorous prey animal he is in the process of hunting, he refrains from sexual intercourse with his wife or girlfriend. Women, notes James Suzman, "after all, are 'like meat' and so also can break the empathetic bond" (2017: 175).

As already noted, the wife-eland identification can also be reified so that an actual eland is the hunter's wife and the chase for the antelope a sexual chase (McCall 1970). This myth-theme—the "animal-wife"—is a universal and possibly age-old notion of hunting people, with respect especially to ungulate prey animals.[14] The Ju/'hoan healer G/aq'o Kaqece rapturous dream (or vision) report to Bradford Keeney is on that theme:

> Sometimes we make love with an animal in a kabi [spirit vision or dream]. I have had special kabis where I made love with the eland, duika [duiker],

[14] See Vol. II, Chap. 2.

> gemsbok, and giraffe. Their songs enable me to own the feelings for them. When I am in a kabi with a special animal, I hold it close to me as I would my wife. We become like husband and wife because the song makes our heart one. It's the strongest love. When we love an animal it is like we are married to it. (Keeney and Keeney 2015: 151)

A more sensuous variation on the theme is described by Jiro Tanaka among Central Kalahari San who reportedly esteem "the mature and curvaceous eland body [which] exemplifies the ideal woman's figure". Also appreciated, notes Tanaka, is "its large rump" as it epitomizes for the Kalahari San "fecundity and easy birth" (1996: 21–2). As observed by John Parkington, the linkage between sex and hunting is a "widely used metaphorical framework that fits the allusive, indirect verbal style of these Kalahari people" (2002a: 47; see also Parkington 2002b), allowing such Ju/'hoan hunters as Dahm N!a'an and G/aq'o Kaqece the widest rhetorical scope for waxing eloquent about their "enchanted" antelope-wives, in registers that range from chaste and maidenly to salacious and zaftig.

The symbolic and mystical basis for utterances and allusions of this sort, about women, sex and antelopes, is the girl's transformation inside the menstrual hut into antelope and prey, which, as noted earlier, is the ontological equivalent to her social status and role transition to wife and sexual partner. The latter—wooing and sex—is all-pervasive and tone-setting also for what happens to people outside the hut, in the same ontological context, of human-antelope mimesis, sufficiently intense and self-absorbing as to border on metamorphosis.

For the participants at the rite, the old kinswomen who attend to the girl and other senior members of her extended family group, the experience of eland transformation is not so much inward and spiritual, the way it is for the girl, as it is outward and physical (including, as we will see below, in the sexual sense). Its manifestation, somatically, is the much-described, iconic and, subject to regional variation, pan-San (Guenther 1999: 170–2) eland dance that is performed by attendants at the menarcheal site, usually around the seclusion hut housing the girl (Guenther 1999: 170–2). The American ethnomusicologist Nicholas England description of the dance among the Ju/'hoansi vividly conveys the eland'esque physicality of the dancers' movements:

> Eland dancing is heavier and more deliberate than any other. It can perhaps be best described as a moderately slow, flat-footed run in which the body weight is allowed to settle firmly on each alternate foot as it is planted on the ground; indeed, at the moment of impact all of the dancer's flesh sags toward the ground, graphically illustrating the direction of the weight. And her body ornaments follow the motion downwards, adding a small but clear clicking sound effect to the movement. The feet land flat and firm on each step, producing a thud in the sand. … The entire effect conjures a picture of the grandly muscular, fleshy eland, trotting along unhurriedly in the veld. (England (1968: 596, cited in Biesele 1993: 197))

The rising intensity of the dance, as well as of its eland mimicry component, was recently described by anthropologist Jenny Lawy, on the basis of observations and participation in a six-day long menarcheal rite (from March to early April 2010[15]) among Naro-speaking San at D'Kar village in Botswana's Ghanzi district (2016: 239–68). And, linked to it, its erotic overtone, as though the women dancers' Rubenesque aspects, as per Tanaka—mature curvaceousness, large rump, fecundity and easy birth—had spurned on the male dancers. The men, in Lawy's account, joined the dance on the fifth day of the rite, the dance having in the first four days been performed only by women. When the first of the "bulls", an elderly man, simulating eland horns with his hands held to the sides of his head, joined the dance behind the last in the dancing row, this "energised" the dance, which "became more jovial and excited than before" (ibid.: 262). Moreover, "the chase had a distinctively sexual element to it", underscored by some of the women dancers lifting their skirts—one of them even "pulled knickers down"—to expose their buttocks (ibid.: 259). The mounting ribaldry—which made Lawy, a participant in the dance, "feel uncomfortable"—came to an abrupt end when a fight broke out between two inebriated participants, a man and a woman.

[15] There are very few first-hand descriptions of the San menarcheal rite. An early description (that includes a valuable photograph) is by Siegfried Passarge, among the Ghanzi // Aikwe (1907: 101–3). Lorna Marshall's detailed account (1999: 287–301) among the Ju/'hoansi is based not on actual observations but on reports that Marshall collected from Ju/'hoan women, and an eland dance some Nharo informants "acted out" for the Marshalls. For ethnographic accounts of the rite among the Kalahari /Gwi and !Kõ, see Silberbauer (1963) and Heinz (1994: 121–6), respectively. Schapera's account in his 1930 classic *The Khoisan Peoples of South Africa* (pp. 118–22) is a summary of the early ethnographic literature on the topic.

The sexual component of the rite's dance is manifested through—and, in terms of Naro propriety, contained within and restrained by—its eland aspect, especially its ludic-mimetic framework. This is eland mating behaviour which pervades the dance's choreography, with respect to both the female and male participants. The women in the past wore only leather aprons in front of their lower bodies leaving their buttocks exposed throughout the dance. Accentuating this feature and linking it to eland courtship was the eland "tail" that dangled between the buttocks of some of the dancers, a strand of beads—also worn by some of the contemporary Naro women—that wiggled when she danced, in tandem with other body ornaments the clicking sounds of which simulate the clicking of eland hooves. Among the Ju/'hoansi, women will also create this distinctively eland sound by linking two adze blades together. Moreover, the intricate rhythmic patterns of the women's clapping that accompanies their singing of the Great Eland Song, in part by means of a steady two-beat pulse, and that women strike when dancing the Eland Dance, is a "beat that represents the slow trot of the eland" (Marshall 1999: 196). As noted, the male dancers simulated eland horns with their hands, in the past they wore horns made of branched sticks and even further back in time, when elands were still abundant in the Kalahari, they might have used actual horns (as did springbok-shamans of the /Xam a century-and-a-half back). Marshall describes the "delightful pantomime" of the men's eland bull courtship:

> [T[hey imitated bull elands approaching the females, sidling up to them, following close behind, and turning and brandishing their horns at other men to ward off rivals. (ibid.: 198)

When "dancing eland", as the Ju/'hoansi refer to this central component of the female initiation rite, participants are "lifted entirely out of ordinary daily experience", observes Lorna Marshall, attributing its extraordinariness to "the nakedness, the clarion singing, and the intensity of the dance" (1999: 199).

To these I would add, as the most important factor in rendering the eland dance an extraordinary experience, the experience, by some dancers and especially the girl, of "becoming eland" and "being eland". They become and are such in the menarcheal rite not so much through transformation, the process undergone by the tranceforming shaman-lion, but through close attunement to and resonance with the animal they dance, with steps and props that mimic its behaviour. Through these kinaesthetic moves they "catch the feeling" of eland, which, in the course of energetic dancing, may then escalate into a transformative experience. Eland behaviour is manifested through an especially intense mode—courtship and mating—that symbolically and somatically, as well as psycho-erotically, as some of the male dancers, as observed by Jenny Lawy, seem to actually become sexually aroused in the course of dancing—emphatically links the Great Eland Dance to an abiding cosmological concept of San myth, belief and life: the equivalence of women and antelope. In addition to bringing this aspect of ontological mutability of San cosmology into experience through spirited mimetic dance and song, "dancing eland" also, notes Marshall, "makes a vivid affirmation of femaleness" (ibid.). It is, among the Ju/'hoansi, a dance primarily by and for women, one of them young the others old, that celebrates the premier game antelope, one especially generous with meat and fat and for this reason all the more linked, mythically and me(a)taphorically, to women, and their potency and power.

Further to the extraordinariness of the menarcheal rite, Lewis-Williams and Pearce consider the possibility if this passage rite, among the /Xam, may also have included trance. That it might have done so is suggested in the accounts of female initiation in the Bleek/Lloyd archive which at times refer to the girl's physical condition as "trembling" (*!kouken*), the same term used to describe a shaman's trance state (2004a: 162–3). Moreover, among the /Xam the "bull" eland dancer was usually an experienced shaman who, through his role in the eland dance, would bring to it elements and associations from the trance dance, most particular that of healing potency. Among the /Xam that potency had a direct metaphorical and magical link to the girl, through her association with rain and !Khwa.

Among the Ju/'hoansi to the north, the menarcheal girl's linkage to healing potency is even more direct: one of the many meanings in the Ju/'hoan language for *n/om* is menstrual blood (Lee 1968: 33). *N/om* is lodged also in eland fat, expanding the metaphorical and metonymic linkage between trance and transformation and the trance dance ritual and the initiation rite: kept in a tortoise shell contained with other substances by shamans to help induce trance at a healing dance, eland fat is rubbed on a girl' skin at her menarcheal rite to induce eland transformation (Lewis-Williams 2015a: 86), activating, in these two ritual contexts the two potencies that inhere in this meaning-laden animal, trance and transformation. In the initiation rite it is the girl, rather than the trance dancer, who is the "focus of potency" (Lewis-Williams and Pearce 2004a: 163), drawing her within the symbolic and mystical sphere of the shamanic dance. As for the dancers at the initiation site, intense dancing, some of it by trance dancers (some of them women), might possibly have triggered in one or the other more energetic participant Altered State of Consciousness.

It was an element, perhaps, of being "lifted entirely out of ordinary daily experience", through dancing eland and, in this process of mimesis, also becoming eland and partaking in the being of this significant other-than-human. This, for the dancers outside the menstrual hut, was very much a bodily experience, as opposed to the girl inside, who "caught the feeling" of eland not by "dancing" but "thinking eland", her imagination guided by myth, symbols and tropes that abound in her culture about this animal. Her merger of identities with the eland, via the imagination, would likely be stimulated by the people outside, whose mimetic eland'ness—stomping and clicking, grunting and rutting as they dance around the hut singing the Eland Song—could be expected to act as a stimulant to her own eland transformation.

The Male Rite

Whereas the foremost social concern of the girl's initiation rite is sex and marriage, that of the boy's is hunting. Symbolically, mythically and experientially, the latter is a replay of the former: ontological transformation of a human into an animal, specifically an antelope, in particular an eland, on a scale from virtual to actual, direct to vicarious, mimetic to metamorphic, as well as social transformation from a child to an adult, sex- and marriage-eligible, as wife- and husband-to-be. And, as regards the latter, as a hunter-provider: what is at the bottom of the male initiation is to "wake up the boy's heart", to his prey animal, the eland antelope, to whom the boy or novice hunter becomes mentally and experientially linked through symbolic and mimetic acts, as described by Lorna Marshall in her account of this protracted rite (1999: 155).

> Merely to rub the mixture on the surface of the skin, I was told later, would not "wake up" a boy's heart. The mixture must be rubbed into the boy through the cuts. The meat of the animal's "arm" (foreleg) in the mixture strengthens the boy's arm for pulling the bow. ... The zau powder makes the boy able to shoot far and accurately. The chest meat makes him say to himself as he sits in the encampment, "Why am I sitting here? Why am I not out looking for meat?" The blood in the second mixture also gives the boy the will to hunt. (Marshall 1999: 155)

The explicit purpose of these features of the rite is to confer to the initiand hunter the marks and skills of the trade, not so much its technical aspects—which he would have already acquired, incrementally since boyhood when he shot dung beetles with miniature arrows and birds and rodents with his slingshot or a scaled-down bow (Lebzelter 1934: 32)—but the hunt's supererogatory, magico-mystical aspects. As shown elsewhere (Guenther 2017), these consist of a set of prescriptions and proscriptions through which the hunter's sympathetic engagement with and control over the animal comes about. This is part of a hunter's skill set, as seen in Chap. 7, for which he is prepared as a boy or young man at initiation.

Part of his initiation rite is hunting an antelope, "shooting the eland" actually, rather than virtually, as does the girl at and through her first menstruation (as well as subsequently and at greater symbolic remove, as "shooting the eland" in /Xam is the circumlocutionary term for menstruation [Lewis-Williams 1981: 51]). This along with such other key elements as mimetic dancing, the strong presence and prominence of *n/om* and, at the base of it all, cosmology-pervading symbolic and somatic links with antelopes (specifically eland) link the male and female initiation rites. Indeed, even the "pan-Bushman or nearly so" (McCall 1970: 18) notion of the equivalence of hunting and mating resonates with the boy's rite, the implicit or explicit purpose of which is to develop and demonstrate a man's readiness for the status and role of husband and, beyond mating and sex, provider. "A boy may not marry until he has killed a big meat animal and had the rite performed", notes Lorna Marshall, writing about the Ju/'hoansi (1999: 154) and, throughout the !Kõ rite, the boys are enjoined "to hunt for their wives, to feed them properly, not to allow them to get thin, to clothe them properly with skins and to work for them" (Heinz 1994: 129).

The rite, referred to what Marshall and Lewis-Williams and Biesele call the "Rite of First Kill" (1999: 153–6) and the "First-Kill Rituals" (1978: 128–30), respectively, is one of two initiation rites Kalahari San people practise for young male members of their society. The other, obtained possibly through San contact with Bantu-speaking neighbours, is the Tshoma (choma) rite which, through generations or centuries of practice by San, has assumed traits that link it conceptually to the other San initiation rite (Marshall ibid.: 205); indeed, some San groups—such as the !Kõ—have amalgamated the two into the one male rite (Heinz 1994: 216–31). Tshoma is more elaborate ritually than the First Kill rite and its duration is weeks rather than days. It is also more restricted in its distribution and there is no evidence for a "bush school"-type of male initiation rite in the /Xam ethnographic record. This suggests an external cultural origin for the Tshoma and a San cultural provenance of the First Kill rite which was evidently also found among the /Xam, as part of a suite of eland-related ritual hunting practices (Lewis-Williams and Biesele 1978). Having described these two rites elsewhere (1986: 275–8, 1999: 167–73[16]), I will here deal with their aspect of transformation.

That aspect is more pronounced and ritually elaborated in the First Kill rite, which is directly focussed on the antelope killed by the teenage boy. This, ideally, is an eland (as in the case of the Ju/'hoan lad Kan//a, Lewis-William's and Biesele's key informant), but may be any large game antelope species (in the rite observed and filmed by the Marshalls it was a female wildebeest, killed by the 13-year-old /Ti!kay[17]). Whereas in the girl's rite such eland'ness as she experiences derives from her inner state, her feelings, thoughts and imagination, spurred through vicarious identification with mimetic eland dancers stomping and snorting outside her seclusion hut, the boys' access to the eland's or wildebeest's being is more physical and corporeal, through substances of the antelope's body that are transferred to the young hunter's body. The central—and universal—feature of the San First Kill rite is the administration of thin cuts to parts of the young hunter's body into which charred bits of boiled and pounded animal meat, froth and liquid from the broth, eland fat, burned eland hair

[16]See also Bleek (1928: 23–5) and Barnard (1980) for descriptions of the Naro case. Heinz (1994: 126–31) and Marshall (1999: 153–6, 203–20; see also Lee 1979: 238–40) have presented accounts of the !Kõ and Ju/'hoan cases, respectively. Schapera presents a summation of early ethnographic accounts from the colonial literature (1930: 122–6). Lewis-Williams and Biesele combine contemporary with historical ethnographic sources in a fascinating comparative paper of the rite among the /Xam and Ju/'hoansi (1978), the latter information elicited in part by the two researchers by posing to them questions about the male rite they drew from the Bleek/Lloyd archive.

[17]/Ti!kay's father orchestrated his son's First Kill rite after the latter had just shot the wildebeest with one of his father's arrows. He participated in the rite as one of the elders, administrating the medicinal cuts to the boy. Marshall identifies the father as Khan//a, along with his home village, Kai Kai. Lewis-Williams and Biesele's key informant, Kan//a, was from the same village and, at the approximate age of 65 at the time of the study. It would be informative to know if this was he the same man as /Ti!kay's father, in Marshall's study a couple of decades earlier. This would be an interesting and serendipitous coincidence of information, on inter-generational transmission of culture, from two separate ethnographic studies complementing each other. It also gives poignancy to more recent ethnographers' reports about male initiation rites having all but disappeared among contemporary San, two to three generations after these studies.

mixed with powered or charred plant "medicine" is rubbed. The hunter initiand is "cut with meat", say the Ju/'hoansi (Marshall 1999: 154); it must be rubbed deeply into the cuts in order to "wake up the boy's heart", leaving, thereby, indelible dark scars or tattoos—"meat marks" (Bleek 1927: 114)—on the hunter's body visible for the rest of his life and attesting to his participation in the rite.

It is at this point in the rite that the people deem the young hunter ready for big game hunting, whom they may see picking up his bow and leaving the ritual site, his heart, "awakened" and bent on hunting the animal and procuring its meat, which, transferred to his body via the "meat marks" (Bleek 1927: 114), is now also in his heart and mind: "His heart is burning hot towards meat;", Kan//a's group told Lewis-Williams and Biesele, "he desires meat. He has become a real hunter and will spend the whole day out and not come back to camp" (1978: 129).

There are several other ways in which a mystical-physiological connection between the young hunter and his prey animal is established. One—which we also saw in the menarcheal rite—is the application of species-defining antelope markings—such as the gemsbok's brow-shield pattern daubed on the young !Kõ intiand's face (Heinz 1994: 128). Two especially potent procedures for transferring essence utilize the animal's two life sustaining substances, its blood and its fat, the one sustaining the animal's own life, and the other the hunter's. The first, referred to as the "Blood Rite" by Lorna Marshall, consists of drinking the hunted antelope's blood; it is a ceremony performed for and by the young hunter both at the First Kill rite and at the subsequent three kills of big game animals (Marshall 1999: 156). The other procedure that transfers not only bodily but also, in terms of San symbolism and cosmology, spiritual essence from antelope to hunter consists of flicking eland fat over the boy's shoulders and smearing his body with this potency substance. The fat is rendered from the eland the young hunter has killed and flicked over him with parts of the same animal's body (forelock, ears, dewlap skin, tail tuft), while at the same time sticking hairs from the eland's ears to the boy's temple. Throughout this ritual procedure the initiand sits on an antelope skin around which the elder men attending to him make imprints with the clicking hooves of the hunted eland's foreleg, surrounding the skin with eland hoof prints. The eland's foot is then placed on the mat beside the boy (Lewis-Williams and Biesele 1978: 129). The antelope—specifically eland—skin was used in another symbolically and ontologically powerful way by southern San: as cloaks for the initiands (Jolly 2015: 209–10).

Some of these were sewn together from several eland skins, enveloping the upper bodies of a dozen or so boys, some of whose legs, as depicted in a north-eastern Cape rock painting, are undergoing antelope-transformation, an outgrowth, perhaps, of the surge of eland'ness that courses through their bodies from the eland skin that enfolds them.

Such ritual actions are explained, by Lewis-Williams and Biesele, in terms of sympathy-based "hunting magic"—eland chest meat to confer the "will to hunt", foreleg meat to strengthen his legs and running speed and endurance, cuts between the eyes to enhance keenness of vision, "flap-flapping" the eland's ears rather than holding them up thus rendering the animal less attentive when stalked by the hunter (Lee 1979: 239), and the like. The end effect and culmination of such ploys of sympathetic magic is the young hunter's readiness for the hunt, spurred on by the "burning" desire, in his heart, for meat.

While this explanation is no doubt on the mark, and, as discussed further later on in the chapter, applies also to hunting in general, outside the context of the ritualized "First Kill", my contention is that the young hunter's physical immersion in the prey animal likely established, at the experiential level, a physical bond with the animal. Its physical components—fat, blood, meat, hair, skin—become, at various points in the rite, placed in, on and over his body, some of it processed through his gastro-intestinal tract or, more salient experientially, rubbed into cuts in his skin or muscle. All of this may instil in the San initiand a sensation of bodily merging with the animal, a sense perception entertained at such moments of mystical-physical ontological boundary dissolution within the cosmological context of a connective cosmology and thus intellectually corroborated. This underscores the mental and emotional bonding he may feel for it through sympathetic magic and ritual at his hunter's initiation, and continue to feel for the remainder of his hunting career, *vis-à-vis* eland or antelope like them (especially keenly when he feel's his prey animal's "tappings", as seen in Chap. 7).

Such is the experiential trajectory also along which cross-species identity bonding moves for the girl, inside the hut, surrounded and lulled by singing eland dancers, whose eland-like clicking ornaments complement the eland-like stomping of their feet. The parallel here, to the boy, sitting on the antelope skin and surrounded by eland hoof prints created by clicking eland feet is especially beguiling. As is the other parallel, already noted above: both initiands "shoot the eland", she in spirit, as she seeks out eland or they seek out her in that fashion, he in person, as he takes is bow and sets out for the veld, a "real hunter", "with the will to hunt".

The second component of the male initiation rite, the Tshoma rite, points to yet another parallel it has with the female rite: mimetic dancing. This is especially elaborate in this rite's performative and mytho-magical aspects. The First Kill rite, more "quietly and simply" performed than the Tshoma rite (Marshall 1999: 205), may not have any dancing associated with it; for instance, Marshall makes no mention of its in her account. Lewis-Williams and Biesele make mention of an all-men "regular eland dance" performance at the beginning of the rite (a ritual occurrence that may accompany any eland kill after the butchering and meat division, in part because of the joyousness of the occasion, in part to assuage possible resentments over meat division). Moreover, at the conclusion of the Tshoma rite, in the presence once again of all camp members, the eland song may be sung by everyone, intriguingly described by one of the two ethnographer's informants as resembling "the wailing of hyenas cheated of the eland meat" (reflecting the human hunters' awareness of the presence of equally aware non-human fellow-hunters, whom they had just managed to get the better of).

Like the healing dance, the Tshoma dance is held at night, around a fire at a special Tshoma site about half a kilometre from the camp and enclosed by a brush fence, to underscore its seclusion and privacy, as a male ritual site from which women and children are excluded. The men and boy dance the all-night long dance every night, days and weeks on end. Its physical strain on the boys is part of their endurance testing, along with meagre rations and a number of food proscriptions that further reduce their food intake. During the day they sleep in flimsy grass huts, on the open ground and without cover, as exposure to cold is part of what the initiands are expected to endure at the Tshoma. Women are not allowed participation in the Tshoma rite and dance as they are in the healing dance.

Whereas the goal of "enhancing the power to hunt" in the initiand is achieved in the First Kill rite by actual hunting and ritual acts of sympathy and transference of animal essence physically, in the Tshoma rite there is no actual hunt and what is transferred, through dancing, is more mystical than it is physical (or a blend of the two). What is transferred is *n/om*. However, the *n/om* here is not a shaman's *n/om* but a hunter's "*Tshoma* n/om" which, according to some of the Marshall's informants—the older though not younger ones—are different potency types, the one conferring healing powers the other enhancing the power to hunt. The way Tshoma *n/om* is dispensed by the older men to the initiates points to another similarity to the healing dance (and, by extension trance dancing and hunting):

by means of sweat they exude through their strenuous dancing, which the older man rubs over the body of the younger (and shaman curer over his patient's body). This transfers the *n/om*—much the same way that eland potency is transferred to him in the First Kill rite, when his body is rubbed with eland fat, one of the most potent repositories of *n/om* available to San from the natural world.

Another such substance is honey which, at the metaphorical level, too, is transferred to the boys at Tshoma (and through it, through its linkage to sex, sexual potency and energy). One way in which this transfer comes about, mytho-magically, is in the use of the bullroarer, the rite's principal ritual prop (called *n≠abbi* in Ju/'hoan and *hisi* in Naro). This somewhat enigmatic and to the San vaguely portentous object, possibly of Bantu provenance[18] (Marshall 1999: 204), is symbolically and physically linked to bees and is used only on that occasion, being kept hidden during the five- to six-year intervals between Tshoma rites. It is used at night, when one of the old men will sound the object, once or twice during a nocturnal dance a distance away in the dark and away from the fire which the object is to be kept away from. Yet, fire is also brought into contact with the bullroarer for one of the rite's most important ritual procedure: shavings from the object were charred and the powder is rubbed into the initiand's skin. While different in substance—vegetal instead of faunal—the procedure here is the same as scarifications administered at the First Kill rite.

Marshall does not indicate any reason for this ritual act; an intriguing explanation was recently suggested by Neil Rusch, in the context of /Xam myth and cosmology (2017) on the basis of /Xam ethnography and rock art imagery. Rusch links the bullroarers to bees, in terms of both a practical and a metaphorical goal: attracting and directing bee swarms and harnessing potency, which bees and honey signify in /Xam belief (see also Hollmann 2005: 172; Lewis-Williams 2015a: 68–9, 92–6). As noted above, this symbolic chain also extends, via fat, to elands and antelopes. Thus, through these metaphorical links, scarification by means of charred bullroarer wood and by antelope substances in the Tshoma and First Kill rites, respectively, are symbolically equivalent ritual operations, both connecting the initiand somato-psychically to his prey animal.

Among the Naro the bullroarer had another function, pointing to yet another dimension of San initiation rite, having to do with spirit and myth beings. The object was used to attract the trickster-god to the initiation

[18] However, it was used also by the /Xam, and actual specimens of bullroarers were obtained from them by Bleek (Rusch 2017).

site. He is drawn to the sound of the bullroarer; indeed, he "talks" like one, as suggested to me by a Naro informant, who also pointed out a related function of sounding bullroarer: to impersonate //Gaũwa (or Hise) to make his presence real to the boys at the */ri kaxu* rite (Guenther 1986: 277–8). Another such ploy was to erect small grass huts out in a remote spot in the veld and tell the lads that this was "//Gaũwa's house". I discussed in an earlier chapter (Chap. 2) Hise's appearance at the initiation site, as per a number of bizarre accounts Dorothea Bleek collected from people who claimed to have seen them (in visions or dreams). Some of the Hise-//Gaũwa beings came to dance with the boys and men (Bleek 1927: 122). Among the Ju/'hoansi, notes Nicholas England, certain Tshoma songs, when "sung under the prescribed conditions, brings about an appearance from the god", who will, once present, "handle" some of the mytho-magical aspects of Thoma dancing. The god's ministrations include the "activation" of the medicines inherent in the dance, conferring this power to the "initiates to have and use later throughout their lives", especially should they become "medicine men" (1963: 4).

There were occasions among some Kalahari San groups at which the trickster-divinity appeared at the male passage rite in his female persona—or, "sometimes in double form as man and woman", as among the Naro (Bleek 1927: 122). Apart from its liminality-enhancing effect, the gender reversal of what is usually a male spirit or myth being, featuring the same as a woman, who dances with the men and boys, gives a sexual overtone to the male rite (which, as seen above, in much more embellished and raunchy tones, is one of the dominant themes in the female rite, especially the eland dance). We find what is perhaps that overtone's most explicit expression in the !Kõ rite, to which the female trickster-god comes to taunt and tempt the young hunters and men until their mothers and wives drive them away with shouts and insults (Heinz and Lee 1978: 114). Here once again we come across a mirror-image parallel to the girl's rite: the male trickster, when appearing at the girl's initiation site as is his wont among the Naro, does so because he is sexually attracted to the young girl-woman in the menstrual hut (as is the /Xam rain divinity !Khwa to a secluded "maiden"). And among the !Kung-speaking San of Central Angola whom Dorothea Bleek visited in 1925, it is //Gaũwa himself who may lead the dancers at the menarcheal rites; moreover, the tattoo marks the girl receives "are made in his honour" (1927: 122).

The initiation rite and site, then, much like the trance dance, is the other ritual sphere, within the real world, into which myth and spirit beings may step and make their presence known to the people active at the rite's ritual space. The trickster from the myth and spirit world has special significance at initiations, of both boys and girls. Apart from the attainment of social and gender maturity and its symbolic and kinaesthetic manifestation in each rite—and its drawing, as just noted, a lustful trickster-divinity to the site—both rites have to do with hunting, that is, "shooting eland", in spirit or body. *Qua* that antelope's—and any other game animal's—Spirit Protector, the trickster-divinity's role is cut out for him (or her) when present at a boy's rite. As noted by Dorothea Bleek, Hise, as "protector of game" was also "giver of good and bad luck to the hunter" (1927: 123). We see him (or her) dispense the last in an account I received from a Naro informant who told me that when attending a male initiation rite Hise or //Gaũwa would sometimes shoot arrows at the young men, in random fashion, which, when hitting their target, would cause a hunter to fall ill. Other hits might cause him to become a trance dancer (rather than hunter). Aware of the spirit being's potentially spoiler effect on the rite's intent—to raise the next generation of successful hunters—vigilant old men would attempt to keep him at bay and, when noticing his approach, drive the trickster-god back out into the veld (Guenther 1986: 277–8). Is this perhaps why the spirit being, when visiting, as a woman, the !Kõ initiation site as temptress, to distract the men from their manly business at hand? Were they insulted and shooed by the women not out of sexual jealousy with regard to their husbands or lovers but concern over the rite's successful outcome, making hunters and providers of their sons?

That outcome is well served by the dancing that takes place at initiation. Like the eland dance at the girl's rite, the dances at men's rite (called *tshxai!go* and */gi* by the Ju/'hoansi and Naro, respectively) are vigorously mimetic, not of only of eland but of a diversity of other animals, including predators such as leopards, jackals and hyenas. Also included in the roster of animals is the lion, who is not only danced mimetically but whose identity may be assumed through metamorphosis: lion transformation is something young initiands also learn about and, according to one of the Ju/'hoan informants in Le Roux and White's oral history project, were also, in the past, taught to do themselves (2004: 57).

The animal dances at the male initiation rite were danced to chants, that, while wordless, were frequently named after animals ("Giraffe", "Long-eared Fox", "Ostrich") or animals' body parts ("hyena Penis"), or a variety of other things, such as rain, honey or sun (Marshall 1999: 213–4). Dorothea Bleek mentions the Naro ostrich dance—"a weird solemn tune with the refrain of 'honk a honk'"—sung by the men while stomping their feet in circular motion and waving their arms. Another component of the ostrich pantomime was wearing ostrich feathers or "the head and beak of the black and white stork as head ornament" (1928: 24, 25). Marshall describes the Tshoma dance chant as "gasp or grunt a pattern of vocables" produced—citing the musicologist Nicholas England—by a "sudden pressure from the diaphragm combined with tension of the vocal folds, and sometimes ... a glottal release" (ibid.: 214). Marshall witnessed one of the "practice dances" that precede an actual Tshoma rite, noting that, when the lads "shouted the sounds that animals make", each of them a different species, the resulting animal babel was a "loud hubbub". John Marshall, a lad himself at the time of her field work, participated in the same dance. Having been told by one of his Ju/'hoan friends and fellow-participant that "they were making special animal sounds so I would 'know' the animals", John Marshall's sense of these dances was that they imparted "some kind of spiritual 'knowledge' or awareness that would help me know what to do when I met the animal and what the animal might be intending to do" (ibid.: 211).

The degree and intensity of absorption of such mimetic dancing, with steps and gestures, vocalizations and adornments attuned to or derived from the danced animal, is similar to what we see in the eland dance. In both cases we can assume that such dancing might elicit ontological wavering within the human dancer between humanness to animalness (more so in some than in others and in some dancers, perhaps not at all as the degree of "musicality" on matters mystical varies amongst individuals). Such an experience would keep the dancer, an actual or novice hunter, attuned to the animal and his hunter's mission in relation to it. So attuned, he might be both inattentive and resistant to the wily Trickster-Animal Protectress's attempts at disrupting that mission.

We turn now to another arena—of play and recreation—in which the San perform dances, many of them mimicking animals and hunters, with much the same effect: of edging, through mimesis, towards metamorphosis, and experiencing, more or less subliminally, ontological mutability.

References

Alexander, James E. 1838. *An Expedition of Discovery into the Interior of Africa Through the Hitherto Undescribed Countries of the Great Namaquas, Boschmans, and Hill Damara. Vol. 2. London: Henry Colburn, Publisher.* Reprint edition. London: Henry Colburn.

Barnard, Alan. 1979. Nharo Bushman Medicine Men. *Africa* 49: 68–80.

———. 1980. Sex Roles Among the Nharo Bushmen of Botswana. *Africa* 50: 115–124.

Bennum, Neil. 2004. *The Broken String: The Last Words of an Extinct People.* London: Viking.

Biesele, Megan. 1993. *Women Like Meat: The Folklore and Foraging Ideology of the Kalahari Ju/'hoansi.* Johannesburg/Bloomington: Witwatersrand University Press/Indiana University Press.

Bleek, Wilhelm H.I. 1864. *Reynard the Fox in South Africa; or Hottentot Fables and Tales.* London: Trübener and Co.

Bleek, Dorothea. 1927. Bushmen of Central Angola. *Bantu Studies* 3: 105–125.

———. 1928. *The Naron.* Cambridge: Cambridge University Press.

———. 1935. Beliefs and Customs of the /Xam Bushmen. Part VII. Sorcerors. *Bantu Studies* 9: 1–48.

Deacon, Janette. 1994. Rock Engravings and the Folklore of Bleek and Lloyd's / Xam San Informants. In *Contested Images: Diversity in Southern African Rock Art Research*, ed. Thomas A. Dowson and David Lewis-Williams, 238–256. Johannesburg: Witwatersrand University Press.

DeMothe, Kimerer L. 2015. When Words Don't Get It: The Challenges of Writing About Ritual Dance. Paper Presented at the Society of Dance History Scholars/ Congress of Research Dance (SDHS/CORD Conference), Iowa City, University of Iowa, November 13–16, 2014, Conference Proceedings, 127–134.

Dornan, Samuel S. 1925. *Pygmies and Bushmen of the Kalahari.* London: Seeley, Service & Co.

Eastwood, Edward B., and Cathelijne Eastwood. 2006. *Capturing the Spoor: An Exploration of Southern African Rock Art.* Cape Town: David Philip.

England, Nicholas. 1963. Fieldwork in the Kalahari. Paper presented at the Symposium on the Current Research in Ethnomusicology, University of Washington, March 7–9, 1963, 1–9. https://www.ethnomusicology.org/page/Resources_Symposium?

———. 1968. Music Among the Zu/'wa-si of South West Africa and Botswana. Ph.D. Dissertation, Harvard University. (Published as *Music Among the Ju/'hoansi and Related Peoples of Namibia, Botswana and Angola.* New York: Garland Publishing, 1992.)

Gentz, Leutnant. 1903/04. Einige Beiträge zur Kenntnis der südwestafrikanischen Völkerschaften in Deutsch-Südwestafrika. *Deutsche Kolonialzeitung* 26: 450–452.

Grinnell, Jon, Craig Packer, and Anne E. Pusey. 1995. Cooperation, in Male Lions: Kinship, Reciprocity or Mutualism? *Animal Behaviour* 49: 95–105.

Guenther, Mathias. 1986. *The Nharo Bushman of Botswana: Tradition and Change*. Hamburg: Helmut Buske Verlag.

———. 1992. 'Not a Bushman Thing': Witchcraft among the Bushmen and Hunter-Gatherers. *Anthropos* 87: 83–107.

———. 1999. *Tricksters and Trancers Bushman Religion and Society*. Bloomington: Indiana University Press.

———. 2005. The Professionlisation and Commoditisation of the Contemporary Bushman Trance Dancer and Trance Dance, and the Decline of Sharing. In *Property and Equality, Volume 2: Encapsulation, Commercialisation, Discrimination*, ed. Thomas Widlok and Wolde Gossa Tadesse, 208–230. Oxford: Berghahn Books.

———. 2017. '…The Eyes Are No Longer Wild, You Have Taken the Kudu into Your Mind': The Supererogatory Aspect of San Hunting. *The South African Archaeological Bulletin* 72: 3–16.

———. 2018. 'I Can Feel My Lion Hair Growing and My Lion Teeth Forming…': San Lion Transformation – Real or Imagined? God's Truth or Hocus Pocus? Paper Presented at the 24th Biannual Meeting of the Society of Africanist Archaeologists, University of Toronto, June 16–21, 2018.

Heinz, Hans-Joachim. 1975. !ko-Buschmänner (Südafrikas, Kalahari): Mädchen Initiation. *Encyclopaedia Cinematographica*, E 1849: 3–15.

———. 1994. *Social Organization of the !Kõ Bushmen*. Cologne: Rüdiger Köppe Verlag.

Heinz, Hans-Joachim, and Marshall Lee. 1978. *Namkwa: Life Among the Bushmen*. London: Jonathan Cape.

Hewitt, Roger. 1986. *Structure, Meaning and Ritual in the Narratives of the Southern San*. Hamburg: Helmut Buske Verlag.

———. 2008. *Structure, Meaning and Ritual in the Narratives of the Southern San*. 2nd ed. Johannesburg: Wits University Press.

Hollmann, Jeremy C., ed. 2005. *Customs and Beliefs of the /Xam Bushmen*. Johannesburg/Philadelphia: Wits University Press/Ringing Rock Press.

Ijäs, Mikko. 2017. Fragments of the Hunt: Persistence Hunting, Tracking and Prehistoric Art. Doctoral Thesis, Aalto University: School of Arts, Design and Architecture, Department of Art.

Jansen, William H. 1959. The Esoteric-Exoteric Factor in Folklore. *Fabula: Journal of Folklore Studies* 2: 205–211.

Jolly, Pieter. 2002. Therianthropes in San Rock Art. *South African Archaeological Bulletin* 57 (176): 85–103.

———. 2015. *Sonqua: Southern San History and Art After Contact.* Cape Town: Southern Cross Ventures.

Katz, Richard. 1976. Education for Transcendence: !Kia-Healing with the Kalahari !Kung. In *Kalahari Hunter-Gatherers: Studies of the !Kung San and Their Neighbours*, ed. Richard B. Lee and Irven DeVore, 281–301. Cambridge, MA: Cambridge University Press.

———. 1982. *Boiling Energy: Community Healing Among the Kalahari Kung.* Harvard: Harvard University Press.

Katz, Richard, Megan Biesele, and Verna St. Denis. 1997. *"Healing Makes Our Hearts Happy": Spirituality and Cultural Transformation Among the Kalahari Ju/'hoansi.* Rochester: Inner Traditions.

Keeney, Bradford. 1999. *Kalahari Bushman Healers.* Philadelphia: Ringing Rocks Press.

Keeney, Bradford, and Hillary Keeney. 2013. Reentry into First Creation: A Contextual Frame for the Ju/'hoan Bushman Performance of Puberty Rites, Story Telling, and Healing Dance. *Journal of Anthropological Research* 69: 65–86.

———, eds. 2015. *Way of the Bushman as Told by the Tribal Elders.* Rochester: Bear & Company.

Kinahan, John. 1991. *Pastoral Nomads of the Namib Desert: The People History Forgot.* Windhoek: Namibia Archaeological Trust.

Lang, Andrew. 1901. *Myth, Ritual and Religion.* Vol. I. 2nd ed. London: Longman, Green and Co. (Both Available as ebooks Through Project Gutenberg, Released November 12, 2009).

Lawy, Jenny. 2016. An Ethnography of San: Minority Recognition and Voice in Botswana. Unpublished D.Phil. Thesis, University of Edinburgh.

Le Roux, Willemien, and Alison White, eds. 2004. *Voices of the San.* Cape Town: Kwela Books.

Lebzelter, Viktor. 1934. *Eingeborenenkulturen von Südwestafrika: Die Buschmänner.* Leipzig: Verlag Karl W. Hiersemann.

Lee, Richard B. 1968. The Sociology of !Kung Bushman Trance Performances. In *Trance and Possession States*, ed. Raymond Prince, 35–54. Montreal: R. M. Bucke Memorial Society.

———. 1979. *The !Kung San Men, Women, and Work in a Foraging Society.* Cambridge: Cambridge University Press.

———. 2003. *The Dobe Ju/'hoansi.* 3rd ed. Fort Worth/Belmont: Wadsworth.

Lewis-Williams, David J. 1981. *Believing and Seeing: Symbolic Meanings in Southern San Rock Paintings.* New York: Academic Press.

———. 1988. The World of Man and the World of Spirit: An Interpretation of the Linton Rock Paintings. In *Margaret Shaw Lecture 2.* Cape Town: South African Museum.

———. 2015a. *Myth and Meaning: San-Bushman Folklore in Global Context.* Walnut Creek: Left Coast Press.
Lewis-Williams, David J., and Megan Biesele. 1978. Eland Hunting Rituals Among Northern and Southern San Groups. *Africa* 48: 117–134.
Lewis-Williams, David J., and David G. Pearce. 2004a. *San Spirituality: Roots, Expression, and Social Consequences.* Walnut Creek: Altamira Press.
———. 2004b. Southern African Rock Paintings as Social Intervention: A Study of Rain-Control Images. *African Archaeological Review* 21: 199–228.
Low, Chris. 2013. The Role of the Body in Bushman Healing Dances. Paper Presented at the 10th International Conference on Hunting-Gathering Societies (CHAGS 10), Liverpool, June 25–28, 2013.
Marshall, Lorna. 1969. The Medicine Dance of the !Kung Bushmen. *Africa* 39 (4): 347–381.
Marshall, Lorna J. 1999. *Nyae Nyae !Kung Beliefs and Rites*, Peabody Museum Monographs, No. 8. Cambridge, MA: Harvard University.
McCall, Daniel F. 1970. Wolf Courts Girl: The Equivalence of Hunting and Mating in Bushman Thought. *Ohio University Papers in International Studies, Africa Series*, No. 7. Athens: Ohio University, Center for International Studies.
McGranaghan, Mark. 2012. Foragers on the Frontiers: The /Xam Bushmen of the Northern Cape, South Africa, in the Nineteenth Century. Unpublished Ph.D. Thesis, Oxford University.
Parkington, John. 2002a. *The Mantis, the Eland and the Hunter Follow the San….* Cape Town: Creda Communications.
———. 2002b. Men, Women and Eland: Hunting and Gender Among the San of Southern Africa. In *In Pursuit of Gender Worldwide: Archaeological Approaches*, ed. Sarah M. Nelson and Myriam Rosen-Ayalon, 93–118. Walnut Creek: Altamira Press.
———. 2003. Eland and Therianthropes in Southern African Rock Art: When Is a Person an Animal? *African Archaeological Review* 20 (3): 135–147.
Passarge, Siegfried. 1907. *Die Buschmänner der Kalahari.* Berlin: Dietrich Reimer.
Rusch, Neil. 2017. Sound Artefacts: Recreating and Reconnecting the Sound of the *!goin !goin* with the Southern Bushmen and Bees. *Hunter Gatherer Research*, 2015 3 (2): 187–226.
Schapera, Isaac. 1930. *The Khoisan Peoples of South Africa: Bushmen and Hottentots.* London: Routledge & Kegan Paul.
Schmidt, Sigrid. 2013. *A Catalogue of Khoisan Folktales of Southern Africa*, 2 vols. Research in Khoisan Studies 28.1&2. Cologne: Rüdiger Köppe Verlag.
Silberbauer, George B. 1963. Marriage and the Girl's Puberty Rite of the G/wi Bushmen. *Africa* 33: 12–24.
———. 1972. The G/wi Bushmen. In *Hunter-Gatherers Today*, ed. M.G. Bicchieri, 271–326. New York: Holt, Rinehart and Winston.

Skotnes, Pippa. 1996. The Thin Black Line: Diversity and Transformation in the Bleek and Lloyd Collection and the Paintings of the Southern San. In *Voices from the Past: /Xam Bushmen and the Bleek and Lloyd Collection*, ed. Janette Deacon and Thomas A. Dowson, 234–244. Johannesburg: Wits University Press.

Solomon, Anne. 1997. The Myth of Ritual Origin? Ethnography, Mythology and the Interpretation of San Rock Art. *South African Archaeological Bulletin* 52: 3–13.

———. 2011. Writing San Histories: The /Xam and 'Shamanism' Revisited. *Journal of Southern African Studies* 37: 99–117.

Sugawara, Kazuyoshi. 2015. On the G/ui Experiences of Being Hunted: An Analysis of Oral Discourse on the Man-Killings by Lions. Paper Presented at the 12th International Conference of Hunting-Gathering Societies (CHAGS 12), University of Vienna, Vienna, September 7–12, 2015.

Suzman, James. 2017. *Affluence Without Abundance: The Disappearing World of the Bushmen*. New York: Bloomsbury.

Tanaka, Jiro. 1996. The World of Animals Viewed by the San. *African Studies Monographs*, Supplementary Issue, 22: 11–28.

Thorp, Carolyn. 2013. 'Frog People' of the Drakensberg. *Southern African Humanities* 25: 245–262.

Turner, Victor. 1970. *The Forest of Symbols*. Ithaca: Cornell University Press.

Van Gennep, A. 1908. *The Rites of Passage*. Trans. M.B. Vizedom and G.C. Caffee. Chicago: Phoenix Books.

Vinnicombe, Patricia. 2001 [1976]. *People of the Eland: Rock Paintings of the Drakensberg Bushmen as a Reflection of Their Life and Thought*. Johannesburg: Wits Press.

Wittenberg, Hermann. 2014. The Story of //Kabbo and *Reynard the Fox*: Notes on Cultural and Textual History. In *Courage of //Kabbo: Celebrating the 100th Anniversary of the Publication of* Specimens of Bushman Folklore, ed. Janette Deacon and Pippa Skotnes, 93–98. Cape Town: UCT Press.

CHAPTER 6

Animals in San Dance and Play: Between Mimesis and Metamorphosis

It is a circular dance in which they use their bodies to mimic the cumbersome movements of the galloping gnu, while with one hand sketching the length of the beard-like head hair of the wildebeest. Suddenly the circle opens. One of the participants plays the role of the wildebeest as it defends itself, lunging about with its horns. Others, barking loudly, enact dogs that follow and bring to bay the game animal and try to bite and hold on to the thigh of its hind leg – all of it a performance that at times is in no way lacking in realism. The rest of the people play the hunters, who throw their spears at the animal, which finally collapses and is delivered the coup de grâce *by jubilant hunters, to the shrill singing of hand-clapping women.*

Gentz 1904: 157, my translation

The /gae-ǂnab *or gemsbuck* [sic] *dance represents a gemsbuck bull, cow and calf being chased by two men with three or four dogs. The bull and cow carry horns but the calf does not. The gemsbuck lead and are followed by the men and dogs, who dance to and fro on the flanks until the animals are brought to bay. Then the men pretend to shoot and one of the buck goes down wounded while the others make off pursued by some of the dogs. The wounded animal fights in realistic fashion, sweeping at both man and dogs with its horns. The dance is continued until all of the animals have been accounted for. Bows and arrows are represented by sticks in this dance.*

Fourie (1925/26 61)

© The Author(s) 2020

M. Guenther, *Human-Animal Relationships in San and Hunter-Gatherer Cosmology, Volume I*,
https://doi.org/10.1007/978-3-030-21182-0_6

This is a very exciting drama. The girls act as springboks with little children as the kids. Two or three men act as lions. The girls fold up and secure their skin petticoats [karosses] *till they look like "shorts", the better to imitate animals. Then they imitate in cries and actions the graceful springboks with wonderful skill. They prance along plucking at the grass with their hands and pretending to feed; every now and then uttering whistles of alarm, they bunch together with the kids between them; then run off again bounding over imaginary tracks and obstacles. The lions stalk with consummate skill and crouch at strategic spots. Every now and then, one of them makes a rush and a bound, catching a kid and pretending to rend and devour it. Soon all the kids are lying about dead, and the real fun commences when the lions catch the plump springboks, which they do with much tantalizing on the part of the girls … In one depiction of this play, which I witnessed, the end came dramatically. One springbok had long evaded two lions till eventually both sprang at once from opposite sides, but instead of catching the girl, they caught one another and then was depicted a wonderful imitating of a lion fight with mighty roaring, snarling, and contortion of faces. One lion eventually emerged with nose bleeding furiously.*

Doke (1937: 90)

Non-ritual dancing is a much-practised component of San recreational life. Children perform some of it, as part of the culture's "child's play" repertoire, but adults do as well and, like storytelling, dancing is an established feature of adult down-time. As noted by Isaac Schapera, in a valuable section of his classic *Khoisan Peoples* that deals with this underreported field of San expressive culture,[1] "dancing is often indulged in purely for pleasure, especially in times of plenty, and may take place any fine moonlit

[1]Apart from specialist studies on child development and socialization, or socio-biology (e.g. Sbrzesny 1976), recent and contemporary general San ethnographies hardly ever include discussions of recreational dance and play (except for a Ju/'hoansi competitive "traditional dance" troupe described by Richard Lee [2003: 201–5]; see also Hermans 1998). Early ethnographic accounts included this information (viz. Stow 1905: 97–102; Fourie 1925/26: 30–3; Bleek 1928: 18–23, 1927: 119–22; Gentz 1904 156–7; Wilhelm 2005: 172–4; Kaufmann 2005: 72–4; Doke 1937). I note as well that, with the exception of Vinnicombe's section on "Dance, Mime and Music" in her *People of the Eland* (2001: 299–310) and some of the German socio-biological work on the !Kõ, San dancing is almost exclusively viewed in religious rather than expressive terms in more recent San ethnography, as an element of the "trance hypothesis".

night, when the people have had a good meal, and there are enough of them to make it festive" (1930: 202). Doke, who studied "games, plays and dances" among the ≠Khomani in the 1930s, refers to this ludic aspect of San culture as "the lighter side of Bushman life" (1937: 90, cited in McCall 1970: 6). The bucolic tenor of such spontaneous, for-the-joy-of-it-all dancing is reflected also from the /Xam story teller /Han≠kasso's account of a dance that followed in the after-glow of a successful honey-gathering outing from which the men return, laden with bags full of cut honey, to their women at home camp who begin to "frolic" on their arrival, by clapping and chanting and getting out the drum for the men who, after a bit of a rest, join in the dance. "While they feel that the women are satisfied with food", they dance until sunrise, when fatigue sets in, because "they have danced strongly", kicking up dust that "covers the women's faces" (Bleek and Lloyd 1911: 356–57; see also Rusch 2017).

Adding to the prominence of dance and play in the lives of people is the fact that ludic components can also be found in ritual dancing and that ludic dancing may itself include ritual elements (in a system of religious belief and practice that draws no clear distinction between sacred and profane or the numinous and the frivolous[2]). We see it in play in the early, warm-up phase of an all-night trance dance when children, from toddlers to teens, may join the dance circle and play at trancing and dancing, at times with great hilarity triggered by the at times exaggerated clumsiness and drollness of their dance steps and moves (Guenther 1986: 256–7, 1999: 183; Biesele 1993: 74–5; Marshall 1999: 85). Moreover, the trance dance may itself be a game, for children (Heinz 1975: 7), as may the Eland Dance, including the part of the secluded girl played at by one of the little girls (Sbrzesny 1974). In its adult version, the Eland Dance, along with its male counterpart, the *tshxai !ŋo* and */ŋi* dances, both performed in a ritual

[2] This is a corollary to what my book *Tricksters and Trancers* is about: an examination of an open, ambiguous, fluid, resilient and tolerant religion in which myth and belief feature a trickster divinity who, as "*Ur*shaman", has a presence also in ritual. This sort of ambiguity and fluidity would allow for dances that combine playful, diversionary ease (even, at moments, irreverence) with earnest ritual intent. It would also obviate moot speculation about "hidden meanings" on the "ritual character" of San ludic dancing, which, so it is reasoned, might, in the past, have been ritual dancing (Schapera 1930: 204). Antje Krog sees this conjoining of play and ritual as part of the Khoisan "interconnected world view": "an all-encompassing philosophy … with 'the wholeness of life' – religious and secular, spiritual and material – which can never be compartmentalized or understood in isolation from one another" (2009: 184, quoted in Wessels 2012: 187).

context charged at moments with mytho-magical portent and aquiver with liminality, the performance contain a strong measure of drama and mimetic histrionics, along with, in the male dances, a dose of playfulness. Other adult recreational dances may be tempered with ritual overtones; for instance, the Ju/'hoan Drum Dance derived from the Bantu-speaking Mbukushu (Lee 2003: 136–7). As to be discussed below, some ludic dances may actually generate a trance state in the dancer while another ludic dance, the Gemsbok Dance, among the Central Kalahari San, may evidently be performed as both a "hunting" and a "healing" dance, giving different intents and purposes to one and the same dance (Tanaka 1996: 22). All of this obscures the distinction between recreational and ritual dancing. Ritual and ludic dancing, as well as singing that is more or less directly associated with it, while different in style and purpose—"potency filled" the one, "purely for pleasure" the other (Lewis-Williams 2015: 191)—do also overlap, with respect to content and intent.

The common element in San dances—and much of their children's games (Sbrzesny 1976: 128–47) and adult play (McCall 1970: 7)—is mimicry of animals (Schapera 1930: 203–5). The animals mimicked cover the widest range of species; bees, frogs and snakes, baboons and vervet monkeys, ostriches and vultures, steenboks, duikers, kudus, wildebeest and elands, honey badger, dogs, jackals and hyenas are the ones I have come across in my reading and my own field work. What I have rarely if ever come across and what is conspicuous through its almost-absence from the inventory of animals mimed in ludic dancing is the lion; presumably this is because of the respect relationship people have towards this predator and because of its ominous-numinous role in the trance dance (although, as seen earlier, the lion gets its due share of lampooning in myths and stories; moreover, it is included in the roster of animals mimed in Tshoma dancing).

This wide cast of animal characters that people engage with mimetically in dancing that, as seen from the two examples featured in the section epigraphs, may be spirited and strenuous through its complicated choreography, is drawn primarily, it seems, from "real-life experiences" with animals people encountered in the veld, as suggested by Fourie (1925/26: 33), writing about the Hai//om (Hei//kum) of the then South West Africa (Namibia). What made these "actual animal experiences" remarkable and memorable and warranting re-enactment and re-experience (or, in the case of stories, retelling) is that they were either uncanny or dangerous, and thus all the more salient mentally and affectively.

An example is the Hei//kum woman's encounter out in the veld with an albino monkey whose subsequent report to her people at home about seeing this "wonderful thing" developed, probably after many retellings, into a dance (Schapera 1930: 204–5). Fourie describes the dance's mimetic aspects of the experience:

> The part of the monkey is taken by a woman who stands with her hand covering the sides of her face, leaving only a small opening for peeping through. She sways her head from side to side. The other women form up in a row opposite to her. One goes forward to look at the monkey, gazes at it intently and, when the latter starts moving its head from side to side, beckons and calls to the others to come and see the wonderful thing. This they do one by one, each in turn beckoning to the others to come and see. (Fourie 1965: 33)

Another Hai//kom women's dance was based on a highly dangerous experience one of them had had with a ringhals snake, who had spat into her eyes while she was "trying to drive mice out of a shrub by means of a stick" (Fourie 1965: 33). This is how the experience played out in the dance, again with high drama and mimetic choreography:

> One woman acts as a snake, standing with her hand and forearm raised above the forehead. As the other women approach her she hisses and strikes with the raised hand after the manner of a snake, causing them to scream and retire. (ibid.)

Another experience—spiced as it is with an element of danger, risk and excitement—that may yield a mimetic dance is the emptying of a beehive. Among other /Xam groups the replay of this event included a vivid pantomime of a bee swarm into which the dancers "transformed themselves ... with a buzzing chorus" (Schapera 1930: 204; see also Gentz 1904: 157, for an example from the ≠A//eisi). The latter sound may also have been produced with a bullroarer,[3] and as noted above in the account of male initiation, the other ritual-ludic context for the employment of this object.

[3] See Rusch (2017) for an examination of the use of the bullroarer in the context of the ethological and subsistence, as well as mystical and mythical aspects of honey-gathering and bee imagery and symbolism. He also deals with the vibroacoustic dimension of bullroarer sounding and its connection to bee buzzing, linking this feature of San ritual and ludic culture to San music-making. This linkage is evident in Bushmen, when hearing an organ at a church service to which they were taken in Cape Town by the missionary Johannes Kicherer, "mistak[ing] the sound of the organ with the swarming of bees" (Jolly 2015: 78).

Other real-life animal encounters, experienced by only men, are dangerous hunts, giving rise to the sorts of dances described by Gentz and Doke in the epigraph above, each of which was likely triggered by an actual hunt, or represents an idealized, dramatized composite of several hunts, by hunters over decades and generations. The agonistic element of the hunter-prey relationship is sometimes embellished by the dancers, what Heunemann and Heinz (1975a), in one of their remarkable films of !Kõ dances refer to as the "*Wettstreit between Jäger und Tier*" ("competition between hunter and animal"). This is a carefully choreographed, ritualized game with hand gestures that indicate agonistic actions between a hunter and his prey animal (such as ostrich or steenbok) that includes such "explosive vocalizations" as grunting, panting and wheezing. What the performance especially appears to highlight is the "killing moment", the dance's—and its referent, the hunt's—*grand finale*.

A further example of an energetic ludic animal dance that dramatizes the agonistic element, albeit in a different manner and sense, was described by the German colonial army doctor and explorer H. Werner in some detail, along with photographs he took in the field of the performance (1906: 249–51). It features two men, crouched on the ground and growling at each other with fierce grimacing. The dance, Werner was told by his Hai//om informants, depicts the antagonists not as hunter and prey but as two dogs or monkeys engaged in a food fight! Fourie describes a very similar dance among the same people, in which the two animals featured are hyena and jackal. All of the performers of the vividly mimetic dance are men, one of whom mimes the hyena trying to feed on a carcass, while being harassed by two men miming the jackals, getting ever closer to the carcass, dodging the hyena's charges, who manages to chase them off and return to the carcass, when the same action resumes (1927: 32). Thus, not all of the men's agonistic animal dances were about the hunter-prey relationship; some of these beguiling ludic performances had their sight set exclusively on the animal and its actions and antics, the other-being so prominent in San myth and cosmology and the collective imagination.

The most widespread hunter-prey dance appears to be the Gemsbok Dance, perhaps because of the exceptional danger of this antelope species whose spear-like horns can eviscerate a hunter or his dog. The /Xam practised a dance in which the dancers—young men and young women and girls—"imitate the ways of gemsbok with dogs", as described by Diä!kwain to Lucy Lloyd, with reference to one of Stow's rock art "cartoons" he was shown by Lloyd and asked to comment (LV. – 10. 4744–50; also see Bank

2006: 315–16). Among the southern ≠Khomani the Gemsbok dance, labelled by C. M. Doke "Gemsbok as Men" is on the pan-San cosmological theme of transformation. It mimics—similar to the Kalahari San eland dance—a "gemsbok as a human being" copulating with women, who are part of the dance troupe (McCall 1970: 8–9). Turning to Kalahari San, variants of an evidently similar dance, as per Fourie's description (1965: 32, see epigraph) featuring gemsbok, hunters and (usually) dogs, were performed by most of the groups.[4] Ilse Schatz provides an especially lively description of the dance among the Hai//om (1993: 17) which is similar to Fourie's scenario, featuring one gemsbok and one hunter and a pack of dogs, all enacted by male dancers and, as an additional twist in what is usually an all-male cast of characters, women. Dancing, a group of them—the vultures—encircle the scene, to throw themselves, at the end of the dance, on the dead gemsbok, for their vulture's own "*Festschmaus*" (royal feast). This element gives this performance of the Gemsbok Dance a bit of the sexual—and gender reversal—flavour we find in the Eland Dance at the menarcheal rite (a flavour added also to the ≠Khomani lion and springbok dance, when, near its end, the "tantalizing" springbok girls are hunted down and eaten by the lions). Among the /Gui (G/wi) the Gemsbok Dance is a trance curing dance—requiring from the dancer an "intense degree of emotional involvement", along with "great physical exertion" (Silberbauer ibid.: 177). As noted above, according to Tanaka it is also a ludic-mimetic "hunting dance", which, having diffused into the Central Kalahari from Namibia in the 1950s, displaced its predecessor, the Wildebeest Dance (wildebeest having at that time been a widely hunted species whose numbers have sharply declined, leading to its displacement by another hunting dance, of another game species). The dance is performed at night, after a successful gemsbok hunt, to women's clapping and singing and men dancing around them, arms hanging forward, bent down, holding sticks and gemsbok tails,[5] all of it "requiring much energy" and "quite tiring". The women's "high-toned, shrill chorus" contrasts with the men's "low-toned humming" by which they "simulate the gemsbok groaning" (Tanaka 1996: 22).

[4] Other Kalahari San amongst whom this dance is described are !Kõ (Heunemann and Heinz 1975b), /Gui (G/wi) and //Gana (Tanaka 1996: 26, Silberbauer 1981: 176–7), Ju/'hoansi (Marshall 1999: 76), =Au//eisi (Metzger 1990: 34–5), Hai//om (Schatz 1993: 17).

[5] Metzger reports that =Au//ei dancers actually attached gemsbok tails to their loincloths (1990: 34).

The Central Kalahari Gemsbok dancers, leaning forward and supporting themselves with sticks, invite comparison with a /Xam dancer photographed by Dorothea Bleek at Prieska in the Northern Cape around 1910 who is seen in the same dancing posture. In his interpretive analysis of the photographs Pieter Jolly (2006) suggests that, by virtue of his "quadrupedal" stance, the dancer was mimicking an antelope. Jolly goes further, moving from mimesis to metamorphosis: pointing to parallels between rock art depictions of such figures, especially the ones in which the sticks appear to "have fused with the arms of the person to produce a four-legged therianthropic being" (ibid.: 177); he suggests that the dancer on Bleek's photograph might have been undergoing transformation (or performed a dance which, in the recent past had that feature, before the eroding effects on their culture and its enchantment-bearing myth and beliefs, cosmology and ontology all in the context of an exceptional traumatic contact experience with colonial settlers). Depictions of such scenes from an earlier, pre-contact age are drawn in by Jolly to buttress his interpretation of the photographed /Xam dancers, such as to Orpen's Melikane rock art copy published in the *Cape Monthly Magazine* July 1874 issue (ibid.: 176), featuring, in the image on the upper left panel, three bent-over therianthropic dancers with antelope heads, two of them leaning on sticks (see above, Chap. 3, Fig. 3.7). Jolly does not attempt to identify the antelope species as it is not clearly discernible from the image; however, I suggest that it could be young gemsbok, on the basis of Diä!kwain's above-mentioned explanatory comment to Lucy Lloyd of two of Stow's rock art copies depicting antelope headed therianthropes, in terms of a gemsbok dance performance in which dancer wear caps "cut ... out from the gemsboks' horns, the young gemsbok horns. The long horns being a little too long, they take horns from the young gemsbok, which are short" (LV.—10. 4744 rev-4745). It is tempting to assume a similar ontological dynamic, in—the mind and feelings of—the G/wi gemsbok dancer, from a stick-arm-leg, forward-bent human into the antelope itself—the gemsbok tail, transposing itself in the transforming process from the dancing human's hand to the danced animal's rump!

The use of animal paraphernalia is not uncommon for antelope/hunting dances, which link dancers mimetically and materially to both the animal and to the hunt. Apart from sticks for antelope legs, such objects may—and likely did in the past—hold a much more direct and intimate association with their animal referent. Like the G/wi dancer's gemsbok tail, or the Ghanzi trance dancer's gnu tail which, 50 years or so back, I

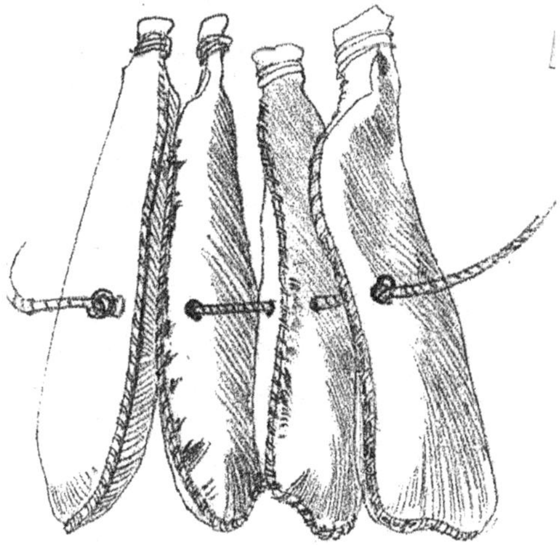

Fig. 6.1 /Xam dance rattles made out of springbok ears. (Source: Bleek and Lloyd 1911: 244)

found in the hands of every other dancers I observed in action, connecting the Ghanzi healing dance culturally to its one-time wildebeest counterpart in the Central Kalahari, these objects were derived directly from the animal, from the carcasses of previously hunted and butchered prey animals. To the eared springbok caps of /Xam springbok shamans mentioned earlier may be added springbok-ear dancing rattles (Fig. 6.1) made by women to be worn by the men who tied them to the instep of their feet—"in fours or fives"—as they danced to the drumming of the women[6] (Bleek and Lloyd 1911: 351–3). The depiction on rock art panels of human-like figures wearing animal skin cloaks (karosses) as scenes of therianthropes or of transformation was discussed in the previous chapter.

[6] William Burchell observed such rattles in use by a Bushman dancer near Klaarwater in the Northern Cape in 1812: "four ears of spring-buck, sewed up and containing a quantity of small pieces of ostrich eggshell, which at every motion of the foot produced a sound that was not unpleasant or harsh, but greatly aided the general effect of the performance" (1824: 65; page 45 of his *Travels in the Interior of Southern Africa*, vol. II depicts his drawing of one of the rattles). Moreover, among the same group of Bushmen he found that a number of people carried around jackal tails which they "frequently drew across their eyes, for the purpose, as I was told, of improving their sight, agreeably to their belief that it possesses a virtue of that kind" (ibid.: 57). Stow, in his description of these springbok rattles, mentions other variants of this musical device (which he refers to as "Bushman bells"), worn not only on the dancers' feet but also on their shoulders and upper arms and as belts around the waist (1905: 10–1).

Such a reading would suggest, albeit with little supporting evidence from contemporary ethnographic sources, that mimetic animal dances, too, may, in addition to their play and dance element—of kinaesthetic titillation—have contained also an element of transformation. In his consideration of this notion—of the use of ritual animal dress in dances and their possible transformative effect—Jolly cites a late nineteenth-century source from north-eastern Bechuanaland, about the use of animal skins from a wide diversity of species—including an "elephant head with trunk attached" (2002: 92), an over-sized and no doubt unwieldy animal accoutrement! These props were reportedly employed by dancers in their wildly histrionic—bordering on hysterical—dances. However, it is not clear if the information reported applies to San or to their Bantu-speaking Tswana neighbours. If such animal substances were in fact used in animal/hunting dances in so elaborate and extensive, body-enveloping and, perhaps also mind-absorbing a fashion, rather than in the much more attenuated way they are seen to be employed among contemporary Kalahari San, ludic dancing might well have constituted yet another phenomenological arena within which human-animal mutability played itself out.[7]

And with the much the same degree of intensity and identity-altering effect as in the arena of ritual—and, as seen below, also of hunting. And, with respect to the last, also on occasion by means of the same device and technique, the donning of animal skins. This San hunters may do, with animal-attuning, sympathy-inducing effect. Indeed, as to be discussed in the next chapter, being worn by pre-contact Bushmen and women throughout their lives, from cradle (an infant's mother's soft steenbok baby carrier) to grave (a deceased person's softened skin burial shroud), skin clothing, quasi-animate and animated, connected all San humans to animals, bodily—skin-to-skin—and, as I will suggest, also mentally.

Given the intensity such mind- and being-altering animal dances were capable of reaching, it is conceivable that on occasion and for some dancers—ones with "easier access to a rich fantasy life" than others and thereby more "predisposed to kia" (Katz 1982: 236)—a trance state could have been brought on. This would happen in a dance, of animal mimesis bordering on metamorphosis, that is not as such focussed on trance (*!kia*) and

[7]Thea Skaanes reports much the same about the nocturnal *epeme* dances of the Hadza of Tanzania, which may include the use of animal skins "The cape or cloak enlarges the dancer's body and obscures the human form. When animal hide was used, the human shape was substituted with an animal presence" (2017: 172).

on activating through its means healing potency (*n/om*). The trigger and trajectory here are both different than in the healing dance; instead of altered states of consciousness it is the intensity of the dancer's animal experience, as they become "fused with the animals that they imitated so closely" (Jolly 2002: 23). Just as a healing dance may channel mind-altering trance into being-altering transformation, an animal transformation dance is capable of eliciting trance. In both cases "tranceformation" may be at play, intensifying, through an altered state of consciousness, an alternate state of being.

Ludic Dancing, Myth and Storytelling

In addition to their connection to ritual, ludic animal dances also connect with the world of myth. Indeed, among the /Xam, a favoured dance called *≠gebbi-gu*, that imitated the "roar" of the ostrich was "taught" to them by the First Race—specifically "baboons that were once people" (Hollmann 2005: 24). Diä!kwain told Lloyd how this happened in Myth Time in a lengthy *kum*, featuring Lion and Ostrich First Race people (Bleek and Lloyd 1911: 126–35). He supplements this story with tangential narratives that feature baboons and people and explain the ontological connection between them (Hollmann 2005: 24–26). These are set in either First or Second Order times, or both, as so often timeframes are not clearly delineated by /Xam story tellers, especially this one. I note that this overlap of myth and real time is evident also in an interesting twist regarding the story's central feature, the *≠gebbi-gu* dance: today's baboons, having in the Second Order apparently forgotten that their kind originally were the "owners" of this dance but somehow sensing some special connection to it, will come to watch humans when they are playing the dance and the humans dancing and singing.

While it is risky if not altogether unwarranted to attribute the two ostrich dances that are found among the Kalahari Naro in any way to the *≠gebbi-gu* of the /Xam of the First and Second Order of Creation, the narrative and choreographic similarities between them are nevertheless intriguing. The /Xam play dance appeared to mime a fight between two Early Race "men", Ostrich and Lion, kicking and inflicting wounds each other with their claws. The issue of contention was the ostrich's "roar" which the lion, given his "wont to hold his tail in his mouth when he calls", was incapable of producing, feeling diminished and jealous of the ostrich, all the more so as the lions' wives "only praised the Ostrich for calling finely; the women did not praise them" (Bleek and Lloyd 1911: 129).

All this may suggest a tenuous connection of these animal mimicry dances to the myth, the former "dancing out" the latter. However, as is the case with regard to San art, it seems unlikely that these performances are danced out stories or scenes from Myth Time featuring therianthropic characters from that time's First Order. We might agree with writers like George Stow that some dance or other, like some San rock paintings, might have, or at one time been linked to a "hidden myth" and had "mythical significance" (1905: 117), indeed, may have been received from Myth Time beings, including Cagn (//Kaggen) himself, as Orpen was informed by his San informant Qing about the *Mo'koma* dance (ibid.: 119).

However, because of the virtual absence of supporting evidence for such an Eliade'an notion—of ritual enacting myth (Eliade 1984), thereby transporting the performer of a mimetic baboon or ostrich dance back into myth time and transforming him into a therianthropic Baboon or Ostrich Man—it seems more likely that the dancer assumes such simian or struthian qualities as the dance might impart to his being through dancing the animal out with reference to its real rather than mythological signifier (all the more so as the connection of such a dancer to myth time is through "hidden" myths, long forgotten). As noted earlier, San animal mimicry dances are more than anything about real animals and animal encounters, usually in a hunting context. Regarding the *≠gebbi-gu* and its Naro cognates, the fighting of real cock ostriches out in the veld—who, rush and jump at each other front on with kicks, apparently trying to knock the opponent down and who do in fact roar "like lions"[8]—provides real-life models from real ostriches for stories or dances that deal with these arresting aspects of ostrich ethology. Indeed, Diä!kwain's *≠gebbi-gu* narrative, set in both myth and historic time, might be an aetiological tale.

This sort of mundane explanation suggested itself to me when I questioned people about one of the play dances I observed among the Ghanzi Naro, which they called the "Fight Game"[9] (Guenther 1986: 253). This then highly popular dance-game among teenage boys and young men consisted of the two opponents, sometimes standing, sometimes squatting, facing each other, all the while kicking and dodging kicks, faces set in

[8] All this information on ostrich behaviour I just found out on YouTube through fascinating video footage after googling "cock ostriches fighting".

[9] The dance-game I observed is similar to D. Bleek describes as "the war game" in her account of Naro children's games. She surmises this ludic dance "may originally have been the imitation of some animal fighting" (1928: 21).

fierce grimaces, swinging and lurching their upper bodies as one player leaps over the other—all of it skilled, synchronized and quite strenuous (reminiscent of Western urban kids' breakdancing). In its choreography it evokes the /Xam First Race's *≠gebbi-gu*, as do the hoarse groans and hisses of the player-dancers. However, as I gather from informants' comments, the dance-game I observed had no such mythic provenance—the lads' histrionic antics simulated some of the birds' movements, in particular that of fighting of two cock ostriches

While myths connect people primarily to the animals and were-animals of the First Order in the mythological past, these dances connect them to the animals of the veld. Both do so by temporarily dissolving the human-animal divide, the one by mimetically acting out animals—ones to be encountered out in the veld or to be hunted on the hunting ground—through dance and play, in ways that link them to real animals, along with their skins, tails and horns on the dancer's body, inadvertently injecting a dose of "hunting magic" into his ludic animal pantomime (a point to return to in the next chapter). The other way dissolves the division between humans from real time with animals from myth time, to be imagined or dreamt or be told stories about. These stories are captivating because they so frequently feature beings that are therianthropes which, *sui generis*, display what the dancer's mimetic performance brings about in his own being: ontological hybridity. Or they are about the process behind it, transformation. And they do so not only through the story's plot but also through its performance.

A number of writers on San expressive culture have remarked on the highly performative element of San story telling[10] (Hewitt 1986: 235–46; Biesele 1993: 51–62; Guenther 1999: 137–8, 2006; Keeney and Keeney 2013: 75–78, 2015: 156–60). It contains, very much like ludic dancing, a strong dose of "dramatic enactment" (Keeney and Keeney 2013: 75), the creativity and artistry of which includes mimetic elements when the story is about animals, of either the First or the Second Order. These "animal folktales … included animal songs and bird songs", the Bugakhwe narrator Peter Goro reminisces about the "olden days" (Le Roux and White 2004: 57) and, as observed by Jiro Tanaka among Central Kalahari San, in

[10] As well as on its earthy, Rabelaisian flavour, in the context of traditional storytelling, as opposed to the clinical, restrained, puritanical context of the Bleek/Lloyd household in which the /Xam storytellers narrated, or dictated, their tales to their two interlocutors (Wittenberg 2012: 677–8; see also Guenther 1996).

telling animal stories, "story tellers use abundant gestures that deftly capture the characteristics of the animals in the story" (1996: 24). Accounts of memorable hunts were dwelled on and drawn out, "step-by-step, in microscopic and baroque detail" (Lee 1979: 205), becoming, in each retelling also a re-enactment.

In their discussion of "the n/om of storytelling" among Ju/'hoansi Bradford and Hillary Keeney expand on the elements of performativity—"tonal shifts, rhythm alteration, dynamics, musicality" (2013: 75)—which brings drama and excitement to storytelling. Generating "energized immediacy and creative possibility", these textural elements of Ju/'hoan storytelling "fire up the performance" for both story teller and story listeners:

> Story, storyteller, and audience all interact in ways that serve the deliverance of n/om. In this ecstatic theater, everyone aims to enter First Creation, the source of renewal and healing. We move from a changing story, to a changed storyteller and a transformed audience. (ibid. 77)

All this contributes to a story's "embodiment", through its narration, through the story teller's "kinesthetic attentiveness" and "somatic sensitivity" (Rusch 2016: 877), rather than the actual narrative itself; that is, using folklorist Alan Dundes's classic terms and concepts, through the story's texture rather than its text (Dundes 1980). Performance in this way gains its own plot, as a "meta-communicative" subplot to the story's own plot (Bauman 1992: 49). This to the Keeneys, as seen in the quote, is the "deliverance" of *n/om* through trance dancing, especially the shaman's bodily shaking. The latter is an element of histrionics also of an animated story teller, who may him/herself be also a shaman, underscoring the link between trance dancing and storytelling, both of them means for gaining entry into First Creation.

The other subplot of storytelling, the one more explicitly conveyed than trance, is transformation. It is seen at play, in a mimetic narrative device employed widely by /Xam story tellers: the special clicks they employed when impersonating animal characters from the First Order (Hollmann 2005: 231–85). Wilhelm Bleek explained these phonetically difficult and idiosyncratic clicks in onomatopoeic terms, as vocalizations created by the story teller by trying to imitate aspects of the referenced animal's vocal operations. This "remarkable attempt to imitate the shape or position of the mouth of the kind of animal to be represented" (1875:

6, cited in Hollmann ibid. 231) is a striking instance of human-animal connectedness, mimetically and somatically expressed through the human vocal apparatus. Because the animal so voiced is from the Second Order, this mimetic ploy brings about the story teller's subtle linking or merging of the therianthropic character from myth with its counterpart from the hunting ground, each conferring to the other aspects of its own ontological being and temporal state. It is one of many instances we find in San expressive culture of "aesthetic mimesis"—"fiction's ability to form reality" (Hawkes 2017: 25).

It is here, in a story's at times highly dramatic-mimetic performance that may absorb the narrator and draw in the listener of the stories to and into myth time, that the similarity between storytelling and ludic dance is most striking: both are animal pantomimes that may reach moments when mimesis borders on metamorphosis. There is also a profound difference, in cases where story tellers tell ***kukumi*** about the First Order. Here the mimetically transformed human actor, the story teller, subliminally identifies primarily with the therianthropes and animals from myth time, whereas in that of the ludic dancer the primary identification is with the animal from real time. Such is the case also with respect to the dancers at the initiation rites, when they, along with the initiands, merge identities with the eland (as well as, albeit more obliquely, in the trance dance where, in addition to human-focussed, potency- and healing-generating altered states, animal transformation, too, may come about).

For all of their mystical and preternatural aspects, all of these dance-triggered transformations are linked also to hunting, informing this domain from the techno-economic infrastructure of San culture with elements of San symbolic and expressive culture. This, as seen in the previous chapter, is most direct in the initiation rite, where both the female and male initiand "shoots eland", metaphorically or actually, and in the ludic dances that enact antelopes and men hunting them, the most arresting and absorbing dances of the large repertoire of animal-imitating dance numbers that make up San non-ritual dances. With regard to the trance dance, the animal the dancer mimics and metamorphoses into may be an eland or another game antelope, which the trancing shaman in spirit may shoot or catch. Or it may be a predatory animal, who may prey on the same animal as the hunter, or even on the hunter himself, all of it broadening the scope of human-animal (and hunter-prey) identity merging that comes about in San ritual and recreational dancing.

Hunting brings us to the topic of the next chapter in which we leave the realm of ritual and of the preternatural and mythic beyond and the direct or vicarious, real or virtual human-animal transformation we've seen coming about in various ways in that surreal, at times ludic context, that range from more or less subtle mimesis, sympathetic stirrings of ontological resonance, to strongly felt metamorphosis, in which human and animal identities morph and merge. Yet, for all its tacit linkages to ritual and myth, hunting, as an instrumental subsistence practice, places us squarely within the real Second Order, the veld's hunting ground, where humans meet animals eye-to-eye and the animal-other is not imagined but real and such mimetic or metamorphic ontological boundary crossing as might take place between hunter and prey occur with reference to real rather than imagined animals.

How stable is this real animal's ontological identity and integrity at such moments of intersubjectivity? How much of it is assumed by the hunter? How much of the ontological hybridity that attaches to the human and to the animal in myth, art and ritual remains in the prey animal stalked, shot, tracked and killed in a hunt? Is hunting and killing an animal something that occurs within or outside the relational sphere of human-animal sociality and personhood? Does it undermine or even negate the ontological and social inclusivity of the two species, humans and other-than-humans or does it undergird and reinforce the same? These questions define the tone and substance of the next chapter.

Moreover, they anticipate ontological and phenomenological questions that are the concern of the core chapters of the following volume: Does the "reality" of the veld-encountered animal eclipse that of the mystically experienced animal of myth and art, ritual and dance? Or, rephrased differently and more broadly, how is the reality of the veld and the hunting ground and its natural and actual "doings" affected by doings of a mythical, mystical, preternatural kind? Do the latter enhance the former, giving the veld and landscape the feel not only of the "true real, coarse and subtle" (Taussig 1993: xvi, xvii; see also Solomon 2014: 714) but also, through the linkage with myth and ritual, and thereby with the supernatural, "the really real"? How is that "really real's" "aura of utter actuality" (Geertz 1972: 175) linked to "the real in its wonder" and the "spell of the sensuous" that envelops it (Abram 2010: 312, 1997)?

References

Abram, David. 1997. *The Spell of the Sensuous: Perception and Language in a More-than-Human World*. New York: Vintage Books.

———. 2010. *Becoming Animal: An Earthly Cosmology*. New York: Vintage Books.

Bank, Andrew. 2006. *Bushmen in a Victorian World: The Remarkable Story of the Bleek-Lloyd Collection of Bushman Folklore*. Cape Town: Double Storey Books.

Bauman, Richard. 1992. Performance. In *Folklore, Cultural Performances, and Popular Entertainments*, ed. Richard Bauman, 41–49. New York: Oxford University Press.

Biesele, Megan. 1993. *Women Like Meat: The Folklore and Foraging Ideology of the Kalahari Ju/'hoansi*. Johannesburg/Bloomington: Witwatersrand University Press/Indiana University Press.

Bleek, Wilhelm H.I. 1875. *A Brief Account of Bushman Folk-Lore and Other Texts*. Cape Town/London/Leipzig: J. C. Juta/Trübner & Co./F. A. Brockhaus.

Bleek, Dorothea. 1927. Bushmen of Central Angola. *Bantu Studies* 3: 105–125.

———. 1928. *The Naron*. Cambridge: Cambridge University Press.

Bleek, Wilhelm H.I., and Lucy Lloyd. 1911. *Specimens of Bushman Folklore*. London: George Allen & Co.

Burchell, William J. 1824. *Travels in the Interior of Southern Africa*. Vol. II. London: Longman, Hurst, Rees Orme, Brown & Green.

Doke, Clement M. 1937. Games, Plays and Dances of the ≠Khomani Bushmen. In *Bushmen of the Southern Kalahari*, ed. John D. Reinhallt Jones and Clement M. Doke, 89–100. Johannesburg: University of the Witwatersrand Press.

Dundes, Alan. 1980. Texture, Text, and Context. In *Interpreting Folklore*, ed. Alan Dundes, 20–32. Bloomington: Indiana University Press.

Eliade, Mircea. 1984. Cosmogonic Myth and 'Sacred History'. In *Sacred Narrative: Readings in the Theory of Myth*, ed. Alan Dundes, 137–151. Berkeley: University of California Press.

Fourie, L. 1925/26. Preliminary Notes on Certain Customs of the Hei-kom Bushmen. *Journal of the South-West African Scientific Society* 1/2: 49–63.

Fourie, L. 1963/64–1964/65. Preliminary Notes on Certain Customs of the Hei-kom Bushmen. *Journal of the South-West African Scientific Society* 18/19: 18–34.

Geertz, Clifford. 1972. Religion as a Cultural System. In *The Interpretation of Cultures*, ed. Clifford Geertz, 87–125. New York: Basic Books.

Gentz, Leutnant. 1903/04. Einige Beiträge zur Kenntnis der südwestafrikanischen Völkerschaften in Deutsch-Südwestafrika. *Deutsche Kolonialzeitung* 26: 450–452.

Guenther, Mathias. 1986. *The Nharo Bushman of Botswana: Tradition and Change*. Hamburg: Helmut Buske Verlag.

———. 1996. Old Stories/Life Stories: Memory and Dissolution in Contemporary Bushman Folklore. In *Who Says? Essays on Pivotal Issues in Contemporary Storytelling*, ed. Carol L. Birch and Melissa A. Heckler, 177–197. Little Rock: August House Publishers.

———. 1999. *Tricksters and Trancers Bushman Religion and Society*. Bloomington: Indiana University Press.

———. 2006. Contemporary Bushman Art, Identity Politics and the Primitivism Discourse. In *The Politics of Egalitarianism: Theory and Practice*, ed. Jacqueline Solway, 159–188. Oxford: Berghahn.

Hawkes, David. 2017. Many-Headed Multitude How Incorporation Can Be a Good Thing. *TLS* 24 (2017): 25.

Heinz, Hans-Joachim. 1975. !ko-Buschmänner (Südafrikas, Kalahari): Mädchen Initiation. *Encyclopaedia Cinematographica*, E 1849: 3–15.

Hermans, Janet. 1998. Basarwa Cultural Preservation in Contemporary Botswana: The Dance. In *The Proceedings of the Khoisan Identities & Cultural Heritage Conference*, ed. Andrew Bank, 281–288, South African Museum, Cape Town, July 12–16, 1997. Cape Town: InfoSource.

Heunemann, D., and Hans-Joachim Heinz. 1975a. !ko-Buschmänner (Südafrika, Kalahari): Wettstreit, Jäger und Tier (Gestenspiel). In *Encyclopaedia Cinematographica*, Film E2105, ed. G. Wolf, 3–12. Göttingen: Institut für den wissenschaftlichen Film.

———. 1975b. !ko-Buschmänner (Südafrika, Kalahari) Spiel Oryx-Antilope, mit einer Auseinadersetzung zwischen zwei Spielgruppen. In *Encyclopaedia Cinematographica*, Film E2107, ed. G. Wolf, 3–12. Göttingen: Institut für den wissenschaftlichen Film.

Hewitt, Roger. 1986. *Structure, Meaning and Ritual in the Narratives of the Southern San*. Hamburg: Helmut Buske Verlag.

Hollmann, Jeremy C., ed. 2005. *Customs and Beliefs of the /Xam Bushmen*. Johannesburg/Philadelphia: Wits University Press/Ringing Rock Press.

Jolly, Pieter. 2002. Therianthropes in San Rock Art. *South African Archaeological Bulletin* 57 (176): 85–103.

———. 2006. Dancing with Two Sticks: Investigating the Origin of a Southern African Rite. *South African Archaeological Bulletin* 61 (184): 172–180.

———. 2015. *Sonqua: Southern San History and Art After Contact*. Cape Town: Southern Cross Ventures.

Katz, Richard. 1982. *Boiling Energy: Community Healing Among the Kalahari Kung*. Harvard: Harvard University Press.

Kaufmann, Hans. 2005 [1910]. The =Auin. In *Kalahari and Namib Bushmen in German South West Africa: Ethnographic Reports by Colonial Soldiers and Settlers*, ed. Mathias Guenther, 37–96. Cologne: Rüdiger Köppe Verlag.

Keeney, Bradford, and Hillary Keeney. 2013. Reentry into First Creation: A Contextual Frame for the Ju/'hoan Bushman Performance of Puberty Rites,

Story Telling, and Healing Dance. *Journal of Anthropological Research* 69: 65–86.

———., eds. 2015. *Way of the Bushman as Told by the Tribal Elders*. Rochester: Bear & Company.

Krog, Antje. 2009. *Begging to Be Black*. Cape Town: Random House Struik.

Le Roux, Willemien, and Alison White, eds. 2004. *Voices of the San*. Cape Town: Kwela Books.

Lee, Richard B. 1979. *The !Kung San Men, Women, and Work in a Foraging Society*. Cambridge: Cambridge University Press.

———. 2003. *The Dobe Ju/'hoansi*. 3rd ed. Fort Worth/Belmont: Wadsworth.

Lewis-Williams, David J. 2015. *Myth and Meaning: San-Bushman Folklore in Global Context*. Walnut Creek: Left Coast Press.

Marshall, Lorna J. 1999. *Nyae Nyae !Kung Beliefs and Rites*, Peabody Museum Monographs, No. 8. Cambridge, MA: Harvard University.

McCall, Daniel F. 1970. Wolf Courts Girl: The Equivalence of Hunting and Mating in Bushman Thought. *Ohio University Papers in International Studies, Africa Series*, No. 7. Athens: Ohio University. Center for International Studies.

Metzger, Fritz. 1990. *Naro und seine Sippe*. Windhoek: Namibia Wissenschaftliche Gesellschaft/Scientific Society.

Rusch, Neil. 2016. The Root and Tip of the *//Kwanna*: Introducing *Chiasmus* in Three /Xam Narratives. *Critical Arts* 30 (6): 877–897.

———. 2017. Sound Artefacts: Recreating and Reconnecting the Sound of the *!goin !goin* with the Southern Bushmen and Bees. *Hunter Gatherer Research*, 2015 3 (2): 187–226.

Sbrzesny, Heide. 1974. !ko-Buschleute (Kalahari) – Der Eland Tanz. Kinder spielen das Mädchen. Initiationsritual. *Homo* 25: 233–234.

———. 1976. *Die Spiele der Ko-Buschleute unter besonderer Berücksichtigung ihrer sozialisierenden und gruppen-bindende Funktionen*. Munich: Piper.

Schapera, Isaac. 1930. *The Khoisan Peoples of South Africa: Bushmen and Hottentots*. London: Routledge & Kegan Paul.

Schatz, Ilse. 1993. *Unter Buschleuten auf der Farm Otjiguinas in Namibia*. Tsumeb: Ilse Schatz.

Silberbauer, George B. 1981. *Hunter and Habitat in the Central Kalahari Desert*. Cambridge: Cambridge University Press.

Skaanes, Thea. 2017. Cosmology Matters: Power Objects, Rituals, and Meat-Sharing Among the Hadza of Tanzania. Ph.D. Dissertation, Aarhus University.

Solomon, Anne. 2014. Truths, Representationalism and Disciplinarity in Khoesan Researches. *Critical Arts: South-North Cultural and Media Studies* 28: 710–721.

Stow, George W. 1905. *The Native Races of South Africa: A History of the Intrusion of the Hottentots and Bantu into the Hunting Grounds of the Bushmen, the Aborigines of the Country*. London: Swan Sonnenschein.

Tanaka, Jiro. 1996. The World of Animals Viewed by the San. *African Studies Monographs*, Supplementary Issue, 22: 11–28.

Taussig, Michael. 1993. *Mimesis and Alterity: A Particular History of the Senses.* London/New York: Routledge.

Vinnicombe, Patricia. 2001 [1976]. *People of the Eland: Rock Paintings of the Drakensberg Bushmen as a Reflection of Their Life and Thought.* Johannesburg: Wits Press.

Werner, H. 1906. Anthropologische, ethnologische und ethnographische Beobachtungen über die Heikum- und Kung-Buschleute. *Zeitschrift für Ethnologie* 3 (1): 241–268.

Wessels, Michael. 2012. The Khoisan Origin of the Interconnected World View in Antje Krog's *Begging to Be Black. Current Writing: Text and Reception in Southern Africa* 24 (2): 186–197.

Wilhelm, Joachim Helmut. 2005. The !Kung Bushmen. In *Kalahari and Namib Bushmen in German South West Africa: Ethnographic Reports by Colonial Soldiers and Settlers*, ed. Mathias Guenther, 91–184. Cologne: Rüdiger Köppe Verlag.

Wittenberg, Hermann. 2012. Wilhelm Bleek and the Khoisan Imagination: A Study of Censorship, Genocide and Colonial Science. *Journal of Southern African Studies* 38: 667–679.

CHAPTER 7

Transformation and Hunting

In … merging somatically with the animal as the hunter "runs its track" and mentally as he looks into its eyes after it has stopped running, eyes that reflect the hunter and that "are no longer wild", identity and otherness of the human and other-than-human dissolve into each other.
Guenther (*2017*: 4)

The epigraph that open this chapter attests to the relevance of its topic—hunting—to the theme of this book—transformation and ontological mutability. It is an excerpt from my recently published essay on the "super-erogatory" aspects of San hunting (2017) and its allusions are to two elegant statements on this matter, the first by a Western researcher (Neil Rusch), and the second a !Kõ (!Ko) San hunter (Karoha):

> The chase leads beyond the narrow boundary of personal identity and familiar consciousness. Identifying with the animal he effectively enters its path; his identity connects with that of the animal as he runs its track. (2016: 23)

> What you will see is that you are now controlling its mind. You are getting its mind. The eyes are no longer wild. You have taken the kudu into your mind. (quoted by Liebenberg 2006: 1024)

© The Author(s) 2020

M. Guenther, *Human-Animal Relationships in San and Hunter-Gatherer Cosmology, Volume I*,
https://doi.org/10.1007/978-3-030-21182-0_7

The chapter epigraph frames the argument I presented in the essay from which it is excerpted, in terms of the phases of a hunt, of the sort—stalk, wound, track, chase, kill—that may instil, in the San hunter, an appreciation of the "humanness and sociality of prey" (McNiven 2013: 99–100). The same argument runs through this chapter: that the animistic cosmology of the San is the ontological and experiential foundation for the non-utilitarian aspect of San hunting. Pervasive ontological blurring of human-animal ontologies, and its resultant inconstancy of human and animal identities and permeability of ontological species boundaries, all have a bearing on San hunting. When the hunter's prey animal is perceived, more or less tacitly, as a person an extra-utilitarian, non-rational, supererogatory element comes to inform this in so many ways practical subsistence activity, marked by relationality and intersubjectivity, communication and sociality (Ingold 2000: 48–52). And, among the San, for whom, as stated by the !Xõ hunter Karoha, "tracking is like dancing—because your body is happy" (Guenther 2017: 3) hunting, so enlivened and embodied, may fleetingly render the hunter into a quasi-therianthrope, marked as well with mystical portent and mythic resonance.

Regarding the latter aspects, the substance of the preceding chapters, we've already come across hunting repeatedly as it has cropped up in everything discussed thus far. Myth, story and art, trance dance, initiation, ludic dancing all link up with hunting at some point or other, either indirectly, by ubiquitously featuring elands and other prey animals, or predators that compete with hunters and even prey on them. Or they are linked to hunting directly—as when male and female initiands "shoot eland" or a gemsbok hunt is danced out. Or when the trickster-divinity of myth, art and ritual himself hunts, as he on occasion is told in stories or painted in rock art images to do, notwithstanding the Spirit Keeper of Animals mandate he paradoxically (as is his wont) may also hold and impose on a hunter.

A more fundamental link of San symbolic, expressive and ritual culture to hunting is through therianthropes. As I suggest in this chapter, hunters, too, may at certain moments during the hunt be or become such therianthropic beings, their being's "theri"-component derived from the hunt and the intersubjective, relational dynamics it generates between hunter and prey. Through this dynamic human and animal identities come close to being merged, or may even be merged. At such moments the ontological inchoateness of the First Race may become palpable in a member of the Second Race, when the hunter, much like the shaman-lion and maiden-eland, merges in his being human and animal. Hunting is thus, *pace* Parkington, a "key context" for therianthropes:

> Hunting is a thrilling and exhausting occupation that involves total immersion in the contest between hunter and prey. The hunter identifies completely with the animal he is pursuing. He learns to think and act like his prey so that by becoming it he can overcome it. This makes hunting a key context for the conflations of human and prey that we see in therianthropes. (2002: 44)

Along with trance and initiation rites, as well as perhaps some ludic dances, hunting is an occasion at which these exemplars of the mythic First Order's Early Race and paradigmatic embodiments of ontological mutability that underlies San cosmology are mediated not through the mind and imagination but through direct experience.

In leading off this account on the mimetic and metamorphic aspects of San hunting,[1] I once again turn to the eland, the San's blue-ribbon prey animal whose meat and fat feed people's bodies and minds. Docile and tractable, its massive bulk, much of it consisting of fat, weighs down on the antelope and slows its run, making it all the more suitable for hunting. Wounded with a poison arrow an eland, weighing as much as one ton, might take several days to expire, while being tracked and stalked and dispatched at close quarters, possibly even near the hunter's home base (Lewis-Williams 2015a: 83), given the animal's tractable disposition (Fig. 7.1). Close to eland as valued quarry are such premier, high-yield game antelopes as gemsbok, kudu, hartebeest and wildebeest, as well as giraffe.

There are two ways whereby San hunt such big game species, both of them ways that include the relational dynamics of sympathy and identity merging. In the one, which Richard Lee refers to as "the mobile hunt" (2003: 48) and deems the Ju/'hoansi's "classic hunt"[2] (1979: 207), hunt-

[1] Having described the relational dynamics of San hunting in some detail elsewhere, in the above-mentioned article (2017: 4–6, 10–11; see also Guenther 2015: 293–98, 1999: 72–76, 1988), I will here limit myself to recapitulating the main points from these other discussions which the reader can refer to for elaboration. Moreover, I will expand on, or add some further points these discussions have dealt with only tangentially or not at all. In this I also draw on James Suzman's discussion of "hunting and empathy", which appears as a chapter (2017: 161–74) in his recent book on the lifeways of traditional and contemporary ≠Au//eisi Ju/'hoansi in the Omaheke region of eastern Namibia and is couched in terms of intersubjectivity and relationality.

[2] This hunting method was made famous by John Marshall's classic film—a North American intro anthro classes' staple—"The Hunters", which chronicles a protracted giraffe hunt by a handful of Ju/'hoan hunters. For accounts of this method of San hunting (along

Fig. 7.1 Eland hunt, driving the eland to home base. (Source: Harris 1852: 75)

ers, hunting alone or in a small group, having stalked and wounded their quarry with a lethal but slow poison arrow, track and kill the animal with a spear (or club, in the case of smaller game).[3] The other, less frequently practised method is "persistence hunting"[4] (Liebenberg 2006). Each

with others), see Schapera (1930: 132–5), Marshall (1976: 134–8), Lee (1979: 214–9, 2003: 48–51) and Silberbauer (1981: 208–14). For an account that also includes the element of sympathy see Suzman (2017: 161–74).

[3] In a recent study on the origins of hunting, the American evolutionary anthropologist Travis Pickering considers the sort of hunting the San practice, which he calls "ambush hunting" (2013: 100–2, 156–7). He traces the same back to Homo erectus horizons and deems it a "milestone" event in human evolution for its "decoupling of predation from aggression". He sees this stemming from a new hunting technology, projectile weapons (spears, atlatls and, later on, bows and arrows), that allow hunters to dispatch animals from a distance rather than at close range, in "hand-to-hand combat". Instead of impulse and affect—"raw emotion"—the hunter's stance towards the prey animal, argues Pickering, changed significantly. It became marked with restraint and reflection (ibid.: 104–5)—providing, it might be argued, the basis for a mindset capable of entertaining notions of sympathetic engagement with a hunter's quarry.

[4] An ethnographic film on this hunting method, by !Xo hunters, is available as well: "The Great Dance: A Hunter's Story" (2000), by the South African documentary filmmakers Craig and Damon Foster. The film complements Liebenberg's *Current Anthropology* article on the subject (2006), featuring some of the same hunts and hunters the author features in his article. The very same hunters are featured in the film clip "The Persistence Hunt" which is part of David Attenborough's "life of Mammals" television series (BBC Worldwide 2003).

attunes the hunter to the hunted animal, in experientially different ways, the one along predominantly psychic, the other along primarily somatic lines.

Underlying these manifestations of sympathy throughout these two hunting modes is what appears to be a pan-San notion, what Kazuyoshi Sugawara, writing about the G/wi of the central Kalahari, refers to as "invisible agency" and that the G/wi refer to as *n!àrè*[5] (2015: 2), that there is a mystic-psychic connection of persons to other persons or to animals. It is also believed by the San to confer to humans, *qua* hunters or shamans, a measure of control over the prey or preying animal to which he is mystically and psychically connected. Regarding animals, some, the big game species, possess this connection and power as well, which runs across species lines. Moreover, this link and control—called *n!ow* by the !Kung—may connect a person to the wind or the rain (which, through this link becomes person-alized[6]). A big game animal's *n!ow* may interact with that of the stalking hunter, felicitously or unfavourably, bringing him either luck or failure in his hunting endeavour. That this form of invisible agency is phenomenologically linked to hunting is suggested by the fact that *n!ow* is not possessed by animals with which Ju/'hoan hunters do not experience any identity-blurring encounters, such as predators, small non-prey species, or the Cape Buffalo—an animal too dangerous to hunt (Marshall 1999: 169).

Tappings: "I Feel the Springbok Sensation"

> The Bushmen's letters are in their bodies. They (the letters) speak, they move, they make their (the Bushmen's) bodies move. They (the Bushmen) order the other to be silent, a man is altogether still, when he feels that his

[5] The Naro have much the same concept, transcribed as *nqàre* (*n!àre*) in Hessel Visser's *Naro Dictionary* and glossed as "'expect'(something to happen) Example: *khóè nerko nqare, dao qgomrko*: I expect people to come, it is itching" (Visser 1997: 34).

[6] To such an extent that the human connected to the rain would engage in conversation with the same, as the Namibian Hai//om healer Alfons /Gao-doseb, who, taking shelter from a thunderstorm in his hut, might be told by the rain that "she would get me out from there. She further said, if her time was ready, she would start to work with me. We understand each other. I you belong to her she will tell you everything that she will do with you" (Le Roux and White 2004: 118). On the personification of the rain, among the /Xam, as a gendered being, appearing either as male or female, with hair and legs, likes and dislikes, sentience and emotions—one of them "quick to take offence"—see Vinnicombe (2001: 332).

body is tapping (inside). A dream speaks falsely, it is (a thing) which deceives. The presentiment is that which speaks the truth; it is that by means of which the Bushman gets (or perceives) meat, when it has tapped. (//Kabbo, 1873 (Bleek and Lloyd 1911: 331))

//Kabbo opens his lengthy, nostalgia-tinged description of a group springbok hunt by his people on their */Xam-ka !au* hunting ground he presented to Wilhelm Bleek over a number of days from February through March, 1873 (Bleek and Lloyd 1911: 330–9) with an account of "Bushman presentiments". As his narrative unfolds it reveals, vividly and engagingly, the operation of invisible agency.

It operates through what //Kabbo refers to as "the Bushmen's letters" (*!gwe*[7]) and we see from his opening comments that they are lodged in the body, which stirs and moves when the letters "speak". And when they do—as "tappings" in a Bushman's body—the person "tapped" listens to them intently, commanding silence from others around him lest he be distracted. Unlike dreams, the other category of presentiments the /Xam recognize, people also assign truth and efficacy to these "letters", specifically in "getting and perceiving meat", that is to say detecting and killing prey animals.

Impending events about people, too, fall within the purview of presentiments. After describing two of these—the arrival of a grandfather through a sensation in the perceiving person's body at the same spot at which his grandfather has a wound scar and of a woman whose pressure sensation in her shoulder from the thong of her baby carrier is perceived by the other person—//Kabbo's discussion turns from human examples back to animals. He transitions to that topic via ostriches, as the illustrative case // Kabbo provides to Bleek in an aside when in his narrative he talks of "a tapping Bushmen feel when other people are coming":

The Bushman, when an ostrich is coming and is scratching the back of its neck with its foot, feels the tapping in the lower part of the back of his own neck; at the same place where the ostrich is scratching. (Bleek and Lloyd 1911: 333, footnote)

[7] Manuel De Prada-Samper argues that a more appropriate gloss for *!gwe* is "pictures" and thereby links /Xam presentiments to rock art (2014). So glossed, as something that is not only felt but also seen, and representable, as well as valorized, through painted or engraved images, presentiments become all the more direct and accessible, as well as salient when they are experienced.

The delicate precision of this tapping, through an ostrich's scratching an itch on its neck, indicates how sensitive and receptive to an animal a human could be. This sensitivity was capable of picking up not only bodily sensations—"beatings of the flesh"—from the animal but also its moods and intentions. An example, again featuring the ostrich, comes from one of Diä!kwain's narrative about an uncle of his who "possessed" ostriches (i.e. was mystically linked to them) and who, when he was angry at people, transmitted that anger to ostriches, causing them to run away from people before they were able to get close enough to shoot them (McGranaghan and Challis 2016: 590).

After the ostrich example //Kabbo's account turns to his favourite animal and one of his favourite topics for stories, springboks. On these he now dwells expansively, for the rest of his narrative. I will pick that narrative up at this point and follow it through, in abbreviated, annotated fashion as the text is too lengthy to give in full (as is usually the case with this most prolix of /Xam narrators). The reason for going so deeply into this text at this point is because of its effectiveness in conveying the affective and somatic aspects of invisible agency, emitted by a prey animal. Moreover, it provides the social context and emotional impact of an animal on a Bushman "when it has tapped", as well as the people around him.

As yet, in the story, the springbok is far off, beyond the hills and concealed amongst trees rendering the hunter's connection to the animals as yet incipient. One of them, the first unnamed character in //Kabbo's story, instructs the children to confirm the antelope's somatically sensed presence.

> He feels a tapping, (at) his ribs; he says to the children: "The springbok seem to be coming, for I feel the black hair (on the sides of the springbok). Climb ye the Brinkkop standing yonder, that ye may look around all the places. For I feel the springbok sensation". (ibid. 333)

The sentiment behind these instructions to the children is corroborated by another hunter at the camp, in a lengthy comment that dwells on the features of the landscape and the springbok's place within it:

> The other man agrees with him: "I think (that) the children should do so; for the springbok come in the sun; for the Brinkkop standing yonder is high; they shall look down upon the ground. And then they can see the whole ground. They can therefore look inside the trees; for the springbok are wont

to go hidden inside the trees. For the trees are numerous. The little riverbeds are also there. They are those to which the springbok are wont to come (in order) to eat in them. For the little river beds have become green." (Bleek and Lloyd 1911: 335)

In giving this bucolic account of springbok milling about and grazing on vegetation in verdant river beds, the other speaker is now seen himself to come into the thrall of "springbok sensation". However, it is not of the live springbok but of a springbok hunted and killed, not as an animal feeding in the green riverbed below the Brinkkop but as meat, bodily sensed and anticipated, from the hunt to-be:

> For I am wont to feel thus; I feel a sensation in the calves of my legs when a springbok's blood is going to run down them. For I always feel blood, when I am about to kill springbok. For I sit feeling a sensation behind my back, which the blood is wont to run down, when I am carrying a springbok. The springbok hair lies behind my back. (Bleek and Lloyd 1911: 335)

The other hunter, the first speaker, now seemingly on the same wavelength, perhaps because like tappings, about springbok hunting and butchering have perhaps become astir in his own body at this point, "agrees with him (saying): 'Yes, my brother'" (ibid.).

//Kabbo's narrative now turns inwards, away from the two hypothetical hunters, to his own musings and reminiscences about springbok, drawn from his own sympathy encounters with them over his hunting career, as both animals alive on the hunting ground and as hunted and butchered prey.[8]

> Therefore, we are wont to wait quietly; when the sensation is like this, when we are feeling the things come, while the things come near the house. We have a sensation in our feet, as we feel the rustling of the feet of the springbok with which the springbok come, making the bushes rustle. We feel in this manner, we have the sensation in our heads, when we are about to chop the springbok's horns. We have a sensation in our face, on account of the black-

[8] In another lengthy //Kabbo narrative (Lewis-Williams 2000: 52–77), we learn that springbok hunting was a prominent subsistence activity in the life of this narrator and his people. He provides a detailed account of one of their group springbok hunts, consisting of young people running behind and beside the springbok herd and driving the animals towards the hunters.

ness of the stripes of the springbok.[9] We feel a sensation in our eyes, on account of the black marks on the eyes of the springbok. (Bleek and Lloyd 1911: 335)

At this point //Kabbo's musings about animal tappings and the hunter's cross-species sensors in his feet and face and eyes and head corresponding to those in the springbok, once again—as if to frame his story, a narrative technique this and other /Xam story tellers used to good effect—turn briefly to the ostrich. As in its previous appearance, the animal is scratching itself, eliciting in the narrator the "sensation of a louse, as it walks, scratching the louse" (ibid.: 335–6).

After the ostrich vignette //Kabbo turns to animals in general—"things"—whom Bushmen detect not only by the spoors they see when they leave early on a day's hunt, but also feel connected to, through their tracks (as do contemporary Kalahari San, as seen below). Bushmen at such an extra-sensory tracking moment, "when the things are walking – feel them coming, feel them as they walk, moving their legs." They do so with the alertness of hunters whose presentiments of the walking animal include not only the perception of the approaching animal but also an anticipated visceral perception of the same as prey, hunted and butchered:

We feel a sensation in the hollows under our knees, upon which blood drops, as we go along, carrying (the game), Therefore we feel this sensation there. (Bleek and Lloyd 1911: 335, 337)

//Kabbo's account now turns back to humans again, to the boys sent out to scout the springbok beyond the Brinkkop hills, and the hunters who now figure out the appropriate strategy for hunting the springbok, in terms of speed, lines and angles of approach. Mystical hunting now turns to actual hunting in //Kabbo's narrative.

Neil Bennum's summative statement on //Kabbo's narrative makes explicit the element of sympathy between hunter and prey, as well as the transformative effects of the latter on the former:

[9] At this point in the narrative //Kabbo offers an explanation (entered as a footnote in the text)—"a black stripe that comes down in the centre of the forehead, and terminates at the end of the nose"—that clarifies a detail on the mechanics of this ontological transference of the springbok's identity traits to the hunter attuned to the springbok's tappings.

> The "Bushmen's letters" so eloquently described by //Kabbo were a kind of transformation. .... Hunter and quarry were linked long before a poisoned arrow had struck its target. Through a web of ritual hunting observances called */nanna-sse* ... the hunter came to embody the creature he was hunting. We could think of it as a kind of "tuning" in which the hunter prepared for the hunt by developing a kind of complicity, or at least sympathy, with the animal he was tracking. It is a hunting technique that allows the hunter to make decisions based on the most subtle information he has ever learned and it works extremely well. (Bennum 2004: 358–9)

Moreover, returning to a matter discussed in the previous chapter, what the hunter here experiences is an instance of "secular" transformation, rather than ritual-linked tranceformation: "//Kabbo did not", notes Bennum, "need to be in a state of trance produced by hyperventilation, concentration and movement to syncopated rhythm to feel himself springbok" (ibid).

Moving from the /Xam of the past to present-day Kalahari San hunters, some of these also appear to experience bodily palpitations that connect them to the prey animal before an impending hunt or during the hunt, such as the G/wi (see below) and the Ju/'hoansi. Like //Kabbo, the Ju/'hoan hunter /Ui Debe allegedly referred to these as "tappings" (Keeney and Keeney 2015: 171), which he "learned to feel" as a boy during hunting outings with his grandfather. Tappings can be strong, making the hunter's heart "quiver and tremble when the animal is near" (ibid.). Like //Kabbo's tappings, they are specific and precise, enabling /Ui Debe to hone in on the animals' size, sex and age:

> The tapping can also be along the chest or the wrist. The tappings tell us whether it is a young or old animal, male or female. It also tells us if the animal is small or large. The tapping is stronger when it is a big animal. (ibid.)

For the ≠Au//eisi hunter /I!ae the bodily sensations he experienced when hunting were less discrete:

> It was something that he felt and so could not be easily translated into words. When he discovered fresh tracks of something he wished to hunt, he shivered briefly, as if tickled by this trace of the animal's presence. It was a sensation, he explained, that he felt in the back of his neck and sometimes his armpits. (Suzman 2017: 170)

As the hunt progresses the hunter's state of attunement to and sympathy with his quarry intensify and the state of attunement and animation they transport him into once again manifests itself specific bodily sensations throughout the animal's death from the arrow's poison, their intensity fading as the animal gets closer and closer to death:

Tappings are the somatic manifestations in the hunter's body of his state of attunement to and sympathy with his quarry, which starts, as we will see in the next two sections, when he sets out to hunt a big game animal and persists throughout the tracking, stalking and chasing phases and ends with the killing of the immobilized, wounded or winded animal. They are sensations that make the human-animal bond palpable and render it an instance *par excellence* of Merleau-Ponty's "phenomena of incarnate significance" (1962: 166; quoted in Csordas 1990: 15), more so, perhaps, than the other two scenarios in which humans experience this bond, the trance dance and initiation rites.

The Classic Hunt

A sympathy bond links a San hunter—or, rather, some San hunters as not all individuals have such sensitivities, a point to which I will return again—to his prey animal at every phase of this mode of hunting. It starts with "presentiments", which a /Xam or Ju/'hoan hunter like //Kabbo and /Ui Debe experienced when sensing the nearby presence of springbok or kudu, and ends "when the fatal spear thrust" dispatches the quarry and ends the hunt (which, as we will see later, may have social, mystical, mythic and ritual repercussions that continue beyond the animal's death and the conclusion of the hunt). Here is James Suzman's account of this sort of hunting that he observed recently among the ǂAu//eisi.

> The practice of hunting is a process in which their bodies and senses progressively merge with those of the prey up to and beyond the moment that the fatal spear thrust is delivered. … Once they locate their prey, each movement they make is dictated by an amplified awareness of the prey's immediate sensory universe: what it sees, what is smells, and what it hears. Moving silently, knees bent and shoulders stooped, bows in hand, they experience the world around them from the perspective of both the hunter and the hunted. As the hunted they are alert to every sound and movement the hunters make. As the hunters they are singularly focused on dissolving into the landscape and emerging undetected close enough to get off a good shot. (2017: 171–2)

Sympathetic engagement with his prey animal, on which the hunter is now intently focused, continues and grows in intensity, enclosing him, notes Suzman, in a "hypersensory bubble" (ibid.: 173) throughout the tracking and stalking phases, its taxing, attention- and concentration-demanding logistics keenly connect the hunter to the animal. And vice versa, as the chased animal may on occasion—when the hunter is near, or his sounds and scent reach the animal's senses—be as aware of the chasing hunter as he is of it.[10]

Magico-mystical links emanate, some San believe, from the animals tracks themselves, bringing about a connection between hunter and prey. As the track "unfolds", notes Neil Rusch, hunters become profoundly intimate with the creature at the end of the path, such that they may already "feel the presence of particular animals before they can see them" (Rusch 2016: 6, quoting Low 2007: 576).

As noted above, a ≠Au//ei hunter may shiver when he encounters animal tracks, "as if tickled by this trace of the animal's presence", while his G/wi counterpart may feel a burning spot on his forehead when crossing the track of a prey animal (Low 2014: 51)—the G/wi variant of the tapping presentiment (something experienced also by the ≠Au//ei hunter /I!ae). Kalahari San informants told Edward and Cathelijne Eastwood that San hunters sometimes place sweat from their armpits on the spoor of a prey animal they are tracking (2006: 16), establishing a sympathetic link through this potency-charged bodily substance.

Even walking along the animal's trail and path and "running its track" may, as implied in the Rusch chapter epigraph, connect the hunter to the prey that has walked there before him. For Arctic hunters, amongst whom

[10] The Danish-Norwegian anthropologist Thea Skaanes has recently described eland hunting among the Hadza of Tanzania in much the same relational-ontological terms. The eland appears to be, for the Hadza, the same *bon à penser* as that animal is for San and hunted in like sympathy-informed fashion (Skaanes 2017a: 212–13, 2017b: 177–84). Relationality is seen to be reflexive and somatically expressed when the hunters, on the basis of clues they gain from the eland's spoor, carry out the same actions the eland did at the same spot hours earlier.

> They do not speak, only whisper. They get up and walk and walk. The eland was tired because of the poison that was killing it. It stood still to rest. Where the eland rested the old men will rest. And smoke. Twice. And walk again. Three times. The older men will say: 'Now it will die.' They get up: they will see it. (Skaanes 2017a: 213)

animal trails, carved by millions of migratory caribou into the landscape over millennia, are more abundant, prominent and permanent than in the Kalahari, the notion of a hunter's connection to his prey through its tracks is more defined and explicit in their animistic world view than for San hunters in whose landscape animal trails are less marked and more ephemeral. As suggested by the Arctic archaeologist Peter Whitridge (2013), in using such paths, human and animal movements become "imbricated" and through this overlap, in sharing the same action, walking, with the same purpose, to change locale, migrate, forage, stalk prey, evade or flee predators, seek shelter, a bond is established between humans and animals that is both kinetic and conative.

Elias Canetti's take on the connection, among Bushmen, between hunter and prey through the latter's tracks and through the act of tracking is both more subtle and more direct. He sees tracking an animal as an incipient form of miming the same, subconsciously and delicately, "by the rhythm of their movements" which the tracks entail. Tracks provide "a rhythmic notation imprinted on the soft ground [that] connects the hunter, as he reads it, with the sound of its formation" (1984: 31). As noted by anthropologist and literary critic Rosalind Morris in her analysis of Elias Canetti's discussion of /Xam presentiments and transformation (1984 [1960]: 337–41), this can affect "a transformation in the person in whose body the tracks make themselves felt" (2011: 181).

Mystical links between hunter and prey derive also from a suite of hunting observances premised on sympathetic magic, consisting of such things as food proscriptions or prescriptions and refraining from certain actions that might alert or invigorate the stalked or wounded prey animal[11] (which, as seen in the previous chapter, boys are introduced to at their First Kill initiation rite and which, for the rest of their hunting careers, will constitute the symbolic component of their hunting labours). As the hunter closes in on the ever more exhausted game animal for the kill, especially in the persistence hunt, when he is as exhausted as it after a chase that matched the hunter's strength and wits with those of the animal, the sympathy bond may assume its most reflexive moment, when the hunter sees himself—not just figuratively but, as we will see below, quite literally—in the eyes of the now immobile animal whose gaze is fixed on him.

[11] See Marshall (1999: 143–61), Suzman (2017: 173–4) and Bleek and Lloyd (1911: 270–83) and Vinnicombe (2001: 172) for accounts of these beliefs and practices among the Ju/'hoansi, ≠Au//eisi and /Xam, respectively. See also Barnard (1992: 58–9) and Le Roux and White (2004: 178–81).

The Persistence Hunt

> Therefore I chase it, in the sun, that the sun may, burning, kill it for me, that I may eat it, dead from the sun: while I feel that I was the one who chased it, while it went along in fear of me. It, in fear, lay down to die from the sun; because it had become dry (while running about) in the sun; because it saw me when I followed it. It did not stop to walk, that it might look backwards. For it had run about, when it was tired. It seemed as if it were about to die; because fatigue had killed it; while it had run about in the heat; for, (it) was the summer sun, which was hot. The ground was hot which was burning its feet. (//Kabbo (Bleek and Lloyd 1911: 311–2))

> One appreciates the stamina of the tough Namib denizen when considering how he will sometimes hunt springbok. Having no hunting weapon on hand, */Ga-.burob*, an old Namib hunter, fills two gemsbok stomachs with water ... divests himself of his greasy leather clothing and heads out. Sighting his first springbok, he gets down on the ground and stalks the animal, naked but for the strap of the water bag hanging from one of his shoulders. As soon as the antelope takes flight he jumps up and follows the animal in a steady jog for an hour and longer, as dog chases hare. Once he has started his pursuit he cannot pause for even a minute, his excrement soiling his thighs. He grabs the animal, collapsed from exhaustion, and smashes the back of its skull with a rock. (Schultze 1907: 99, my translation)

Once again I turn to //Kabbo to introduce the other mode of hunting San employ in which the hunter engages with his prey through a bond of sympathy, persistence (or endurance) hunting. The animal hunted in //Kabbo's first-hand account is not a springbok—"for, the springbok were gone away"—but a hare.[12] The /Gainin hunter /Ga-burob in the second epigraph is seen hunting a springbok in this fashion (albeit, as evident from Leonhard Schultze's starkly matter-of-fact account, with less sympathy on the hunter's part than the /Xam hunter //Kabbo, as well as his !Kõ and Ju/'hoan counterparts Karoha and /Ui Debe whom we will meet below).

The account graphically conveys what is this style of hunting most distinctive aspect—the physical exhaustion it brings about in the hunter, and the prey, both as intent as the other in running—through the heat of the summer sun—to catch the animal prey or to get away from the human predator. The fact that such a hunt may last for hours, with little let-up,

[12] A brief account of a persistence hunt for springbok is given by //Kabbo's son-in-law /Han≠kasso (Bleek and Lloyd 1911: 387–9).

increases its arduousness and the stamina it requires of the hunter, as well as the prey he is focused on intently and unrelentingly. Indeed, it is this intentness, and intensity of sustained effort, that, Louis Liebenberg suggests, elicits the hunter's identification with the pursued animal.

In his ethnographic study of persistence hunting of the Central Kalahari !Kõ—that included himself participating in such hunts which, in the case of a large antelope such as a Kudu bull, may last up to five hours and cover over 30 kilometres—Liebenberg describes this as "a subjective experience of projecting oneself into the animal", in which "one attempts to think like an animal … and, in looking at its tracks, one visualizes the motion of the animal and feels that motion in one's own body" (2006: 1024). And, once the kudu can run no further the hunter having "taken its energy", the spent animal, standing, or dropped to the ground, "looks at the hunter with glazed eyes". In addition to creating a moment of reflexive self-awareness, as it did for the Ju/'hoan hunter /Ui Debe, who, when looking into the eyes of a kudu, "will sometimes see my own eyes reflected in the kudu's eyes" (Keeney and Keeney 2015: 172), for the !Ko hunter Karoha this instance of eye-to-eye encounter also created a moment at which the hunter is attuned to, or one with, the kudu's mind. The hunter has now "taken the kudu into [his] mind", his ontological merging, psycho-kinetic throughout the chase now, at the end and culmination of the hunt, entirely psycho-mental.

In a recent analysis of /Xam persistence hunting, supplemented with Liebenberg's !Kõ-based research on the topic, Neil Rusch (2016) suggests that the state of extreme physical exertion, intensely felt by the running or jogging hunter, may bring on not only transformation but also an altered state. The hunter-runner's "motility-dependent embodied activity", Rusch argues, "sets in motion transformation—chemical-physical, neuropsychological, psycho-phenomenological—which in turn produce changes in bodily state and consciousness" (ibid: 22). Rusch's point here is expressed succinctly, and in a different idiom, by the !Xo hunter Karoha, when asked about this experience: "tracking is like dancing—because your body is happy."

Karoha's observation resonates with Rusch's point, about persistence hunting and its similarity to that other "motility-dependent embodied activity" San carry out regularly, trance dancing.[13] And regarding the lat-

[13] Indeed, as suggested by Harvard evolutionary biologist Daniel Lieberman in an interview with *Scientific American* science writer Thea Singer, the connection between endurance running and dancing is even more basic, experienced not only neuro-psychologically, as sug-

ter, not only is the trancer's body and mental state altered but ontological transformation, too, takes place in both the dancer and hunter, as they merge identities with an animal, the first with a lion (and occasionally also an antelope), the second an antelope.

That transformation is brought on through the effects of an increasingly intensifying experience of mental and physical identification of the hunter with the chased prey animal, after hour upon hour of sympathy-based "speculative tracking" (Liebenberg 2006: 1025) of the unceasingly pursued quarry. Both are maximally exhausted and at the limit of their endurance, doggedly or desperately intent on catching up with or getting away from the other, unrelentingly, through the "summer which was hot" and the midday sun that burned and boiled the chased animal's feet[14] and blood—as well as the hunter's—and "killed it with fatigue". Rusch, struck by the vividness of //Kabbo's account of his hare chase, "feels compelled" to ask: "is he describing the animal's condition or his own?" (2016: 892). Hunting an animal in this fashion may bring about a deeply felt cross-species merging of identities of hunter with quarry as "his identity connects with that of the animal as he runs its track" (Rusch 2015: 23).

The bodily merging of hunter with the expiring prey animal is at its most intense—reaching its consummation—whenever an animal is killed not with a spear—or a rock (!)—but is suffocated by the hunter. Elizabeth Marshall Thomas describes the intimate physicality of this killing mode:

> Still the hunter is coming. The eland must force himself to keep running. Surely the hunter will give up soon, after coming so far in so much heat, so the eland stops again to look back, but still the hunter is coming, even gain-

gested by Rusch in the context of endurance hunting (see also Ijäs 2017: 33–4, 105–9, 117–31 and Bramble and Lieberman 2004), but also motor-physically, with two million years-deep roots in our species's evolutionary past. It is an "outgrowth" of endurance running, in a certain manner of "coordinated movement of torso and limbs" that, Lieberman suggests, "could have grown out of our ability to run—as opposed to just walk—on two legs" (Singer 2017: 70).

[14] The severe heat stress on the chased animal's feet due to the extreme temperatures of the sand in the summer months was noted by Wikar as the decisive factor in bringing down steenbok when they were hunted by the /Xam of the Northern Cape. After the third or fourth chase along the spoor of the targeted buck, with intervals in-between where the hunter might rests in the shade of a bush "for as long as it takes to smoke a pipe", "you can catch it with the hand as it jumps up—then its feet are burnt through". This mode of persistence hunting, noted Wikar, "is the regular custom of the Bushmen of the plains in the hot season" (Wikar 1935: 175, quoted in Iläs 2017: 115).

> ing on him. Eventually, the eland can run no more. He falls to his knees, or just stands still, head low, legs side apart, spent, dying. Then the hunter, with the last of his strength, can catch up and grab him by the tail, then kill him with a spear if he brought one, or he can push the eland over and lie on his neck to keep him from struggling and clamp his hand over the eland's nose and mouth to stop his laboring breath. (2006: 32)

A number of Drakensberg rock paintings suggest that this transformative moment, of a hunter closely attuned to the antelope before him, immobilized from dying through his arrow's poison, may progress in intensity, from a psychosomatic experience on the part of the hunter to its expression, through ritual—even ludic—dancing. Two of the earlier, pre-"trance hypothesis" writers interpret certain pictures along such sympathy/hunting lines. One is as Bert Woodhouse who depicts an image titled "playing with eland" of a dying eland surrounded by four hunters one of whom appears to be leaping over the animal's back (1984: 35, Fig. 6). Another is Patricia Vinnicombe, who presents and describes two paintings (2001: 173, Fig. 101, 307, Fig. 223, 320, Fig. 234) that show stupefied, dying, or dead, eland surrounded by dancing, leaping and soaring figures, some of them hunters with spears, poised to dispatch the eland, others in various stages of transformation, from human—one of them a hunter, his bow still in hand, with hooved feet or fetlock heels (Fig. 234)—to antelope-headed and bodied therianthropes (Fig. 223).

Whether what is depicted here is transformation, of a sympathy-stricken hunter, or trance, by a hunter-*qua* metaphorical shaman who experiences trance through *trance*formation is debatable (and debated) amongst rupestrian scholars. Whatever it might have been—as already suggested, both altered states of consciousness and of being might have been experienced by the men in the hunters in the presence of the dying eland—such images attest to the intensity of the hunter's mimetic-metamorphic, trans- or trance-forming experience face-to-face with the dying prey animal. His feelings may spill over, from something felt and inside the hunter, and contained as he proceeds with his hunter's business, to outward enactment, of the identity merging he may have felt towards the animal with mounting intensity, after hours or days of hunting the same, as well as feeling and thinking antelope.

The extreme demands this hunting technique makes on the hunter physically—something especially evident from the second epigraph—may explain why evidently it was not used very often by San hunters (except as

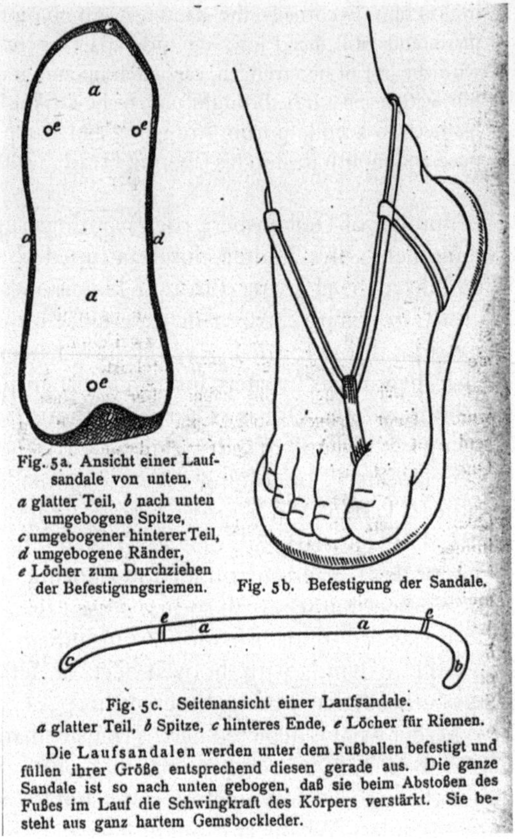

Fig. 7.2 ǂAu//eisi running sandal. (Kaufmann 1910: 1423)

part of stalk and track hunting which may, at its conclusion, involve the hunter chasing the prey animal in its final stage of exhaustion to its point of collapse). Persistence hunting is mentioned much more in the ethnohistorical literature of the nineteenth and early twentieth century (such as Schultze in the section epigraph) than it is by recent or contemporary ethnographers.[15] Kaufmann, who describes this hunting technique among the ≠Au//eisi (1910: 145; see also 2005: 64–5) also describes, by means of annotated illustrations (Fig. 7.2), what he calls *Laufsandalen* (running

[15] See Passarge (1907: 71–3) on ≠Au//eisi, Schultze (1907: 99) and Trenk (2005: 239–40) on "Namib Bushmen", Bleek and Lloyd (1911: 285, 309–13), and Wikar (1935: 115) on /Xam, Schapera (1930: 133), on Kalahari San, Metzger (1990: 20–2) on ≠Au//eisi.

sandals) which, unlike normal sandals worn by the people, were especially constructed—with convexly bent soles made from "very hard gemsbok leather"—"to increase the momentum of the body when the foot is lifted off the ground" (ibid.: 142, 2005: 59). They were worn by men, presumably when chasing game. Wilhelm reports that among the !Kung of north-eastern Namibia this style of hunting is "unlikely" because of its "slight chance of success" (2005: 121). Jodtka reports its use by !Kung hunters for antelope grazing on new shoots in areas of the veld previously burned by them, in order to destroy scorpions, snakes and vermin and as a high-yield hunting technique: to drive game running from the veld fire, set in a semi-circle around the veld area, along a predictable course (1902: 495).

The reason for the higher incidence of persistence hunting among San of the past—apart from dubious ethnographic reportage based not on observation but repetition of other published accounts—may be that the antelope species ideally suited for the chase hunt because of its relative docility, slowness and exhaustibility (Thomas 2006: 31–2; Ijäs 2017: 103, 111), the eland, was more abundant then, in both the Kalahari and the Karoo (and even the Namib). The Naro I worked amongst reported it as something that people practised "in the old days" ("in the rainy season when people ran after animals with their bare feet"). Le Roux and White's San interviewers were told by one of their Ngamiland Bugakhwe informants that their forebears hunted warthog and duiker in this fashion. This, in the case of a duiker endurance hunt might have taken "a few hours", provided the hunt was not aborted because the quarry outran the hunter, an outcome that might transpire when the chased duiker disappeared into a thicket and chased from it another duiker that happened to be at the spot (2004: 147). Liebenberg found a noticeable decline in persistence hunting among the !Kö hunters he worked with over the course of a generation, due to hunters' loss of the skill, along with the required stamina, due to ageing (2006: 1021–2). Silberbauer reports that, among the Central Kalahari G/wi, some men may occasionally chase down an antelope such as eland, kudu or hartebeest that has strayed near their home encampment, but that G/wi hunters would seldom hunt animals in this fashion, in part because of its strenuousness and in part because of its slender chances of success (1981: 213–4). Tanaka, on the other hand reports that, in the 1960s and 1970s, G/wi still practised "hunting by chasing", noting that "hunters can catch up with even a large antelope after a one to two kilometer chase, and fell it with a spear." Yet more remarkable is Tanaka's second-hand report of a cheetah—"the world's fastest mammal"—having been "chased and beaten to death with a club" (1996: 16). I am aware of two other accounts of contemporary San practising endurance hunting, among the Ju/'hoansi (Thomas 2006: 26–7) and, as featured above, the !Kõ (Liebenberg 2006[16]). This hunting technique, which is capable of effecting ontological identity merging with exceptional intensity, is thus something more of the past than present, practiced before the onset of disenchantment in San lives, culture and imagination.

The last hunting method to be examined, one also potently charged with ontological mutability that may affect the hunter, is the ritual or mystical use of animal body parts and substances and, more practically, the use disguises, consisting primarily of antelope skins and animal appendages. Like persistence hunting, it is something from the past with little ethnographic evidence among recent and contemporary San people.

Hunting Ritual and Hunting "Medicine"

> After I have discovered a kudu track we walk on the same very quietly, and when we have reached the spot where the kudu has stood we have to be very careful. One walks on ahead and the others follow on. Once I sight the Kudu I take out the poison arrow and I shoot. Then I wait to see what the kudu will do next, or I shoot once again. Then I follow him to see whether the arrow has fallen out of the kudu; if it has I pick it up and look it over to see how deeply it has entered the kudu. Then I have to return to the other hunters and tell them just what has occurred. Then we walk home. You are not allowed to follow the kudu; it has to be left on its own. The one who made the shot has to go home right away and he must not talk, not eat and not drink because if the shooter talks to others it takes a long time to find the kudu and if he drinks water the poison will become watery and cannot take effect and if he eats the kudu will not move which will prevent the poison from working properly. The next morning we have to get up early to pick up the spoor again. The kudu is already dead. (unnamed !Kung informant (Lebzelter 1934: 33, my translation))

> With the North-Western tribes these [charms] most commonly take the form of scarifications serving to endow the hunter with the qualities of certain animals and enabling him to hunt them successfully. The cuts are made as a rule on the arms, sometimes on the cheeks, occasionally also on the back, chest, or

[16] For a stinging critical discussion of the latter account, which includes the suggestion that the hunt was staged and not representative of how !Kõ would hunt today, see Pickering (2013: 99–102, 156–7).

> belly. A tiny piece of meat is then burned to ashes, which are rubbed into the wounds. The meat is that of an animal whose qualities are especially desirable, e.g. that of the springbok for swiftness. (Schapera 1930: 138)

The performance of certain ritual practices, including the application of cuts into which animal substances were rubbed leaving permanent scars, was discussed above, in a ritual context, namely the male initiation rite in which a boy and hunter-to-be was "cut with meat" in order to "wake up the boy's heart" so that, on awakening, it would be a hunter's heart. The sympathy component of this act was dealt with, along with other ritual observances, mostly food taboos that also were informed with sympathy, reiterating the boy's connectedness, spiritually, and, as I would submit, ontologically.

Such ritual practices continue on in the grown-up man's career as hunter. Viktor Lebzelter describes some of these in the context of a kudu hunt, in the words of his !Kung informant he interviewed in 1926 (first epigraph) and, three generations later, the Naro "chief" Willem Ryperdt, in his reminiscences about traditional Naro big game hunting practices, details sympathy-based food proscriptions for a successful hunter.

> When you have shot an animal and it has run away and you return home, you must not eat anything. You must go to sleep without eating. It is part of the tradition to prevent that animal from running far. If you eat, the animal that you have shot will run away. If you do not eat, the animal will also not eat and will then be powerless. So, if you eat you give the animal power to run away with the arrow, because your saliva is also in the mixture that you have put onto the arrow. (Le Roux and White 2004: 178)

Vinnicombe reports a number of such pre- and proscriptions—"extremely intricate rules and taboos ... which closely identified the hunter with his prey"—among southern San (2001: 171–2).

Sympathetic magic also underlies the "charms" Schapera deals with in his summary of the earlier ethnographic information, from colonial sources and a time when such practices appear to have been practised more extensively than among more recent San (1930: 138–40). This—"hunting medicines"—is the aspect of !Kung ritual hunting dwelled on by Wilhelm, in his remarkable account of one such item consisting of an amalgam of substances from a diversity of game animals.[17]

[17] These accounts can be supplemented with similar information among Ju/'hoansi, who continued with such supererogatory hunting practices into the 1950s and 1960s (Marshall 1999: 157–60).

> A hunting medicine which is in my possession is decorated with bits of skin and pieces of every game animal found in the sand *veld*. It includes eland tail, pieces from the head of gemsbok, duiker and steenbok, pieces of wart-hog skin and various other things. As far as I was able to find out, medicine is made from these parts which are then burned. The charred section is scraped off with a knife and is kept in the tortoise shell. To ensure luck in the hunt, the Bushman will apply several closely spaced cuts between both eyes above the base of the nose, into which this medicine is rubbed, along with soot and fat. (Wilhelm 2005: 159)

These early writers are responsible for the standard view about such animal substances as "good luck charms" needed by hunters going after elusive game, difficult and at times dangerous to hunt, causing the hunters anxiety. "Fully aware of the uncertainty attendant upon their efforts", notes Schapera, Bushman hunters "seek to relieve their apprehensions and obtain confidence by the aid of magic and taboo" (Schapera 1930: 140). Writing in the same vein—"hunting is difficult and exceedingly uncertain in the Nyae Nyae region"—Lorna Marshall has Ju/'hoansi hunters, "in the difficulties and uncertainties … look to the supernatural for help" (1999: 143).

I have a different take on these ritual components of San hunting, seeing it not just, *pace* Malinowski, as a coping mechanism to deal with the vagaries of hunting and killing game that is elusive and difficult to get to and at by an anxious hunter who hopes for a lucky break. For starters, my impression from the ethnohistorical literature of the past and the ethnographies of the present is that San hunters were—and still, on balance, are[18]—not all that anxious but, instead, quite confident when they set out to hunt, animated and upbeat by an activity highly valued culturally and carried out in the context of a subsistence economy that, thanks to the considerable quantities of plant food provided by the women, affords, on the whole, a life of relative "affluence" (albeit "without abundance, the Zen style", to use Marshall Sahlins much cited *bon mot* and *mot juste*[19]).

[18] My impression here, about confident hunters, is borne out by both my reading of early and contemporary San ethnography and by my own field work among the Ghanzi farm San. This admittedly was limited—a couple of hunts I participated in and conversations with hunters setting out on and returning from hunts—and not representative of San traditional lifeways.

[19] Sahlins (1972: 2); see Suzman (2017) for a recent re-consideration of past and present San lifeways and socio-economic conditions in terms of Sahlin's classic portrayal of the "stone age economics" of immediate return hunter-gatherers.

Moreover, the fact that divination is not widely employed by these southern hemisphere hunters (Guenther 1999: 88–9, 227)—as opposed to northern ones, in polar and circum-polar regions, who employ this ritual device much more extensively in hunting (e.g. Tanner 1979: 108–35; see Vol. II, Chap. 5)—further attests to the relative confidence of San hunters about the more or less successful outcome of their hunting endeavours.

While the "rabbit foot", magical charm explanation may apply in some instances, my sense is that a San hunter—or some San hunters—may have employed such sympathy-based ritual measures not as much for luck in getting inaccessible, or hard to access, game but more in order to sharpen or substantiate a connection he has with the prey animal, at the present hunt or from previous hunts, all the way back to his First Kill rite at his hunter's initiation. When sensing the presence of a game animal at camp or out in the veld, either spontaneously through presentiments that create "tappings" in parts of his body and translate into "springbok sensations", or by coming across an animal's spoor or scat or trail, sympathy stirrings may become activated. Once sighted, shot and tracked, such ritual practices as food taboos or small, stereotyped behavioural routines, all of them premised on sympathetic magic—much of it of the "contagious" variety—as per the Old Animists Frazer and Tylor, add to the sympathetic bonds the hunter feels with his quarry at the time of special focus and intensity. Or it may somatically substantiate or "incarnate" that bond through the insertion of animal substances—parts of the animals flesh previously rubbed into hunting scars, or re-applied in a new hunt, as in the Wilhelm excerpt above, where the hunter draws on a plethora of "animal medicines" from an array of game species that all comingle in his tortoise medicine box,[20] which becomes thereby a material isomorph of the chimera encountered at times in San myth and art. This merger, of animal with human flesh, is a more potent and ontologically absorbing version of what I will discuss in the next section, the hunter's wearing animal disguises—skin on skin—when stalking game.

[20] The "charms" of /Xam shamans and hunters, some of which comprised animal parts such as hartebeest hooves, baboon hair, burnt "snake powder"—may also have connected the owner mysto-physically—through Frazer's "law of contagion"—to an animal, as a source of potency or to secure luck in hunting or ward off danger (Lewis-Williams 2015a: 70–1). See also Biesele (1993: 94) on the use of animal materials in ritual context by Ju/'hoansi.

After describing the employment and efficacy of animal disguises in the Bushman section of his comparative survey of such devices by African hunters (1955: 178–89), the German anthropologist Helmut Straube notes that their use by Bushman and other African hunters is more than instrumental, viewing them as such, somewhat gratuitously, as "survivals" of a myth-magical past, when, Straube surmises, "profane hunting"—*Profanjagd*—had a ritual counterpart in ritual hunting.[21] He focuses specifically on the practice of tattooing, suggesting that, by virtue of having animal flesh rubbed into his skin, the hunter inwardly undergoes ontological transformation (*Wesenswandel*), so much so that, having now assumed "animal quality" (*Tierqualität*) and become ontologically transformed, "the [prey] animal cannot perceive him as he is himself 'animal'" (ibid.: 197), that is to say, "the hunter becomes invisible to the animal". Straube, aware of the dubious chances of success of such a hunting strategy—as well as, one might add, a potentially high hunter mortality rate, with respect to certain species such as Cape buffalo or gemsbok—points out that all of this is for the hunter a "religious matter" ("*religiöses Anliegen*"), a ritual obligation harkening back to a mytho-magical past when such beliefs were operationalized in the hunt in some way or other. Today's hunters are too skilled and practised for so "foolish" ("*töricht*") an assumption that tattooing would confer invisibility.

Yet, all the same hunters do, Straube contends, still today (i.e. the "ethnographic present") hold to some notion of ontological transformation, of *Wesenswandel*. For lack of evidence from the San he adduces evidence from a southern Bantu-speaking group, the Thonga of eastern Zululand, amongst whom Junod in 1936 described a set of complicated ritual practices through which a hippopotamus hunter, after killing a hippo, ingests morsels of flesh from different parts of the animal's carcass or has them rubbed into his body. By virtue of "his newly acquired animal quality" the hunter, now invisible to the hippo, is able to hunt "ontologically equivalent game" ("*wesensgleiche Wild*"), unrecognized as human by the animal quarry.

[21] Oswin Köhler some years later described such a hunt among the Kxoe of north-eastern Namibia (1973), a complicated ritual several days in duration, Kxoe hunters carry out, for large game, after a succession of failed hunts ("when hunters find only animal tracks but no animals"). It is a propitiatory rite, Köhler contends, addressed through prayer and dance to Kxyani (the Kxoe god), along with //awa, who is asked to mediate. With the exception of a somewhat similar account from Gusinde among the !Kung of north-eastern Namibia (1966: 30–2), I am unaware of this kind of hunting among San, whose hunts seems to be carried out on a "profane" footing, albeit not without a varying dose of mytho-magical, somato-mystical elements.

Straube does not suggest that the San, to whom he ascribes much the same notion, acquired the same through contact with these Bantu-speakers. While unlikely in this case, the impact of acculturation on San from Bantu-speaking neighbours cannot be discounted, especially, as we will see in Vol. II (Chap. 6), as regards beliefs and practices about animal transformation.

Animal Skins and Disguises

> Removing the skin from the neck of a previously hunted ostrich and leaving the head attached, he pulls this tube over a bent stick; he then fashioned a sort of saddle, covered all over with feathers and at the sides with the killed ostrich's wings. This saddle fully covering the bent-over torso of the hunter; his legs painted grey, one hand held behind him holding the deadly weapon, the other, held in front, the stick covered with the pulled-over ostrich neck, the hunter approaches. He looks ever so harmless, pecks the ground, grazes, moving forward once again with the bird's undulating gait—in short, he brings into being, in full and with amazing verisimilitude, *Struthio africanus*. ... The clever birds, partly out of curiosity, partly from an innate "caste chauvinism", approach to turn away the intruding stranger. This is the critical moment—the saddle and ostrich neck, cast off and hurled onto the ground, the mumming ostrich shoots his poison missiles, into the heart ... of four of the boldest birds. Staggering they fall to the ground. (Hahn 1870: 105, my translation)

> [Bushmen] decoy game with an ostrich skin, which they put on and then imitating the walk of the bird, they go into the region where the game are grazing. As soon as they are well in sight of the game they begin to show fear, jumping here and there as an ostrich does when it becomes aware of anything. The zebras, wild horses etc., seeing this, approach the sham ostrich to have a look, since he has been playing the trick below the wind and they have not been able to get his scent. Now that they themselves come toward him he can shoot it at will. (Wikar 1935: 179, quoted by Dowson et al. 1994: 12)

> On the head of one man I remarked an unusually large fur cap. It was made of spring-buck skin, of a shape extending far behind the head, and intended to have as much as possible the appearance of that animal's back. This was for the purpose of deceiving the game and of enabling the wearer, as he creeps along between the bushes, to approach the animal within reach of his arrow. It is called a be-creeping cap (*Bekruip-muts*[22]); and is only worn when in pursuit of game. (Burchell 1824: 56)

[22] As noted by literary critic of Afrikaner literature Helize van Vuuren in her recent work on the San (specifically /Xam) impact on Afrikaans literature, Gideon von Wielligh, author of the popular *Boesman-stories*, also repeatedly mentions such an item—"*bekruipmusse*"

Fig. 7.3 Stow's Bushman putative rock art painting of a disguised ostrich hunter. (Source: Stow 1905: 96)

Even though Stow's iconic rock art painting (Fig. 7.3) that he concocted on the basis of illustrations in the early travel literature (Fig. 7.4; see Dowson et al. 1994) and presented and published as his copy of a rock art painting he allegedly discovered in the north-eastern Cape (Stow 1905: 96) is "almost certainly a fake" (Lewis-Williams 2015a: 176), what the image depicts, a San bowman in ostrich disguise, stalking a group of ostriches, is something San hunters did evidently practise at the time. A number of eighteenth and nineteenth-century sources report this San

("stalking caps")—made from the dried head and neck of an ostrich, in the context of stories he obtained from Afrikaans-speaking /Xam descendants he interviewed in the late nineteenth through early twentieth century. Van Vuuren notes that this is an "unusual Afrikaans word" applied to an item of the people's material culture from "long forgotten indigenous knowledge" (2016: 70).

Fig. 7.4 Stow's source for above painting. (Source: Moffatt 1844: frontispiece)

technique of ostrich hunting, at least three or four of which—by Hendrik Jacob Wikar, Sir James Alexander, Robert Moffatt and Samuel Dornan—appear to be based on first-hand observations[23] (Dowson et al. 1994: 10–4). One of von Wielligh's key informants, the Afrikaans-speaking "very aged /Xam man Old Bles", provided his interlocutor with a vivid hunting narrative when he visited him in 1882 in Calvinia in the Northern Cape, in which he reportedly wore an "ostrich suit" when hunting hartebeest. The narrator adds the interesting detail how, in the course of stalking the hartebeest, he managed to ward off a young lion that "was out on the same business" by gripping his ostrich head "stalking cap" with his teeth, rearing up and "growl[ing] loudly" (van Vuuren 2016: 68–9). The illustration in Farini's—a.k.a. William Leonard Hunt's—travel narrative through the Kalahari to Lake Ngami (Fig. 7.5) is allegedly based on a photograph taken by the explorer's protégé, travel companion and expedition photographer Samuel "Lulu" Wasgatt, in the context of a nocturnal ostrich hunt their Bushman companion Suku had successfully undertaken the night before (ibid.: 330–2).

[23] Given this evidence, Wilmsen's categorical statement—"this so-called hunting method should be considered to be nonexistent" (1997: 177, footnote 143)—is not tenable.

Fig. 7.5 Ostrich hunter's disguise. (Source: Farini 1886: 330)

Another early first-hand report to be added to the list, from pre-colonial Namibia, was by the missionary Reverend Tindall who, as reported by Theophilius Hahn, in his travels through Greater Namaqualand, came across one such hunt in progress: on sighting what he thought was an ostrich some distance from his wagon and from a flock of ostriches yet further off, he had the wagon stopped and instructed his wagon driver to stalk and shoot "the supposed game". The latter, itself too intent on stalking the real ostriches, was unaware of the danger he was himself in "and, wounded in the arm by the wagoner, the Bushman hunter jettisons his disguise, leaving the party of travelers in a state of shocked surprise" (Hahn 1870: 105).

Such disguises were evidently in use in various regions of the Kalahari[24] (Harris 1852: 260; Richters 1886: 80; Kaufmann 2005: 66; Dornan

[24] Passarge reports that, despite repeated questioning of San informants in the Middle-Kalahari, not a single person knew anything about it. One had heard about such a practice amongst Khoe Khoe in Damaraland to the west (1906: 77–8).

1925: 104) as well as by Namib San (Schapera 1930: 135) and Dornan describes the same technique among San of north eastern Botswana. The latter also provides details on the mask's manufacture and dwells in his account on the mimetic deployment of these disguises that included mimicking the twittering of chicks in a nearby egg to attract the anxious mother hen who, "rushing up to protect the nest is easily killed" (Dornan 1925: 105; see also Schapera 1930: 135). Kaufmann describes ostrich disguises among the ≠Au//eisi in eastern Namibia, including the somewhat complicated manufacture of the grass saddle—"a kind of flat, double-cushion stuffed with straw" (Moffatt 1844: 53)—a stalker would wear to which were stuck ostrich feathers—"attached to small pegs" (ibid.)—and wings. The bird's neck and head into which a stick had been stuck[25] was held upright with one hand, the other holding his bow concealed in the costume, legs were daubed grey with limey soil (2005: 66).[26]

In his account of ≠Au//eisi hunting practices, Kaufmann notes that this was the only animal disguise hunters would use, venturing the opinion that an ≠Au//eisi hunter would not have been capable of "utilizing other animal masks because he is unable of softening the skins of large animals

[25] Other sources describe this component of the disguise differently, as a long stick neck with a roughly carved ostrich head (Schapera 1930: 135; for other descriptions see Dowson et al. 1994: 12–5).

[26] Being closely similar in its details to Hahn's description a generation earlier (first epigraph), Kaufmann's source for this detailed information might have been this writer's published article, readily accessible in the widely read monthly *Globus*, a popular item in the early colonial literature. This surmise is supported by Kaufmann's statement that he had never himself actually seen such ostrich costumes, which were described to him by his San informants "with such accuracy that I cannot doubt that they knew how to make the same" (ibid.). Nolte's description of the ostrich disguise hunt of southern Kalahari San, whom he visited in 1881 (Richters 1886: 79–81), is so similar to Hahn's as to suggest borrowing from same source. Nolte description this style of ostrich hunting is second-hand, as opposed to ostrich hunting from horseback, which he appears to have witnessed (ibid.: 80). Hahn's account, in turn, is similar in some details to a yet earlier, and seemingly original source, based, in part, on first-hand observation, namely Moffatt (1844: 53; see Dowson et al. 1994: 13–4). Dowson, Tobias and Lewis-Williams examine 11 of the early accounts, in an effort to separate first-hand from the more prevalent second-hand descriptions (or "borrowings"), with reference to Stow's "blue ostriches" painting (Fig. 7.3) which, the authors conclude, is itself "part of an intricate web of 'borrowing', a kind of pictorial paraphrase" (1994: 33). Yet, as noted above, they do also authenticate some first-hand accounts of this ostrich hunting practice. While the practice evidently did exist in the past, this did not warrant its depiction by South West Africa's postal service in the 1970s on a national stamp series that, along with other primordialist culture tableaux, showcased the country's "Bushmen" (Fig. 7.6).

without altering the animal's natural appearance"[27] (2005: 66). Thackeray, in one of the few twentieth-century examinations of animal disguises as hunting props (1983) which at the time was a notion superseded by the more influential trance hypothesis which explained such animal aspects of San humans in terms of potency (Dowson et al. 1994: 323–4; Jolly 2002: 89; Lewis-Williams 2015a: 176), notes that there is less evidence for use of antelope skin than for ostrich disguises. The fact that the latter technique, as seen from the section epigraph, was used to hunt not only ostriches but also quaggas and zebras, species that mingle with ostriches, perhaps suggests a further reason (along with Kaufmann's point about the technical difficulties of manufacturing animal skin disguises): given the effectiveness for hunting prey other than ostriches with the ostrich disguise, the making and employment of which seemed to require some effort, there was possibly a reduced incentive for investing yet further effort into making yet other disguises.

Nevertheless, there is some evidence, notably a first-hand report from the 1880s by the Kalahari traveller Izak Bosman who describes a Bushman hunt he observed in north-eastern Bechuanaland in which the hunter, in stalking his quarry, "hung animal skins over themselves", along with "dressing up in ostrich skins so they look just like the ostriches" (Thackeray 1983: 39). Schultze, writing about the /Gaingin of the Namib, whom he visited in 1903, describes hunters stalking ostrich wearing a fur cap made of aardwolf fur the upright-pointing hairs of which blend in with clumps of grass concealing the hunter when he raises his head to orient himself (1907: 99). Burchell's description, presented in the section epigraph, of the use by a hunter of a "be-creeping" cap made of springbok skin—an item reminiscent of hunters' and dancers springbok ear caps dealt with earlier—is also marshalled as evidence by Thackeray (ibid.: 42). However, notwithstanding these three sources from the early literature, the evidence for this hunting technique among San of colonial times is undeniably meagre. So much so that Jolly, while including it in his inventory of interpretive approaches to therianthropes in San art, nevertheless excludes it from

[27] This observation about the limited efficiency of San animal skin processing is at odds with another colonial source, namely Burchell who describes in detail the animal skin processing techniques of southern San (1824: 590–2), producing karosses, of even large antelopes, that were soft and "exceedingly pliable" (ibid.: 591). Ethnographic accounts on recent San groups, too, describe such techniques (e.g. Lee 1979: 124; Silberbauer 1981: 223–4; Le Roux and White 2004: 82–5).

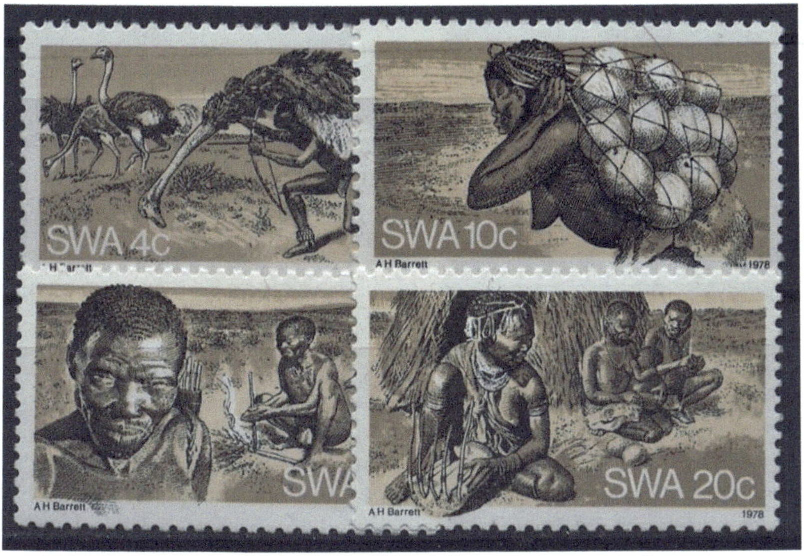

Fig. 7.6 Iconic "Bushman" cultural tableaux depicted on South West Africa stamps. (Source: internet)

further consideration, deeming it not plausible or insignificant, for lack of evidence (2002: 89).

Despite this serious obstacle, I submit that the hunting disguise notion deserves such consideration. Here I also reiterate one of Thackeray's points, even though it is basically moot, that despite there being little evidence among contemporary and recent historical San groups for the use of animal skins in hunting, this might nevertheless have been an established hunting technique in earlier and prehistoric times (1983: 42).

Skin Clothing and Relational Ontology

Thackeray's second point is the reason I give consideration to the matter of hunters' animal skin disguises. It is that "although the adoption of skin disguises could have fallen into disuse, *concepts* which were once associated with animal-skin costumes may have persisted in *belief systems*" (ibid., my emphases). The "concepts" and "belief" that underlie this hunting technique derive from what this book is about, the San connective cosmology and relational ontology. Hunting with the use of animal skins, in the context of a belief system with such cosmological moorings, may well take this hunting technique beyond instrumental, techno-economic practicality—the way it is described by writers like Schapera, Stow or Schultze—and invests it as well with supererogatory significance. Donning an animal skin might gain such significance, in the context of a hunting mode charged with sympathy and relationality the way we saw in this chapter, in which the hunter's mind is intently attuned to the animal and "tappings", "letters in the Bushmen's bodies" that "speak" to the hunter and convey to him bodily equivalent sensations from the quarry's body. This somatic link, sporadically sensed in this or that spot in his body, becomes less fleeting and more absorbing ontologically if the hunted animal's skin—or ostrich's feathers, as in Fig. 7.5—envelops his body. "Getting into a [leather] bag", as aptly put by Lewis-Williams in an insightful exegesis of //Kabbo's classic Mantis *kum* "A Visit to the Lion's House" (Bleek 1923: 15–9), "was thus like 'getting into an animal'" (2015a, b: 121). Especially so if, along with the worn antelope skin *on* his body, the hunter also, *in* his body, holds some of the animal's essence and potency, inserted into cuts at his First Kill hunter's initiation rite, and after successful kills thereafter. And *vice versa*, as in the case of a wounded eland or kudu, whose arrow wound inserts into its flesh not only the arrow's poison but also the hunter's saliva (as per Willem Ryperdt's statement on the matter).

There is precedence in San myth, art and ritual for the notion of a worn animal skin transferring the animal's being, incrementally or in one gulp, mimetically or metamorphically, partially or wholly. We saw the trickster /Xue transform himself into the monkey whose skin he had donned and a lion—and other myth beings—undergoing transformation while enclosed in a kaross, the latter itself transformation-prone and capable, in myth times, of reanimating and reconstituting itself into the antelope it had once been, to bound off into the veld. Other /Xam and Khoisan myths and legends were detailed, on the same theme, of ontological transformation—human to animal and *vice versa*—effected through an enveloping animal skin.

Turning to contemporary San, we detect animatistic vibes in the statement to Willemien LeRoux's and Alison White's San interviewers by the Naro story teller Anna Matikie about jackal and wild dog skin clothing—when you moved about in them and enjoined the wearer to "cover yourself carefully":

> They wore a skin blanket. It was wild dog skins and jackal skins. Such a blanket was used by the grand [old] people, our grand people, who made such things. ... If you had sat down with the skins, they made certain strange sounds, and you covered yourself carefully. (Le Roux and White 2004: 82)

In addition to making "strange sounds", another, implicitly and—arguably—animistic element was raised by two or three informants about old-time skin clothing: the transference of potency and vitality from animal to human, via skin clothing. One informant described how the skin—of a kudu—was "made soft with the brains and fat of other animals" (Le Roux and White 2004: 82). These essence-embodying substances would remain in the skin garments throughout its use by the human wearer as these were not washed—"we did not wash our clothes, only our bodies"—but were oiled with fats, as well as tree gum and a plant powder paste (*n‡'ang*) (ibid.: 84). Skin clothing, in the First Order, reanimating and reconstituting themselves into antelopes and departing the scene leaving a mythic Early Race person unclothed, would seem less fantastical to people in a Second Order, in which Second Race persons wore clothes aquiver with such animistic and vitalistic vibes. Such clothes, when manufactured, do not alter, notes Silberbauer in his ethnography on the Kalahari G/wi, "the essential identity and nature of the material used" and "the act of manufacture does not sunder the link between an artefact and its origin" (1981: 132).

If the latter is an animal, whose skin, when cured, is rubbed with animal fat and brains which is retained throughout its wearing and not washed out, and envelops the wearer's body from birth—in a soft steenbok baby carrier on his mother's back—to death—buried in a skin shroud "when it had been softened, was wrapped around the deceased" (ibid.: 128, 126), that link, between artefact and origin, may well extend also to the user of the artefact, the wearer of the skin to its original animal.

We might surmise that this item of practical, every-day life, skin clothing might, by virtue of its inherent animate and animist qualities, be linked conceptually and phenomenologically to the depictions we saw on rock panels of kaross-wearing humans who blend into the wearer's own skin and legs and arms that become forelegs of an antelope, with hands transforming/transformed to hooves. And linked also to ritual and ludic dancers wearing animal skins and props, in the context of animal transformation dances. Some of these dancers are themselves hunters, either young ones initiated to their trade by killing a large buck—and dancing about and around the event—or mature hunters, solidifying the link between dancing the animals and hunting them. Moreover, a number of older hunters were likely also trance dancers performing a ritual that is itself mythically and symbolically linked to hunting, especially antelope or eland hunting, the culture's ideal antelope and hunt, hunted on occasion as a trance-lion the culture's and hunting ground's ideal hunter-predator.

There are also kinetic and conative links between dancing with skins and hunting with skins: both are dance-like in execution and mimetic in intent. Thus, when a dancer dances out an ostrich—say an ostrich looking for food or preening itself, or two cocks fighting or a hen rushing to her chirping chicks—he plays out ostrich kinetic moves for the benefit of an audience appreciative of his mimetic antics and skill. *Qua* hunter, rather than entertainer, these antics also deepen and broaden the dancer's understanding of his quarry, all the more so as they include moves he may use when he stalks actual ostriches—"show fear, jumping here and there as an ostrich does when it becomes aware of anything", "picks away at verdure, turning the head as if keeping a sharp look-out, shakes his feathers, now walks, and then trots, til he gets within bow shot" (Moffatt 1844: 53). Having raised mirth in the watching humans with these choreographed moves when he dances the mimetic ostrich dance back with his people, he piques curiosity in the stalked animals—"zebras, wild horses etc. … seeing this, approach the sham ostrich to have a look". Whereas the dancer's intention in his mimetic animal dance performance is to amuse his fellow-humans, with respect to the animal-others it is to deceive them. As is known from interpersonal deception theory (Buller and Burgoon 1996), as well as cognitive ethology (Mitchell and Thompson 1986, Ristau 1991: 300–3), intentionally—or, to ethologists, "tactically"—deceiving someone is a ploy that, in terms of its intersubjective dynamics, is as relational and reflexive as is amusing someone, requiring of the performer to project himself, *qua* dancer and hunter, into the human spectator's and prey animal's mind, respectively. This is what the hunter does all along when hunting in the sympathy mode; deception, not only through stealth and guile but also through disguise and mimicry, adds to this mode sharpness of focus and intent.

Moving from the expressive cultural domain to the symbolic—specifically language—and to another, deeper level at which the inherent animate-animal nature of karosses, along with eland armbands, antelope caps, hunting bags, antelope tail whisks, is in evidence, George Silberbauer reflects on the "synonymity [G/wi] language applies to raw materials and derived artefacts", semantically maintaining thereby the above-mentioned "link between an artifact and its origin" (1981: 132). The "linguistic persistence of identity" of a duiker skin cloak, eland armband, springbok cap, steenbok hunting bag or gemsbok tail whisk with respect to animal referent is a symbolic recapitulation of the ontological persistence that also inheres in such objects. These, Silberbauer notes, notwithstanding "the act of manufacture" that has ontologically altered them by rendering them into artefacts, all retain their animalness. Writing about the /Xam, John Parkington restates the point: "an object never ceases to be the material from which it has been made" (2002: 44–5). This material ambiguity would make such objects—caps, karosses or bags—all the more suitable devices for transformation.

As hunting disguises animal skins are indeed "a highly effective hunting technique that elevate chances of success" notes Helmut Straube in concluding his exhaustive comparative study of the same by "*afrikanische Naturvölker*" (1955: 177), cast in terms of the same techno-economic paradigm that we find in Schapera's and Stow's account of the same among San. This insight, while interesting in its own right, hides what goes on below the surface, at the mytho-mystical, neo-animistic level when a hunter employs such gear and may help explain why students of San rock art are dismissive of so materialist and limiting a theory. As suggested in this section and chapter much of the ontological entanglement of hunter and prey generated by San hunting happens when a hunter stalks his quarry, all the more so whenever he is also wearing the latter being's skin, on his own skin tingling with tappings from the animal whose animalness now envelops the hunter.[28]

[28] This sort of ontological, human-animal identity transference, proposes archaeologist Maria Viestad in her fascinating recent study of Bushman dress, can be seen as one of the principal "embedded properties of clothing" (2018: 148), specifically, in the case of the San and other hunting people, skin clothing. San dress, Viestad argues, through its "bodily practice" and as an "embodied practice of social relations between humans, animals and other powerful beings of the Bushman world", "incorporates the perspective of others", citing the example of /Xam shamans' employment of springbok skin caps "to make the springbok follow him/her" (ibid.: 24; see also McGranaghan and Challis 2016: 591–2). Much the same idea was contained, in nascent form, in an essay by Pippa Skotnes's two-and-a-half decades back in which it is argued, in the context of a critique of the explanatory monopoly of the trance thesis, that animal skins, from such animals as cheetah, leopard, cat, dassie, jackal, as well as eared springbok caps, worn by dancers conferred to them varying measures of identification with the animal (1990: 20).

His experience is an especially being-absorbing, somatic replay of a central cosmological and ontological theme reiterated in San myth, ritual, dance and language. In the context of European Palaeolithic cave artists of the ancient past, Lewis-Williams refers to the same as the "concept of enveloping, transforming animality" (2010: 226). In the context of the recent southern African San this concept is as *hautnah*—up-close, skin on skin—an encounter as a hunter can have with his prey.

"Prosaic" Hunting?

We should note that by no means all of San hunting is charged with relational ontology (as is the case arguably, as seen in Vol. II, among circumpolar hunters who hunt primarily large game species, in a taxing environment and in dangerous and arduous fashion that requires them to stay closely focused on and engaged with their quarry). Some San hunts are in fact quite "prosaic" in their operations, carried out soberly and with little if any sympathy stirrings linking the hunter to his prey animal, a hunter who, *qua* hunter, has been described by ethnographers as "a realist, a pragmatist and a proto-scientist" (Heinz 1978: 148; see also Konner and Blurton-Jones 1979), as well as an optimizing strategist (Lee 1979: 209–19, 261–9; Silberbauer 1981: 290–3; Guenther 2017: 2).

As such, "opportunistic" and optimizing, broad-spectrum subsistence hunters, San hunt a variety of small animals, from rodents, through birds and reptiles. Like woman gatherers, men, too, will go after some insects, especially bees, for honey, and termites, for fat, caught seasonally by the kilo in large, nocturnal group "hunts" (Wilhelm 2005: 119–20). Some of the small mammals are burrowing creatures, such as springhares, aardvarks or porcupines, that hunters may impale underground and dig up, a hunting procedure that reduces or altogether obviates any human-animal engagement moments (all the more so in nocturnal hunts that de facto reduce visual hunter-prey encounter).

Moreover, a variety of tools and techniques may be used in hunting such small game which do not generate much by way of hunter-prey intersubjectivity, such as hunting from blinds or snaring. The same applies to the large game, some of the high-yield, mass hunting techniques of some historic (as well as likely prehistoric) San hunters, out of an encircling ring of set veld fires or along previously erected game fences, into pit-falls (Guenther 1999: 14). And then there is the recently acquired hunting technique, on horseback and with spears and dogs (Osaki 1984; Tanaka 1991: 131), or, more recent yet, on the back of a farmer's *bakkie* (pick-up truck) with a hunting rifle (a hunt, of a gemsbok, in which I participated in my field work in the mid-1990s). Such chases are quick and short and the chased-down antelope is swiftly and efficiently dispatched with spear or rifle, once again all but precluding any sympathy encounters between hunter and prey. This new hunting technique is one of a number of manifestations of the "disenchantment" of San people's culture and collective imagination. I will return again later on, in Chap. 4, Vol. II, in a general discussion of the general decline in San culture and experience of not only a relational-ontological perspective on animals and nature but of myth, magic and "shamanimistic" religion in general.

Moreover, even in the context of the two hunts that are capable of eliciting moments of mimetic or metamorphic hunter-prey identification, such moments may not come about in all hunts. Nor in all hunters, some of whose sympathetic sensitivities may be less finely tuned (as in the case evidently of the Namib springbok hunter /Ga-burob, in one of the epigraphs above who smashed the head of the collapsed quarry he had run down with a rock). San hunters are no different than their counterparts amongst animistic hunting peoples generally, of both the past and the present in there being "significant variation in how humans engage with animals," as noted by Arctic archaeologist Erica Hill on this point, "not only between societies, but between individuals in single societies due to difference in age, sex, occupation and life experience" (2013: 5). As seen in the previous chapter, the same can be said of dancers at the healing and initiation rituals not all of whom may be capable of undergoing lion transformation, or willing to attempt to embark on so perilous a course of magico-mystical action.[29]

As for a boy's socialization as hunter, by no means is all of this is acquired during initiation, at his First Buck ceremony or the Tshoma dances, ritual and ludic occasions in a hunter's formative years that lay down in his education as a hunter sympathy-based notions and practices. Small toddlers and pre-teen boys shoot with miniature bows and arrows, "hunting" quarry that ranges in size from beetles to bunnies. Ghanzi farm

[29] As remarked by Michael Jackson and Albert Piette, in their critical remarks on the "ontological turn" in anthropology (specifically as per Vivieros de Castro's perspectivism), it is not possible to "infer individual experience from collective representations, ideologies, mythologies, and cosmologies" (2015: 20). Instead, world view shapes the consciousness of individuals each in her or his own way, in terms of their inner "self-state" (*pace* William James).

San boys use slingshots to kill songbirds (with astounding accuracy, as I witnessed again and again among the Ghanzi San). Boys set snares at an early age and some of their games are play-hunts. Pre-teen and teenage boys may be taken along, singly or in a group, by fathers and uncles, who teach them how to make a bow and arrows, the ins and outs of arrow poisons, how to track, about upwind and downwind and on the varying distances from which effectively to shoot arrows at different game species, on the relative advantages of hunting with bow and arrow—the "silent killing" of animals—over rifle hunting. They also caution them about perils, from such species as Cape buffalo and gemsbok (as well as shoo them from the killing site of a cornered gemsbok, up a tree to wait out the dangerous moments of the hunt). All of these—and many more that San informants from a variety of Kalahari groups related to their interlocutors with alacrity (Le Roux and White 2004: 138–43)—are very much practical skills that stem from, transmit and foster a prosaic mindset about hunting and prey animals.

As for the all-important, iconic poison arrow hunting technique which is instrumental, as noted above, in a number of ways in bringing about protracted moments of intersubjective hunter-prey engagement as it allows the hunter close proximity to the weakened and dying prey animal, that technique may on occasion also altogether obviate such moments. Depending on the poison's strength, the animal's size and where in its body the arrow is lodged, such an arrow may have the effect of killing the animal before the hunter has gotten to it to dispatch it, precluding any of the "killing dialogue" encounters this moment of the hunt may activate. The only human-animal interaction that such a scenario may on the rare occasion bring about would be with lions, hyenas or vultures, who may be at the carcass when the hunters get to it. The first of these animals may elicit an encounter of intersubjectivity, when, as among the Ju/'hoansi, the hunters in such a situation may slowly approach the lions, importuning them, with quiet but insistent respect-discourse, to "share" the spoils of a hunt the human hunters have effected and that rightfully belongs to them, as Richard Lee observed in one of his more tense moments of participant observation when he was along on one such hunt, nervously scanning the landscape for nearby climbable trees (Lee pers. com., 19 Nov. 2016). The encounter with the hyenas and vultures at the scene may, as seen in the previous chapter, get re-enacted *post facto* in a play dance.

And, finally, the two hunting modes, that are most instrumental in imparting a bodily identification with the quarry have all but disappeared.

One is running down the prey animal, in a feat of physical exertion and mental concentration that people seem no longer inclined towards or capable of. The other, stalking game, intently, protractedly, deliberately, by donning an animal skin as a disguise, was evidently practiced only infrequently in the past and not at all in the present, tilting the observed practice of San hunting yet again towards the practical-prosaic side.

So, in light of all of this matter-of-factness, what do we make of the supererogatory, magico-mystical, relational-ontological aspect of San hunting that this chapter is about? I have dealt with this question at some length in the essay I mentioned at the start of the chapter where I argued that the two dispositions towards animals and the environment—prosaic-efficient and mystic-animistic—not only coexist in San hunting but are also complementary.

Regarding the first, I note that, notwithstanding all of the instances of prosaicness of San hunting adduced in this section, San nevertheless do not bracket out the mythic and spirit world altogether. When men hunt and women gather on the hunting and gathering ground, mythic and spirit beings are presences of which people sense an awareness, some individuals more than others and, it appears, some San cultural-linguistic groups, at certain times in their history, more than others. The example for the latter is the /Xam, whose lives and world view appears to have contained more myth and magic than that of the more prosaic Kalahari San (Guenther 2014: 197, 2015: 293–4). Unless the /Xam are deemed culturally unique among Khoisan peoples, a position few if any anthropologists, archaeologists, historians and linguists hold any more (Guenther 1989: 33–6; Barnard 1992; Lewis-Williams 2016: 52–5), the decidedly "animistic" cosmology of this San people must stand as an antidote and corrective to excessively prosaic, "disenchanted" portrayals of the world view of Kalahari San that we come across in some of the ecological, cultural-materialist accounts (Guenther 2015: 293–4, 2017: 11–2).

Regarding the second point, about the prosaic and animistic dispositions by a hunter towards his prey, not necessarily being mutually exclusive—the one mytho-magical feeding only the mind, mystically attuned to the quarry, the other feeding the body, after instrumentally hunting and butchering the same—I note that both approaches to the quarry may in fact be adaptive in a number of ways. The premise of human-animal mythical, ontological and relational connectedness may curb the potential for overexploitation inherent in hunting that is prosaic, functional and efficient by fostering a "conservation ethic" in San hunters (Smith 2016: 10–1). The two approaches to the prey animal may actually be synergistic: empirical-practical efficiency and the sound knowledge about animals it may yield (Blurton-Jones and Konner 1976) may be deepened and sharpened through the "animistic" hunter's focussed awareness of, and attuned engagement with the prey animal. This synergy may actually increase his efficiency *qua* hunter as it sustains and keeps at a sharp edge his interest in animals, their behaviour and their interactions with other animals and the environment, capable of yielding "the most subtle information" to a San hunter, information that may work "extremely well", as noted by Neil Bennum in his commentary on //Kabbo's statement about "Bushman letters" earlier in this chapter.

One of the dividends of the San's—and other hunting peoples'—"fundamental fascination with the wild species they shared their world with" (Kover 2007: 443; also see Shepard 1998)—is that it may translate into sound, grounded understanding of their behaviour (Guenther 1988). The ecological philosopher Richard Kover considers "the ability subjectively to identify with the prey", along with respect towards the same "as the very taskscape of technologically primitive hunting" (2017: 357). In the San instance, a consequence may be, that the hunter's sympathetic bond with the prey animal may actually afford him a measure of "sympathetic control" over the same, a point made by the South African zoologist-archaeologist J. Francis Thackeray (2005a, b). As "extensions of people's senses", suggests Jeremy Hollmann, in observing animals' behaviour with such rapt intent, "the /Xam transcended the limitations of their own sense perceptions and tapped into those of other species" (2005: 66).

"Attentiveness", "attunedness", "subjective identification with the prey", "extension of people's senses", "tapping into sense perceptions of other species": the language of the last two paragraphs of this chapter, which also reverberated throughout the entire discussion of hunting, as it did through that of ritual and ludic dancing, is the language of the body, of perception and experience. The effect of the reiterations of this experience of cross-species intersubjectivity and its transforming effects on the human being's being within these different domains of San culture, and of thought, imagination and action, is that the central theme of San cosmology, ontological mutability, is both mutually corroborated within the people's thought world and grounded, at times bodily, in experience.

Volume II is concerned with this dimension of San cosmology, specifically its ontological component, on the intersubjective human-animal relationship and the porous species divide. In addition, it will consider this relationship in a historical context, amongst "traditional" and contemporary San, again in terms of affect and perception (as mediated through an experience of "enchantment", generated by animals and their natural surroundings *sui generis* or through mytho-mystical beings and doings of the veld and hunting ground that derive from preternatural and supernatural domains). The other comparative context is synchronic rather diachronic and looks at the cosmology and ontology of people and cultures other than the San, preindustrial ones such as the San's agro-pastoral neighbours and an altogether different hunting-gathering people far distant from the San of southern Africa. And it will attempt a "them-us" comparison, not in ontological terms (animals vs. humans), but cultural ones (San vs. Westerners), with a view to understanding and bridging a significant gap in two cosmologies, which, on closer inspection, are as much complementary as they are antithetical.

References

Barnard, Alan. 1992. *Hunters and Herders of Southern Africa*. Cambridge: Cambridge University Press.

Bennum, Neil. 2004. *The Broken String: The Last Words of an Extinct People*. London: Viking.

Biesele, Megan. 1993. *Women Like Meat: The Folklore and Foraging Ideology of the Kalahari Ju/'hoansi*. Johannesburg/Bloomington: Witwatersrand University Press/Indiana University Press.

Bleek, Dorothea. 1923. *The Mantis and His Friends: Bushman Folklore*. Cape Town: T. Maskew Miller.

Bleek, Wilhelm H.I., and Lucy Lloyd. 1911. *Specimens of Bushman Folklore*. London: George Allen & Co.

Blurton Jones, Nicholas, and Melvin Konner. 1976. !Kung Knowledge of Animal Behavior (*or: The Proper Study of Mankind is Animals*). In *Kalahari Hunter-Gatherers: Studies of the !Kung San and Their Neighbours*, ed. Richard B. Lee and Irven DeVore, 325–348. Cambridge, MA: Cambridge University Press.

Bramble, Dennis M., and Daniel E. Lieberman. 2004. Endurance Running and the Evolution of *Homo*. *Nature* 432: 345–352.

Buller, D.B., and J.K. Burgoon. 1996. Interpersonal Deception. *Communication Theory* 6: 203–242.

Burchell, William J. 1824. *Travels in the Interior of Southern Africa*. Vol. II. London: Longman, Hurst, Rees Orme, Brown & Green.

Canetti, Elias. 1984 [1960]. *Crowds and Power*. Trans. Carol Stewart. New York: Farrar, Strauss & Giroux.

Csordas, Thomas J. 1990. Embodiment as a Paradigm for Anthropology. *Ethos* 18 (1): 5–47.

De Prada-Samper, José M. 2014. 'The Pictures of the /Xam People Are in Their Bodies': Presentiments, Landscape and Rock Art in //Kabbo's Country. In *Courage of //Kabbo: Celebrating the 100th Anniversary of the Publication of* Specimens of Bushman Folklore, ed. Janette Deacon and Pippa Skotnes, 225–241. Cape Town: UCT Press.

Dornan, Samuel S. 1925. *Pygmies and Bushmen of the Kalahari*. London: Seeley, Service & Co.

Dowson, Thomas A., Phillip V. Tobias, and David J. Lewis-Williams. 1994. The Mystery of the Blue Ostriches: Clues to the Origin and Authorship of a Supposed Rock Painting. *African Studies* 53: 3–38.

Eastwood, Edward B. 2006. Animals Behaving Like People: San Rock Paintings of Kudu in the Central Limpopo Basin. *South African Archaeological Bulletin* 61 (183): 26–39.

Eliade, Mircea. 1984. Cosmogonic Myth and 'Sacred History'. In *Sacred Narrative: Readings in the Theory of Myth*, ed. Alan Dundes, 137–151. Berkeley: University of California Press.

Farini, G. Antonio [Hunt, William Leonad]. 1886. *Through the Kalahari Desert: A Narrative of a Journey with Gun, Camera, and Note-Book to Lake N'Gami and Back*. London: Sampson, Low, Marston, Searle & Rivington.

Guenther, Mathias. 1988. Animals in Bushman Thought, Myth and Art. In *Property, Power and Ideology in Hunting-Gathering Societies*, ed. James Woodburn, Tim Ingold, and David Riches, 192–202. Oxford: Berg.

———. 1989. *Bushman Folktales: Oral Traditions of the Nharo of Botswana and the /Xam of the Cape*, Studien Zur Kulturkunde No. 93. Stuttgart: Franz Steiner Verlag Wiesbaden.

———. 1999. *Tricksters and Trancers Bushman Religion and Society*. Bloomington: Indiana University Press.

———. 2014. Dreams and Stories. In *Courage of //Kabbo: Celebrating the 100th Anniversary of the Publication of* Specimens of Bushman Folklore, ed. Janette Deacon and Pippa Skotnes, 195–210. Cape Town: UCT Press.

———. 2015. 'Therefore Their Parts Resemble Humans, for They Feel That They Are People': Ontological Flux in San Myth, Cosmology and Belief. *Hunter-Gatherer Research* 1 (3): 277–315.

———. 2017. '...The Eyes Are No Longer Wild, You Have Taken the Kudu into Your Mind': The Supererogatory Aspect of San Hunting. *The South African Archaeological Bulletin* 72: 3–16.

Gusinde, Martin. 1966. *Von gelben und schwarzen Buschmännern*. Graz: Akademische Druck- und Verlagsanstalt.

Hahn, Theophilius. 1870. Die Buschmänner. Ein Beitrag zur südafrikanischen Völkerkunde, IV. *Globus* XVII (5): 120–123.

Harris, William C. 1852. *The Wild Sports of Southern Africa; Being the Narrative of a Hunting Expedition from the Cape of Good Hope, Through the Territories of the Chief Moselekatse to the Tropic of Capricorn*. 5th ed. London: Henry G. Bohn, York Street, Covent Garden.

Heinz, Hans-Joachim. 1978. The Bushmen's Store of Scientific Knowledge. In *The Bushmen*, ed. Philip Tobias, 148–161. Cape Town: Human & Rousseau.

Hill, Erica. 2013. Archaeology and Animal Persons: Towards a Prehistory of Human-Animal Relations. *Environment and Society: Advances in Research* 4: 117–136.

Hollmann, Jeremy C., ed. 2005. *Customs and Beliefs of the /Xam Bushmen*. Johannesburg/Philadelphia: Wits University Press/Ringing Rock Press.

Ijäs, Mikko. 2017. Fragments of the Hunt: Persistence Hunting, Tracking and Prehistoric Art. Doctoral Thesis, Aalto University: School of Arts, Design and Architecture, Department of Art.

Ingold, Tim. 2000. Hunting and Gathering as Ways of Perceiving the Environment. In *The Perception of the Environment*, ed. Tim Ingold, 40–60. London/New York: Routledge.

Jackson, Michael, and Albert Piette. 2015. Anthropology and the Existential Turn. In *What Is Existential Anthropology?* ed. Michael Jackson and Albert Piette, 1–29. Oxford: Berghahn.

Jodtka, Assistenzartzt. 1902. Reise des Assistenzartztes Jodtka nach dem Okavango. *Deutsches Kolonialblatt* 13: 493–495, 590–593.

Jolly, Pieter. 2002. Therianthropes in San Rock Art. *South African Archaeological Bulletin* 57 (176): 85–103.

Kaufmann, Hans. 2005 [1910]. The =Auin. In *Kalahari and Namib Bushmen in German South West Africa: Ethnographic Reports by Colonial Soldiers and Settlers*, ed. Mathias Guenther, 37–96. Cologne: Rüdiger Köppe Verlag.

Keeney, Bradford, and Hillary Keeney., eds. 2015. *Way of the Bushman as Told by the Tribal Elders*. Rochester: Bear & Company.

Köhler, Oswin. 1973. Die rituelle Jagd bei den Kxoe-Buschmännern von Mutsiku. In *Festschrift zum 65. Geburtstag von Helmut Petri. Kölner ethnologische Mitteilungen*, ed. Kurt Tauchmann, vol. 5, 215–257.

Kover, Tihamer R. 2007. The Beastly Familiarity of Wild Alterity: Debating the Nature of Our Fascination with Wildness. *Ethical Perspectives: Journal of the European Ethics Network* 14 (4): 431–456.

———. 2017. Of Killer Apes and Tender Carnivores: A Shepardian Critique of Burkert and Girard on Hunting and the Evolution of Religion. In *Animals and Religion*, ed. Nathan Kowalsky and Tihamer Kover. Special Issue of *Studies in Religion* 46 (4): 536–567.

Le Roux, Willemien, and Alison White, eds. 2004. *Voices of the San*. Cape Town: Kwela Books.

Lebzelter, Viktor. 1934. *Eingeborenenkulturen von Südwestafrika: Die Buschmänner*. Leipzig: Verlag Karl W. Hiersemann.

Lee, Richard B. 1979. *The !Kung San Men, Women, and Work in a Foraging Society*. Cambridge: Cambridge University Press.

———. 2003. *The Dobe Ju/'hoansi*. 3rd ed. Fort Worth/Belmont: Wadsworth.

Lewis-Williams, David J. 2000. *Stories that Float from Afar: Ancestral Folklore of the San of Southern Africa*. Cape Town: David Philip.

———. 2010. *Conceiving God: The Cognitive Origin and Evolution of Religion.* London: Thames & Hudson.

———. 2015a. *Myth and Meaning: San-Bushman Folklore in Global Context.* Walnut Creek: Left Coast Press.

———. 2015b. Text and Evidence: Negotiating San Words and Images. *South African Archaeological Bulletin* 70 (201): 53–63.

———. 2016. The Jackal and the Lion: Aspects of Khoisan Folklore. *Folklore* 127: 51–70.

Liebenberg, Louis. 2006. Persistence Hunting by Modern Hunter-Gatherers. *Current Anthropology* 47: 1017–1025.

Low, Chris. 2007. Khoisan Wind: Hunting and Healing. *Journal of the Royal Anthropological Institute* (N.S.) 13: 571–590.

———. 2014. Locating /Xam Beliefs and Practices in a Contemporary KhoeSan Context. In *Courage of //Kabbo: Celebrating the 100th Anniversary of the Publication of* Specimens of Bushman Folklore, ed. Janette Deacon and Pippa Skotnes, 349–361. Cape Town: UCT Press.

Marshall, Lorna J. 1976. *The !Kung of Nyae Nyae.* Cambridge, MA: Harvard University Press.

———. 1999. *Nyae Nyae !Kung Beliefs and Rites*, Peabody Museum Monographs, No. 8. Cambridge, MA: Harvard University.

McGranaghan, Mark, and Sam Challis. 2016. Reconfiguring Hunting Magic: Southern Bushmen (San) Perspectives on Taming and Their Implications for Understanding Rock Art. *Cambridge Archaeological Journal* 26 (4): 579–599.

McNiven, Ian J. 2013. Between the Living and the Dead: Relational Ontologies and the Ritual Dimension of Dugong Hunting Across Torres Strait. In *Relational Archaeologies Humans / Animals / Things*, ed. Christopher Watts, 97–116. London: Routledge.

Merleau-Ponty, Maurice. 1962. *Phenomenology of Perception.* Trans. Colin Smith. London: Routledge and Kegan Paul.

Metzger, Fritz. 1990. *Naro und seine Sippe.* Windhoek: Namibia Wissenschaftliche Gesellschaft/Scientific Society.

Mitchell, Robert W., and Nicholas S. Thompson, eds. 1986. *Deception: Perspectives on Human and Nonhuman Deceit.* Albany: SUNY Press.

Moffatt, Robert. 1844. *Missionary Labours and Scenes in Southern Africa.* New York: Robert Carter.

Morris, Rosalind C. 2011. Crowds and Powerlessness: Reading //Kabbo and Canetti with Derrida in (South) Africa. In *Demenagerie: Thinking (of) Animals After Derrida*, ed. Anne Emmanuelle Berger and Marta Segarra, 167–212. Amsterdam: Brill/Rodopi.

Osaki, Masakatsu. 1984. The Social Influence of Change on Hunting Techniques Among the Central Kalahari San. *African Studies Monograph* 5: 49–62.

Parkington, John. 2002. *The Mantis, the Eland and the Hunter Follow the San….* Cape Town: Creda Communications.

Passarge, Siegfried. 1907. *Die Buschmänner der Kalahari.* Berlin: Dietrich Reimer.

Pickering, Travis R. 2013. *Rough and Tumble: Aggression, Hunting, and Human Evolution*. Berkeley: University of California Press.

Richters, Ferdinand. 1886. *Bericht über die senckenbergische naturforschende Gesellschaft in Frankfurt am Main*. Frankfurt: Senckenberger naturwissensh-cftlich Gesellschaft.

Ristau, Carolyn, ed. 1991. *Cognitive Ethology: The Mind of Other Animals (Essays in Honour of Donald Griffin)*. Hillsdale: Lawrence Erlbaum.

Rusch, Neil. 2015. The Root and Tip of the *//Kwanna*: Introducing *Chiasmus* in Three /Xam Narratives. Unabridged Version of Rusch (2016a), Accessible at the Centre for Curating the Archive, 1–26. http://www.cca.uct.ac.za/publications/

———. 2016. The Root and Tip of the *//Kwanna*: Introducing *Chiasmus* in Three /Xam Narratives. *Critical Arts* 30 (6): 877–897.

Sahlins, Marshall. 1972. The Original Affluent Society. In *Stone Age Economics*, ed. Marshall Sahlins, 1–40. Chicago: Aldine.

Schapera, Isaac. 1930. *The Khoisan Peoples of South Africa: Bushmen and Hottentots*. London: Routledge & Kegan Paul.

Schultze, Leonhard. 1907. *Aus Namaqualand und Kalahari*. Jena: Verlag von Gustav Fischer.

Shepard, Paul. 1998 [1978]. *Thinking Animals: Animals and the Development of Human Intelligence*. Athens: The University of Georgia Press.

Silberbauer, George B. 1981. *Hunter and Habitat in the Central Kalahari Desert*. Cambridge: Cambridge University Press.

Singer, Thea. 2017. The Evolution of Dance. *Scientific American* 317 (1): 66–71.

Skaanes, Thea. 2017a. Sounds in the Night: Ritual Bells, Therianthropes, and Eland Relations Among the Hadza. In *Human Origins: Contributions from Social Anthropology*, ed. Camilla Powers, Morna Finnegan, and Hillary Callan, 204–223. Oxford: Berghahn Books.

———. 2017b. Cosmology Matters: Power Objects, Rituals, and Meat-Sharing Among the Hadza of Tanzania. Ph.D. Dissertation, Aarhus University.

Skotnes, Pippa. 1990. Rock Art: Is There Life After Trance? *De Arte* 44: 16–24.

Smith, Andrew B. 2016. Why Would Southern African Hunters Be Reluctant Food Producers? *Hunter-Gatherer Research* 2 (4): 1–12. (Online Open Access journal).

Stow, George W. 1905. *The Native Races of South Africa: A History of the Intrusion of the Hottentots and Bantu into the Hunting Grounds of the Bushmen, the Aborigines of the Country*. London: Swan Sonnenschein.

Straube, Helmut. 1955. *Tierverkleidungen der afrikanischen Naturvölker*. Wiesbaden: Franz Steiner Verlag.

Sugawara, Kazuyoshi. 2015. On the G/ui Experiences of Being Hunted: An Analysis of Oral Discourse on the Man-Killings by Lions. Paper Presented at the 12th International Conference of Hunting-Gathering Societies (CHAGS 12), University of Vienna, Vienna, September 7–12, 2015.

Suzman, James. 2017. *Affluence Without Abundance: The Disappearing World of the Bushmen*. New York: Bloomsbury.

Tanaka, Jiro. 1991. Egalitarianism and the Cash Economy Among Central Kalahari San. In *Cash, Commoditisation and Changing Foragers*, ed. Nicolas Peterson and Toshio Matsuyama. *Senri Ethnological Studies* 30: 117–135.
———. 1996. The World of Animals Viewed by the San. *African Studies Monographs*, Supplementary Issue, 22: 11–28.
Tanner, Adrian. 1979. *Bringing Home the Animals: Religious Ideology and Mode of Production of the Mistassini Cree Hunters.* St. John's: Institute of Social and Economic Research Memorial University of Newfoundland.
Thackeray, J. Francis. 1983. Disguises, Animal Behaviour and Concepts of Control in Relation to Rock Art of Southern Africa. In *New Approaches to Southern African Rock Art, Goodwin Series*, ed. David J. Lewsi-Willimas, vol. 4, 38–43.
———. 2005a. Eland, Hunters and Concepts of 'Sympathetic Control' in Southern African Rock Art. *Cambridge Archaeological Journal* 15 (1): 37–34.
———. 2005b. The Wounded Roan: A Contribution to the Relation of Hunting and Trance in Southern African Rock Art. *Antiquity* 79: 5–18.
Thomas, Elizabeth M. 2006. *The Old Way: A Story of the First People.* New York: Farrar Strauss Giroux.
Trenk, Oberleutnent. 2005. The Bushmen of the Namib and Their Legal and Family Conditions. In *Kalahari and Namib Bushmen in German South West Africa: Ethnographic Reports by Colonial Soldiers and Settlers*, ed. and trans. Mathias Guenther, 237–246. Cologne: Rüdiger Köpe Verlag.
Van Vuuren, Helize. 2016. *A Necklace of Springbok Ears: /Xam Orality and South African Literature.* Stellenbosch: Sun Press.
Viestad, Maria V. 2018. *Dress as Social Relations: An Interpretation of Bushman Dress.* Johannesburg/New York: Wits Press/NYU Press.
Vinnicombe, Patricia. 2001 [1976]. *People of the Eland: Rock Paintings of the Drakensberg Bushmen as a Reflection of Their Life and Thought.* Johannesburg: Wits Press.
Visser, Hessel. 1997. *Naro Dictionary.* 3rd ed. Gantsi: Naro Language Project.
Whitridge, Peter. 2013. The Imbrication of Human and Animal Paths: An Arctic Case Study. In *Relational Archaeologies: Humans, Animals, Things*, ed. Christopher Watts, 228–244. London: Routledge.
Wikar, Hendrik J. 1935. *The Journal of Hendrik Jacob Wikar (1779) with an English Translation by A. W. van der Horst and the Journals of Jacobus Coetsé Jansz (1760) and Willem van Reenen (1790) with an English Translation by Dr. E. E. Mossop.* Cape Town: The van Riebeeck Society.
Wilhelm, Joachim Helmut. 2005. The !Kung Bushmen. In *Kalahari and Namib Bushmen in German South West Africa: Ethnographic Reports by Colonial Soldiers and Settlers*, ed. Mathias Guenther, 91–184. Cologne: Rüdiger Köppe Verlag.
Wilmsen, Edwin N. 1997. *The Kalahari Ethnographies (1896–1898) of Siegfried Passarge: Nineteenth Century Khoisan- and Bantu-Speaking Peoples.* Cologne: Rüdiger Köppe Verlag.
Woodhouse, Bert. 1984. *When Animals Were People: A-Z of Animals of Southern Africa as the Bushmen Saw and Thought Them and as the Camera Sees Them Today.* Melville: Chris van Rensburg Publications (Pty) Limited.

Correction to: Human-Animal Relationships in San and Hunter-Gatherer Cosmology, Volume I

Correction to:
© The Author(s) 2020
M. Guenther, *Human-Animal Relationships in San and Hunter-Gatherer Cosmology, Volume I*,
https://doi.org/10.1007/978-3-030-21182-0

This book was inadvertently published with few errors which has been corrected now.

Corrections:

p. 5, l. 18, has been changed as 'with a dozen-odd like ontological oddities'.
p. 107, l. 15, (fig 17) has been changed as (fig. 4.14)
p. 171, 6th line from bottom, Fig. 3.31 has been changed as Fig. 4.31.
p. 197, missing reference has been added now:

The updated versions of the book can be found at
https://doi.org/10.1007/978-3-030-21182-0

© The Author(s) 2020
M. Guenther, *Human-Animal Relationships in San and Hunter-Gatherer Cosmology, Volume I*,
https://doi.org/10.1007/978-3-030-21182-0_8

England, Nicholas 1968. "Music among the Zu/'wa-si of South West Africa and Botswana." Ph.D. diss, Harvard University. (Published as Music among the Ju/'hoansi and Related Peoples of Namibia, Botswana and Angola. New York: Garland Publishing, 1992.)

p. 238, lines 21-22, Lorna Marshall has been replaced with Elizabeth Marshall Thomas.

p. 246, footnote 21, l. 4: 'proprietary rite' has been changed as 'propitiatory rite'.

Appendices

Appendix 1: A Lion Transformation

May 30, 1974, early evening

It took him [Sebetwane] about 1 hour to get ready – he made the rattles, changed into his dance costume (shorts and a singlet). Throughout the preparations he danced in short spurts, always with women around whom he danced beguilingly and with more or less subtle eroticism; both African [Black] and Bushman women. There are two women in the village (one a Herero and one Kgalagadi) who can sing his song. They were joined by four others – M T, S and two other Bushmen women … two of the BU women slightly drunk) – who stood upright, in a line, clapping and singing a refrain.

Then he began dancing, in a preliminary and warming-up fashion and in the process he took the *doeks* [head-kerchiefs] off the heads of all the women, both those that sang for him and spectators. The de-*doek*ed women covered their heads again with caps or rags. He hang the *doek* like trophies – from his belt so that he looked as if he wore a skirt (note: both the erotic and sex reversal elements of this trait)

Then he began dancing: it was effective and skillful, captivating to watch. He danced vigorously, with precise and skillful steps and marvelous coordination – arms held down stiffly, hands stretched out parallel to the ground, moving back and forth, in scissor motion. While dancing he sang his refrain, loudly and harshly. His dance was erotic and scatological –

© The Author(s) 2020

M. Guenther, *Human-Animal Relationships in San and Hunter-Gatherer Cosmology, Volume I*,
https://doi.org/10.1007/978-3-030-21182-0

erotic in that he would dance with women in the song circle for brief spells – around them like a snake around prey and with pelvic thrusts (including kissing one woman on the mouth) and scatological, in one amusing act (which, I think, was intended just to amuse spectators): he danced in a kneeling and squatting position. He held his hands under his buttocks and picked up a piece of cardboard from the ground and held it against his buttocks. Then to his nose and, with a grimace of disgust, he threw it away. Everyone roared with laughter.

His grimacing was another highly effective aspect of his dance: he screwed up his face – to look surprised, sneaky, sexually aroused or otherwise interested (including licking his mouth with his tongue) and angry. All this grimacing was accompanied by contortions of his head and neck. The angry-emotion seemed to become increasingly predominant and at its height he shouted and screamed and charged at some of the spectators, especially the women and children. The latter ran away screaming and laughing; the former stood their ground and half laughingly and half annoyed slapping him or pushing him away.

The children became really afraid after he had caught one of them – a small boy (maybe 2 ½ years old), whom he grabbed, tried to carry on his shoulders, dropped and just caught by the arm, picked up again and held to his chest and dropped again. The poor child was screaming in terror and the mother tried to take him away from Sebetwane; he, however, grabbed the child again and threw him in the air and caught him and tried to hold the squirming child to him while dancing; lost his grip and dropped the terrorized child from his shoulders onto the ground. Everyone was shocked and I rushed forward to intervene but J [a middle-aged Herero man] carried the boy and rushed away some fifty yards, out of S's reach, who now seemed in some kind of wild, vicious pre-trance.

Kennedy [my field assistant and interpreter] became quite apprehensive at this point and wanted us to leave. I stayed and saw S dancing more and more wildly, charging at spectators to glare at them and sing at them, and snarling at people, crouched, at times, on his fours. The child was now back with his mother who, strangely, quietly sat on the ground beside the singing women holding the child and comforting it. The poor boy kept looking at S, with big, fearful eyes, especially whenever S came near him. However, he had lost interest in the child.

Another act, which made Kennedy apprehensive and other people serious, was when he picked up some sand and threw it in one direction and then went to the spot where he had thrown it to scoop it all up in his hand.

This was *moloi* [Tswana sorcery; see Guenther 1992], I was told. Then he motioned J to come and sit on the ground and his dancing was now focused on J, who stared at him continuously.

During the dance there was a drunk Bushman (whom I do not know, the one in the blue overalls and green hat) who insisted on dancing, alongside S, until he was vehemently dragged away by J and given a warning that must have been severe and intimidating as he walked away sullenly and stood at one spot, about 3 yards from where Sebetwane danced, motionless, staring at S.

The dance lasted about one hour and everyone seemed relieved when it was over as, had he gone on, one wouldn't know what other violent or destructive things to expect.

Appendix 2: A Baboon Transformation

From field notes of Renée Sylvain (with permission)

September 8, 1996

Later in the day we got another phone call from Willam – this time to tell us that one of his friends is sick and needs to see a BM doctor, and if we could give him a lift, maybe we would like to 'see the experience' of a BM doctor. We dashed out to Epako.

When we arrived at Willam's place, a young man was being supported and helped toward the bakkie, Ellie on one arm, Willam on the other. This young man was jumping and dancing and generally acting very crazy. E and W assured us immediately that the man, whose name is Oba, was not drunk. Rocky and I agreed that his symptoms resembled someone on a bad acid trip. Jan had to hold Oba with all his strength as Oba was flopping around violently. In the truck on the way to F#2 son's farm, where the BM doctor is, Willem told us that the BM people are very powerful and can turn themselves into animals; Oba had turned himself into a baboon (eerily, just as he was explaining this, a big male baboon, the first we had seen since our first day out here, ran in front of the truck). Apparently, the poor fellow was just sitting in Epako when his hand started turning into a baboon's hand. He begged Willem to hack it off with an axe, which Willem, of course wouldn't do. Then he just continued his transformation.

We had to drive a fair way into the farm to get to the worker's shacks behind the farm house. There were 3 BM doctors, an old man and his wife and one of Willem's uncles. They lived in a metal and plastic shack thrown together with some bush-wood – the most dilapidated old scrap-metal shacks I have seen so far. When we arrived at the workers' quarters (such as they were), two young men who came with us opened the canopy of our bakkie, and Oba bounded out—fully a baboon now, and loped across the veld, bounding up trees, and knuckle-running so convincingly that I need only squint slightly to mistake him for an actual baboon in the dusk. The two young men gave chase, occasionally trying to pull Oba from a tree. Just as I was fixating on the baboon-man, I heard a heart-wrenching wail to my right and turned around to see Hendrik with his head thrown back, howling and stomping the ground with his foot, tears and snot streaming down his face. Wow – what animal is this man turning into???

We had brought Hendrik with us from Epako. Rocky and I didn't know this at the time, but it turns out that Hendrik is the son of the two BM doctors and as soon as he got out of the bakkie he burst into hysterical tears upon seeing his parents again. Hendrik is, by any definition, an incredibly goofy-looking guy: he is short and squat, with a pushed-in round face that wears a constant grin and eyes that stare off in different directions. He has a likeable Quasimodo look about him. Whenever he talks his speech is accompanied by a whistle-hissing from a tracheotomy, required after he sustained a very bad beating in Epako. He had been in jail for about 8 months (for killing one of F#2's goats) and he hadn't seen his parents for a very long time. He flung himself onto the lap of a one-legged woman sitting by the fire and wept pitifully while she stroked his face and stared off into the distance. It was clear that they were family and were very close. I was almost moved to tears at such an emotional display. I am quickly coming to the conclusion that you cannot over-state the importance of family for San people. Not surprising really, they don't have much else and family is probably to only thing saving them from complete isolation and despair that their crippling living conditions would inevitably produce. I watched the intensity of the interaction with some anguish, but this was mixed with fascination as I also followed the activities of the newly formed baboon-man.

After he was caught, the baboon-man had not ceased his violent flailing – flapping his arms, jumping around, head bobbing, tongue hanging out, eyes vacant and rolling back – he eventually just collapsed. Then his legs started twitching, making motions like a dog dreaming. There was nothing 'medically' wrong with him that I could detect – it appeared more like a psychotic episode of some sort. How can anyone from a comfortable background begin to comprehend the psychological effects of living as a

BM here – in poverty, desperation and dirt, being the most oppressed, disenfranchised, dispossessed, and shit-on group of all? Maybe many San live on the edge of psychosis their whole life. Communicating psychological distress engages the expressive resources the BM culture offers – and so Oba turned himself into a baboon. Who could blame him? Life as a human hasn't proven to be such hot shit so far.

While I was trying to comprehend the events around me, Besa run toward Rocky and I and told us, in a very urgent tone, that it was time to leave. Being an obtuse anthropologist, I resisted – there is no way I am leaving while a guy is turning into a baboon! Hell no! Then I saw the angry old man with the big stick coming toward us and I reluctantly agreed that we could consider leaving. We were not allowed to stay around to witness the 'doctoring' – which Willem explained, would involve singing to the sick one and some dancing (described as a smaller version of a traditional trance dance). We were to leave the 'baboon' with the doctors. And we were to leave damn fast! Also, the old man was slightly (or a little more than slightly) pissed and chased us away with a stick. When we were in the bakkie, we weren't moving fast enough for him so he continued to chase us away with his stick and picked up a rock to throw at the bakkie. Then the old man lay down on the ground in front of the truck to make sure we couldn't come back.

Over the next 5 days ... We learned later that the old woman stroking Hendrik's face is his mother and the old man with the stick is his father and they had thought that Rocky and I were police officers (there was also some inebriation involved). They hadn't seen their son for months and thought the police had brought him home. Willem told us that when the doctors sobered up and all was explained, everybody had a good laugh. The doctors offered to help us with trance dancing if we needed it, and so we went to them to help our ex-soldier friend who has a very serious case of TB. We also learned that the BM doctors were unable to completely cure Oba, so they sent him to the white doctors for follow-up treatment.

References

Abram, David. 1997. *The Spell of the Sensuous: Perception and Language in a More-than-Human World*. New York: Vintage Books.

———. 2010. *Becoming Animal: An Earthly Cosmology*. New York: Vintage Books.

Alexander, James E. 1838. *An Expedition of Discovery into the Interior of Africa Through the Hitherto Undescribed Countries of the Great Namaquas, Boschmans, and Hill Damara. Vol. 2. London: Henry Colburn, Publisher.* Reprint edition. London: Henry Colburn.

Bahn, Paul. 2003. Chauvet-ism and Shamania: Two Ailments Afflicting Ice Age Art Studies. *The London Magazine*, August/September: 38–46.

———. 2010. *Prehistoric Rock Art Polemics and Progress*. Cambridge: Cambridge University Press.

Bank, Andrew, ed. 1998. *The Proceedings of the Khoisan Identities & Cultural Heritage Conference*, South African Museum, Cape Town, July 12–16, 1997. Cape Town: InfoSource.

———. 2006a. *Bushmen in a Victorian World: The Remarkable Story of the Bleek-Lloyd Collection of Bushman Folklore*. Cape Town: Double Storey Books.

———. 2006b. Anthropology and Fieldwork Photography: Dorothea Bleek's Expedition to the Northern Cape and the Kalahari; July to December 1911. *Kronos* 32: 77–113.

Barnard, Alan. 1979. Nharo Bushman Medicine Men. *Africa* 49: 68–80.

———. 1980. Sex Roles Among the Nharo Bushmen of Botswana. *Africa* 50: 115–124.

———. 1992. *Hunters and Herders of Southern Africa*. Cambridge: Cambridge University Press.

© The Author(s) 2020

M. Guenther, *Human-Animal Relationships in San and Hunter-Gatherer Cosmology, Volume I*,
https://doi.org/10.1007/978-3-030-21182-0

———. 2013. Cognitive and Social Aspects of Language Origins. In *New Perspectives on the Origin of Language*, ed. Claire Lefebre, Bernard Comrie, and Henri Cohen, 53–71. Amsterdam: John Benjamins Publishing Company.

Bauman, Richard. 1992. Performance. In *Folklore, Cultural Performances, and Popular Entertainments*, ed. Richard Bauman, 41–49. New York: Oxford University Press.

Bennum, Neil. 2004. *The Broken String: The Last Words of an Extinct People*. London: Viking.

Berliner, Paul F. 1978. *The Soul of the Mbira: Music and Traditions of the Shona People of Zimbabwe*. Berkeley: University of California Press.

Berman, Morris. 1999. *Wandering God: A Study in Nomadic Spirituality*. Albany: State University of New York Press.

Biesele, Megan. 1975. Song Texts by the Master of Tricks: Kalahari Thumb Piano Music. *Botswana Notes and Records* 7: 171–188.

———. 1976. Aspects of !Kung Folklore. In *Kalahari Hunter-Gatherers: Studies of the !Kung San and Their Neighbors*, ed. Richard B. Lee and Irven DeVore, 302–324. Cambridge, MA: Harvard University Press.

———. 1993. *Women Like Meat: The Folklore and Foraging Ideology of the Kalahari Ju/'hoansi*. Johannesburg/Bloomington: Witwatersrand University Press/Indiana University Press.

———, ed. 2009. *Ju/'hoan Folktales: Transcriptions and English Translations: A Literacy Primer for Youth and Adults of the Ju/'hoan Community*. San Francisco: Trafford Publishing.

Bird-David, Nurit. 1999. 'Animism' Revisited: Personhood, Environment, and Relational Epistemology. *Current Anthropology* 40 (Supplement): S67–S92.

Bleek, Wilhelm H.I. 1864. *Reynard the Fox in South Africa; or Hottentot Fables and Tales*. London: Trübener and Co.

———. 1874. Remarks by Dr. Bleek. In Orpen, Joseph M. 1874. A Glimpse into the Mythology of the Maluti Bushmen. *Cape Monthly Magazine* [N.S.] 9: 9–13.

———. 1875. *A Brief Account of Bushman Folk-Lore and Other Texts*. Cape Town/London/Leipzig: J. C. Juta/Trübner & Co./F. A. Brockhaus.

Bleek, Dorothea. 1923. *The Mantis and His Friends: Bushman Folklore*. Cape Town: T. Maskew Miller.

———. 1927. Bushmen of Central Angola. *Bantu Studies* 3: 105–125.

———. 1928. *The Naron*. Cambridge: Cambridge University Press.

———. 1929. Bushman Folklore. *Africa* 2: 302–313.

———. 1932. Customs and Beliefs of the /Xam Bushmen. Part II. The Lion. *Bantu Studies* 6: 47–63.

———. 1933. Beliefs and Customs of the /Xam Bushmen. Part V. The Rain. *Bantu Studies* 7: 297–312. *Bantu Studies* 10: 163–199.

———. 1934/53. !kun Mythology. *Zeitschrift für Eingeborenen-Sprachen* XXV (4): 261–283.

———. 1935. Beliefs and Customs of the /Xam Bushmen. Part VII. Sorcerors. *Bantu Studies* 9: 1–48.

———. 1936. Special Speech of Animals and Moon Used by the /Xam Bushmen. *Bantu Studies* 10: 161–199.

Bleek, Wilhelm H.I., and Lucy Lloyd. 1911. *Specimens of Bushman Folklore*. London: George Allen & Co.

Blundell, Geoff, and David Lewis-Williams. 2001. Storm Shelter: A Rock Art Discovery in South Africa. *South African Journal of Science* 97: 1–4.

Blurton Jones, Nicholas, and Melvin Konner. 1976. !Kung Knowledge of Animal Behavior (*or: The Proper Study of Mankind is Animals*). In *Kalahari Hunter-Gatherers: Studies of the !Kung San and Their Neighbours*, ed. Richard B. Lee and Irven DeVore, 325–348. Cambridge, MA: Cambridge University Press.

Boyer, Pascal. 2001. *Religion Explained: The Evolutionary Origins of Religious Thought*. New York: Basic Books.

Bramble, Dennis M., and Daniel E. Lieberman. 2004. Endurance Running and the Evolution of *Homo*. *Nature* 432: 345–352.

Brightman, Marc, Vanessa E. Grotti, and Olga Ulturgasheva, eds. 2012. *Animism in Rainforest and Tundra: Personhood, Animals, Plants and Things in Contemporary Amazonia and Siberia*. New York/Oxford: Berghahn Books.

Buller, D.B., and J.K. Burgoon. 1996. Interpersonal Deception. *Communication Theory* 6: 203–242.

Burchell, William J. 1824. *Travels in the Interior of Southern Africa*. Vol. II. London: Longman, Hurst, Rees Orme, Brown & Green.

Butler, Shelley R. 2000. The Politics of Exhibiting Culture: Legacies and Possibilities. *Museum Anthropology* 23 (3): 74–92.

Canetti, Elias. 1984 [1960]. *Crowds and Power*. Trans. Carol Stewart. New York: Farrar, Strauss & Giroux.

Challis, Sam, Jeremy Hollmann, and Mark McGranaghan. 2013. 'Rain Snakes' from the Sequ River: New Light on Qing's Commentary on Rock Art from Sehonghong, Lesotho. *Azania: Archaeological Research in Africa* 48: 331–354.

Clottes, J., and D.J. Lewis-Williams. 1998. *The Shamans of Prehistory: Trance and Magic in the Painted Caves*. New York: Harry Abrams.

Costa, L., and C. Fausto. 2010. The Return of the Animists: Recent Studies of Amazonian Ontologies. *Religion and Society: Advances in Research* 1: 89–109.

Csordas, Thomas J. 1990. Embodiment as a Paradigm for Anthropology. *Ethos* 18 (1): 5–47.

De Prada-Samper, José M. 2014. 'The Pictures of the /Xam People Are in Their Bodies': Presentiments, Landscape and Rock Art in //Kabbo's Country. In *Courage of //Kabbo: Celebrating the 100th Anniversary of the Publication of* Specimens of Bushman Folklore, ed. Janette Deacon and Pippa Skotnes, 225–241. Cape Town: UCT Press.

———., ed. 2016. *The Man Who Cursed the Wind and Other Stories from the Karoo/ Die Man wat die wind vervloek het en ander stories van die Karoo*. Cape Town: African Sun Press.

Deacon, Janette. 1994. Rock Engravings and the Folklore of Bleek and Lloyd's /Xam San Informants. In *Contested Images: Diversity in Southern African Rock*

Art Research, ed. Thomas A. Dowson and David Lewis-Williams, 238–256. Johannesburg: Witwatersrand University Press.

———. 2002. Relationship Between the San Drawings and Rock Art. In *Der Mond dals Schuh: Zeichnungen der San/The Moon as Shoe: Drawings of the San*, ed. Miklós Szalay, 67–89. Zurich: Scheidegger & Spiess.

Deacon, Janette, and Thomas A. Dowson, eds. 1996. *Voices from the Past: /Xam Bushmen and the Bleek and Lloyd Collection*. Johannesburg: Wits University Press.

Deacon, Janette, and Pippa Skotnes, eds. 2014. *Courage of //Kabbo: Celebrating the 100th Anniversary of the Publication of* Specimens of Bushman Folklore. Cape Town: UCT Press.

DeMothe, Kimerer L. 2015. When Words Don't Get It: The Challenges of Writing About Ritual Dance. Paper Presented at the Society of Dance History Scholars/ Congress of Research Dance (SDHS/CORD Conference), Iowa City, University of Iowa, November 13–16, 2014, Conference Proceedings, 127–134.

Doke, Clement M. 1937. Games, Plays and Dances of the ≠Khomani Bushmen. In *Bushmen of the Southern Kalahari*, ed. John D. Reinhallt Jones and Clement M. Doke, 89–100. Johannesburg: University of the Witwatersrand Press.

Donald, David. 2009. *Blood's Mist*. Auckland Park: Jacana.

Dornan, Samuel S. 1917. The Tati Bushmen (Masarwas) and Their Language. *Journal of the Royal Anthropological Institute of Great Britain and Ireland* XLII: 37–112.

———. 1925. *Pygmies and Bushmen of the Kalahari*. London: Seeley, Service & Co.

Dowson, Thomas A. 2007. Debating Shamanism in Southern African Rock Art: Time to Move on .... *South African Archaeological Bulletin* 62 (183): 49–61.

Dowson, Thomas A., Phillip V. Tobias, and David J. Lewis-Williams. 1994. The Mystery of the Blue Ostriches: Clues to the Origin and Authorship of a Supposed Rock Painting. *African Studies* 53: 3–38.

Dundes, Alan. 1980. Texture, Text, and Context. In *Interpreting Folklore*, ed. Alan Dundes, 20–32. Bloomington: Indiana University Press.

Eastwood, Edward B. 2006. Animals Behaving Like People: San Rock Paintings of Kudu in the Central Limpopo Basin. *South African Archaeological Bulletin* 61 (183): 26–39.

Eastwood, Edward B., and Cathelijne Eastwood. 2006. *Capturing the Spoor: An Exploration of Southern African Rock Art*. Cape Town: David Philip.

Eliade, Mircea. 1984. Cosmogonic Myth and 'Sacred History'. In *Sacred Narrative: Readings in the Theory of Myth*, ed. Alan Dundes, 137–151. Berkeley: University of California Press.

England, Nicholas. 1963. Fieldwork in the Kalahari. Paper presented at the Symposium on the Current Research in Ethnomusicology, University of Washington, March 7–9, 1963, 1–9. https://www.ethnomusicology.org/page/Resources_Symposium?

———. 1968. Music Among the Zu/'wa-si of South West Africa and Botswana. Ph.D. Dissertation, Harvard University. (Published as *Music Among the Ju/'hoansi and Related Peoples of Namibia, Botswana and Angola*. New York: Garland Publishing, 1992.)

Farini, G. Antonio [Hunt, William Leonad]. 1886. *Through the Kalahari Desert: A Narrative of a Journey with Gun, Camera, and Note-Book to Lake N'Gami and Back*. London: Sampson, Low, Marston, Searle & Rivington.

Foster, Charles. 2016. *Being a Beast: Adventures Across the Species Divide*. New York: Metropolitan Books Henry Holt and Company.

Fourie, L. 1925/26. Preliminary Notes on Certain Customs of the Hei-kom Bushmen. *Journal of the South-West African Scientific Society* 1/2: 49–63.

———. 1963/64–1964/65. Preliminary Notes on Certain Customs of the Hei-kom Bushmen. *Journal of the South-West African Scientific Society* 18/19: 18–34.

Geertz, Clifford. 1972. Religion as a Cultural System. In *The Interpretation of Cultures*, ed. Clifford Geertz, 87–125. New York: Basic Books.

Gentz, Leutnant. 1903/04. Einige Beiträge zur Kenntnis der südwestafrikanischen Völkerschaften in Deutsch-Südwestafrika. *Deutsche Kolonialzeitung* 26: 450–452.

Golifer, Ann, and Jenny Egner. 2011. *I Don't Know Why I Was Created: A Biography of Dada, Born Coex'ae Qgam*. Gaborone: Eggsson Books.

Grinnell, Jon, Craig Packer, and Anne E. Pusey. 1995. Cooperation, in Male Lions: Kinship, Reciprocity or Mutualism? *Animal Behaviour* 49: 95–105.

Guenther, Mathias. 1986. *The Nharo Bushman of Botswana: Tradition and Change*. Hamburg: Helmut Buske Verlag.

———. 1988. Animals in Bushman Thought, Myth and Art. In *Property, Power and Ideology in Hunting-Gathering Societies*, ed. James Woodburn, Tim Ingold, and David Riches, 192–202. Oxford: Berg.

———. 1989. *Bushman Folktales: Oral Traditions of the Nharo of Botswana and the /Xam of the Cape*, Studien Zur Kulturkunde No. 93. Stuttgart: Franz Steiner Verlag Wiesbaden.

———. 1990. Convergent and Divergent Themes in Bushman Myth and Art. In *Die Vielfalt der Kultu: Ethnologische Aspekte von Verwandschaft, Kunst, und Weltauffassung*, ed. Karl-Heinz Kohl, Heinzarnold Muszinski, and Ivo Strecker, 237–254. Berlin: Dietrich Reimer Verlag.

———. 1992. 'Not a Bushman Thing': Witchcraft among the Bushmen and Hunter-Gatherers. *Anthropos* 87: 83–107.

———. 1994. The Relationship of Bushman Art to Ritual and Folklore. In *Contested Images: Diversity in Southern African Rock Art Research*, ed. Thomas A. Dowson and David Lewis-Williams, 257–273. Johannesburg: Witwatersrand University Press.

———. 1995. Contested Images, Contested Texts: The Politics of Representation of the Bushmen of Southern Africa. *Critical Arts* 9 (2): 110–118.

———. 1996a. Attempting to Contextualize /Xam Oral Tradition. In *Voices from the Past: /Xam Bushmen and the Bleek and Lloyd Collection*, ed. Janette Deacon and Thomas A. Dowson, 77–99. Johannesburg: Wits University Press.

———. 1996b. Old Stories/Life Stories: Memory and Dissolution in Contemporary Bushman Folklore. In *Who Says? Essays on Pivotal Issues in Contemporary Storytelling*, ed. Carol L. Birch and Melissa A. Heckler, 177–197. Little Rock: August House Publishers.

———. 1997. Jesus Christ as Trickster in the Religion of Contemporary Bushmne. In *The Games of Gods and Men: Essays in Play and Performance*, ed. Klaus-Peter Koepping, 203–230. Hamburg: Lit Verlag.

———. 1998. Farm Labourer, Trance Dancer, Artist: The Life and Works of Qwaa Mangana. In *The Proceedings of the Khoisan Identities & Cultural Heritage Conference*, ed. Andrew Bank, 121–134, South African Museum, Cape Town, July 12–16, 1997. Cape Town: InfoSource.

———. 1999. *Tricksters and Trancers Bushman Religion and Society*. Bloomington: Indiana University Press.

———. 2002. The Bushman Trickster: Protagonist, Divinity, and Agent of Creativity. *Marvels and Tales Journal of Fairy-Tale Studies* 16 (1): 13–28.

———. 2003. *The Kuru Art Project at D'Kar, Botswana: Art and Identity Among Contemporary San*. M.s.

———. 2005. The Professionlisation and Commoditisation of the Contemporary Bushman Trance Dancer and Trance Dance, and the Decline of Sharing. In *Property and Equality, Volume 2: Encapsulation, Commercialisation, Discrimination*, ed. Thomas Widlok and Wolde Gossa Tadesse, 208–230. Oxford: Berghahn Books.

———. 2006. Contemporary Bushman Art, Identity Politics and the Primitivism Discourse. In *The Politics of Egalitarianism: Theory and Practice*, ed. Jacqueline Solway, 159–188. Oxford: Berghahn.

———. 2007. 'The Return of Myth and Symbolism': Articulation of Foraging, Trance Curing and Storytelling Among the San of the Old Way and Today. *Before Farming: The Archaeology and Anthropology of Hunter-Gatherers* 4, article 4: 1–10. (Online Journal).

———. 2014. Dreams and Stories. In *Courage of //Kabbo: Celebrating the 100th Anniversary of the Publication of* Specimens of Bushman Folklore, ed. Janette Deacon and Pippa Skotnes, 195–210. Cape Town: UCT Press.

———. 2015. 'Therefore Their Parts Resemble Humans, for They Feel That They Are People': Ontological Flux in San Myth, Cosmology and Belief. *Hunter-Gatherer Research* 1 (3): 277–315.

———. 2017. '…The Eyes Are No Longer Wild, You Have Taken the Kudu into Your Mind': The Supererogatory Aspect of San Hunting. *The South African Archaeological Bulletin* 72: 3–16.

———. 2018. 'I Can Feel My Lion Hair Growing and My Lion Teeth Forming…': San Lion Transformation – Real or Imagined? God's Truth or Hocus Pocus?

Paper Presented at the 24th Biannual Meeting of the Society of Africanist Archaeologists, University of Toronto, June 16–21, 2018.

Gusinde, Martin. 1966. *Von gelben und schwarzen Buschmännern*. Graz: Akademische Druck- und Verlagsanstalt.

Hahn, Theophilius. 1870. Die Buschmänner. Ein Beitrag zur südafrikanischen Völkerkunde, IV. *Globus* XVII (5): 120–123.

Halbmayer, Ernst, ed. 2012. Debating Animism, Perspectivism and the Construction of Ontologies. Special Issue of *Indiana* 29: 9–169.

Hallowell, A. Irving. 1926. Bear Ceremonialism in the Northern Hemisphere. *American Anthropologist* 28 (Special Issue 1).

———. 1960. Ojibwa Ontology, Behavior, and World View. In *Essays in Honor of Paul Radin*, ed. Stanley Diamond, 19–52. New York: Columbia University Press. (Republished in Stanley Diamond, ed. *Primitive Views of the World: Essays from* Culture in History. New York: Columbia University Press, 1964.)

Hannis, Michael, and Sian Sullivan. 2018. Relationality, Reciprocity, and Flourishing in an African Landscape. In *That All May Flourish: Comparative Religious Environmental Ethica*, ed. Laura Hartman, 279–296. Oxford: Oxford University Press.

Harris, William C. 1852. *The Wild Sports of Southern Africa; Being the Narrative of a Hunting Expedition from the Cape of Good Hope, Through the Territories of the Chief Moselekatse to the Tropic of Capricorn*. 5th ed. London: Henry G. Bohn, York Street, Covent Garden.

Harvey, Graham. 2006. *Animism: Respecting the Living World*. New York: Columbia University Press.

Hawkes, David. 2017. Many-Headed Multitude How Incorporation Can Be a Good Thing. *TLS* 24 (2017): 25.

Heinz, Hans-Joachim. 1975. !ko-Buschmänner (Südafrikas, Kalahari): Mädchen Initiation. *Encyclopaedia Cinematographica*, E 1849: 3–15.

———. 1978. The Bushmen's Store of Scientific Knowledge. In *The Bushmen*, ed. Philip Tobias, 148–161. Cape Town: Human & Rousseau.

———. 1994. *Social Organization of the !Kõ Bushmen*. Cologne: Rüdiger Köppe Verlag.

Heinz, Hans-Joachim, and Marshall Lee. 1978. *Namkwa: Life Among the Bushmen*. London: Jonathan Cape.

Hermans, Janet. 1998. Basarwa Cultural Preservation in Contemporary Botswana: The Dance. In *The Proceedings of the Khoisan Identities & Cultural Heritage Conference*, ed. Andrew Bank, 281–288, South African Museum, Cape Town, July 12–16, 1997. Cape Town: InfoSource.

Heunemann, D., and Hans-Joachim Heinz. 1975a. !ko-Buschmänner (Südafrika, Kalahari): Wettstreit, Jäger und Tier (Gestenspiel). In *Encyclopaedia Cinematographica*, Film E2105, ed. G. Wolf, 3–12. Göttingen: Institut für den wissenschaftlichen Film.

———. 1975b. !ko-Buschmänner (Südafrika, Kalahari) Spiel Oryx-Antilope, mit einer Auseinadersetzung zwischen zwei Spielgruppen. In *Encyclopaedia Cinematographica*, Film E2107, ed. G. Wolf, 3–12. Göttingen: Institut für den wissenschaftlichen Film.

Hewitt, Roger. 1986. *Structure, Meaning and Ritual in the Narratives of the Southern San*. Hamburg: Helmut Buske Verlag.

———. 2008. *Structure, Meaning and Ritual in the Narratives of the Southern San*. 2nd ed. Johannesburg: Wits University Press.

Hill, Erica. 2011. Animals as Agents: Hunting Rituals and Relational Ontologies in Prehistoric Alaska and Chukota. *Cambridge Archaeological Journal* 2: 407–426.

———. 2013. Archaeology and Animal Persons: Towards a Prehistory of Human-Animal Relations. *Environment and Society: Advances in Research* 4: 117–136.

Hoff, Ansie. 2011a. Guardians of Nature Among the /Xam San: An Exploratory Study. *South African Archaeological Bulletin* 66: 41–50.

———. 2011b. *The /Xam and the Rain*. Cologne: Rüdiger Köppe Verlag.

Hollmann, Jeremy C., ed. 2005. *Customs and Beliefs of the /Xam Bushmen*. Johannesburg/Philadelphia: Wits University Press/Ringing Rock Press.

Ijäs, Mikko. 2017. Fragments of the Hunt: Persistence Hunting, Tracking and Prehistoric Art. Doctoral Thesis, Aalto University: School of Arts, Design and Architecture, Department of Art.

Ingold, Tim. 2000. Hunting and Gathering as Ways of Perceiving the Environment. In *The Perception of the Environment*, ed. Tim Ingold, 40–60. London/New York: Routledge.

Jackson, Michael, and Albert Piette. 2015. Anthropology and the Existential Turn. In *What Is Existential Anthropology?* ed. Michael Jackson and Albert Piette, 1–29. Oxford: Berghahn.

Jansen, William H. 1959. The Esoteric-Exoteric Factor in Folklore. *Fabula: Journal of Folklore Studies* 2: 205–211.

Jodtka, Assistenzartzt. 1902. Reise des Assistenzartztes Jodtka nach dem Okavango. *Deutsches Kolonialblatt* 13: 493–495, 590–593.

Jolly, Pieter. 2002. Therianthropes in San Rock Art. *South African Archaeological Bulletin* 57 (176): 85–103.

———. 2006. Dancing with Two Sticks: Investigating the Origin of a Southern African Rite. *South African Archaeological Bulletin* 61 (184): 172–180.

———. 2015. *Sonqua: Southern San History and Art After Contact*. Cape Town: Southern Cross Ventures.

Katz, Richard. 1976. Education for Transcendence: !Kia-Healing with the Kalahari !Kung. In *Kalahari Hunter-Gatherers: Studies of the !Kung San and Their Neighbours*, ed. Richard B. Lee and Irven DeVore, 281–301. Cambridge, MA: Cambridge University Press.

———. 1982. *Boiling Energy: Community Healing Among the Kalahari Kung*. Harvard: Harvard University Press.

Katz, Richard, Megan Biesele, and Verna St. Denis. 1997. *"Healing Makes Our Hearts Happy": Spirituality and Cultural Transformation Among the Kalahari Ju/'hoansi*. Rochester: Inner Traditions.

Kaufmann, Hans. 2005 [1910]. The =Auin. In *Kalahari and Namib Bushmen in German South West Africa: Ethnographic Reports by Colonial Soldiers and Settlers*, ed. Mathias Guenther, 37–96. Cologne: Rüdiger Köppe Verlag.

KDT. 1995. *Kuru Conquers the Brandberg*. Dekar: Kuru Development Trust.

Keeney, Bradford. 1999. *Kalahari Bushman Healers*. Philadelphia: Ringing Rocks Press.

Keeney, Bradford, and Hillary Keeney. 2013. Reentry into First Creation: A Contextual Frame for the Ju/'hoan Bushman Performance of Puberty Rites, Story Telling, and Healing Dance. *Journal of Anthropological Research* 69: 65–86.

———, eds. 2015. *Way of the Bushman as Told by the Tribal Elders*. Rochester: Bear & Company.

Kinahan, John. 1991. *Pastoral Nomads of the Namib Desert: The People History Forgot*. Windhoek: Namibia Archaeological Trust.

———. 2010. The Rock Art of /Ui-//aes (Twyfelfontein) Namibia's First World Heritage Site. *Adoranten* 8: 39–51.

Köhler, Oswin. 1973. Die rituelle Jagd bei den Kxoe-Buschmännern von Mutsiku. In *Festschrift zum 65. Geburtstag von Helmut Petri. Kölner ethnologische Mitteilungen*, ed. Kurt Tauchmann, vol. 5, 215–257.

———. 1978/1979. Mythus, Glaube und Magie bei den Kxoe-Buschmännern. *Journal of the South-West African Scientific Society* 23/24: 9–50.

———. 1989. *Die Welt der Kxoé Buschleute im südlichen Afrika: Eine Selbstdarstellung in ihrer eigenen Sprache*, Band 1: Die Kxoé-Buschleute und ihre ethnische Umgebung. Berlin: Dietrich Reimer Verlag.

Kohn, Eduardo. 2013. *How Forests Think: Towards an Anthropology of Nature Beyond the Human*. Berkeley: University of California Press.

Kover, Tihamer R. 2007. The Beastly Familiarity of Wild Alterity: Debating the Nature of Our Fascination with Wildness. *Ethical Perspectives: Journal of the European Ethics Network* 14 (4): 431–456.

———. 2017. Of Killer Apes and Tender Carnivores: A Shepardian Critique of Burkert and Girard on Hunting and the Evolution of Religion. In *Animals and Religion*, ed. Nathan Kowalsky and Tihamer Kover. Special Issue of *Studies in Religion* 46 (4): 536–567.

Krog, Antje. 2009. *Begging to Be Black*. Cape Town: Random House Struik.

Kruiper, Vetkat R.B. 2014. *Mooi Loop: The Sacred Art of Vetkat Regopstaan Kruiper*. Pretoria: Unisa Press.

Kruiper, Belinda, and Elana Bregnin. 2004. *Kalahari Rain Song*. Pietermaritzburg: University of Natal Press.

Kuru Art Project. 1994. *Qauqaaua, a San Folk Story*. Johannesburg: The Artists' Press.

Kuru D'Kar Trust. n.d. *Ncoa ne Khoe ne di Hua ne: San Stories in Naro and English.* D'Kar: Kuru D'Kar Trust/Naro Language Project.

Lang, Andrew. 1887. *Myth, Ritual and Religion.* Vol. II. London: Longman, Green and Co.

———. 1901. *Myth, Ritual and Religion.* Vol. I. 2nd ed. London: Longman, Green and Co. (Both Available as ebooks Through Project Gutenberg, Released November 12, 2009).

Lange, Mary E. 2011. Rock Art Research in South Africa. *Rozenberg Quarterly The Magazine.* Open-access Online Journal http://rozenbergquarterly.com/rock-art-research-in-south-africa, 10 pp. Excerpt from Mary Lange *Water Stories and Rock Engravings: Eiland Women at the Kalahari Edge.* Amsterdam: Rozenberg Publishers, SAVUSA Series.

———. 2015. *Water Stories: Original !Garib Narrations About the Water Snake.* Pretoria: Unisa Press.

Lange, Mary E., and Lauren Dyll-Myklebust. 2015. Spirituality, Shifting Identities and Social Change: Cases from the Kalahari Landscape. *HTS Teologiese Studies/Theological Studies* 71 (1), Art. #2985, 11 pp. http://dx.doi.org/10.4102/hts.v71i1.2985

Lange, Mary E., Miliswa Magongo, and Shanade Barnabas. 2014. Biesje Poort Rock Engravings, Northern Cape: Past and Present. In *Courage of //Kabbo: Celebrating the 100th Anniversary of the Publication of* Specimens of Bushman Folklore, ed. Janette Deacon and Pippa Skotnes, 363–382. Cape Town: UCT Press.

Lawy, Jenny. 2016. An Ethnography of San: Minority Recognition and Voice in Botswana. Unpublished D.Phil. Thesis, University of Edinburgh.

Le Roux, Willemien, and Alison White, eds. 2004. *Voices of the San.* Cape Town: Kwela Books.

Lebzelter, Viktor. 1934. *Eingeborenenkulturen von Südwestafrika: Die Buschmänner.* Leipzig: Verlag Karl W. Hiersemann.

Lee, Richard B. 1968. The Sociology of !Kung Bushman Trance Performances. In *Trance and Possession States*, ed. Raymond Prince, 35–54. Montreal: R. M. Bucke Memorial Society.

———. 1979. *The !Kung San Men, Women, and Work in a Foraging Society.* Cambridge: Cambridge University Press.

———. 2003. *The Dobe Ju/'hoansi.* 3rd ed. Fort Worth/Belmont: Wadsworth.

Lee, Richard B., and Irven DeVore, eds. 1968. *Man the Hunter.* Chicago: Aldine Publishing Company.

———, eds. 1976. *Kalahari Hunter-Gatherers: Studies of the !Kung San and Their Neighbors.* Cambridge, MA: Harvard University Press.

Lewis-Williams, David J. 1980. Remarks on Southern San Religion and Art. *Religion in Southern Africa* 1: 19–32.

———. 1981. *Believing and Seeing: Symbolic Meanings in Southern San Rock Paintings.* New York: Academic Press.

———. 1988. The World of Man and the World of Spirit: An Interpretation of the Linton Rock Paintings. In *Margaret Shaw Lecture 2*. Cape Town: South African Museum.
———. 1990. *Discovering Southern African Rock Art*. Cape Town: David Philip.
———. 1996. 'A Visit to the Lion's House': The Structure, Metaphors and Socio-political Significance of a Nineteenth-Century Bushman Myth. In *Voices from the Past: /Xam Bushmen and the Bleek and Lloyd Collection*, ed. Janette Deacon and Thomas A. Dowson, 122–141. Johannesburg: Wits University Press.
———. 1998. *Quanto?*: The Issue of Many Meanings in Southern San Rock Art Research. *South African Archaeological Bulletin* 53: 86–97.
———. 1999. 'Meaning' in Southern African San Rock Art: Another Impasse? *South African Archaeological Bulletin* 54: 141–145.
———. 2000. *Stories that Float from Afar: Ancestral Folklore of the San of Southern Africa*. Cape Town: David Philip.
———. 2002. *The Mind in the Cave*. London: Thames & Hudson.
———. 2010. *Conceiving God: The Cognitive Origin and Evolution of Religion*. London: Thames & Hudson.
———. 2015a. *Myth and Meaning: San-Bushman Folklore in Global Context*. Walnut Creek: Left Coast Press.
———. 2015b. Text and Evidence: Negotiating San Words and Images. *South African Archaeological Bulletin* 70 (201): 53–63.
———. 2016. The Jackal and the Lion: Aspects of Khoisan Folklore. *Folklore* 127: 51–70.
Lewis-Williams, David J., and Megan Biesele. 1978. Eland Hunting Rituals Among Northern and Southern San Groups. *Africa* 48: 117–134.
Lewis-Williams, David J., and Sam Challis. 2011. *Deciphering Ancient Minds: The Mystery of San Bushman Rock Art*. London: Thames & Hudson.
Lewis-Williams, David J., and Thomas Dowson. 1989. *Images of Power Understanding Bushman Rock Art*. Johannesburg: Southern Book Publishers.
Lewis-Williams, David J., and David G. Pearce. 2004a. *San Spirituality: Roots, Expression, and Social Consequences*. Walnut Creek: Altamira Press.
———. 2004b. Southern African Rock Paintings as Social Intervention: A Study of Rain-Control Images. *African Archaeological Review* 21: 199–228.
Liebenberg, Louis. 2006. Persistence Hunting by Modern Hunter-Gatherers. *Current Anthropology* 47: 1017–1025.
Lloyd, Lucy. 1889. *A Short Account of Further Bushman Material Collected. Third Report Concerning Bushman Researches, Presented to Both Houses of Parliament of the Cape of Good Hope by Command of His Excellency the Governor*. London: David Nutt.
Low, Chris. 2007. Khoisan Wind: Hunting and Healing. *Journal of the Royal Anthropological Institute* (N.S.) 13: 571–590.

———. 2013. The Role of the Body in Bushman Healing Dances. Paper Presented at the 10th International Conference on Hunting-Gathering Societies (CHAGS 10), Liverpool, June 25–28, 2013.

———. 2014. Locating /Xam Beliefs and Practices in a Contemporary KhoeSan Context. In *Courage of //Kabbo: Celebrating the 100th Anniversary of the Publication of* Specimens of Bushman Folklore, ed. Janette Deacon and Pippa Skotnes, 349–361. Cape Town: UCT Press.

Lowie, Robert H. 1948. *Primitive Religion.* New York: Grosset & Dunlap.

Lyons, Andrew P. 2018. The Two Lives of Sara Baartman: Gender, 'Race', Politics and the Historiography of Mis/Representation. *Anthropologica* 60: 327–346.

Maingard, L.F. 1937. The ≠Khomani Dialect, Its Morphology and Other Characteristics. In *Bushmen of the Southern Kalahari*, ed. John D. Reinhallt Jones and Clement M. Doke, 237–276. Johannesburg: University of the Witwatersrand Press.

Malan, J.S. 1995. *Peoples of Namibia.* Pretoria: Rhino Publishers.

Manguel, Alberto, and Gianni Guadalupi. 1980. *The Dictionary of Imaginary Places.* New York: Macmillan Publishing Co.

Marrett, Robert R. 1914. *The Threshold of Religion.* 2nd ed. London: Methuen & Co.

Marshall, Lorna. 1962. !Kung Bushman Religious Beliefs. *Africa* 32: 221–252.

———. 1969. The Medicine Dance of the !Kung Bushmen. *Africa* 39 (4): 347–381.

Marshall, Lorna J. 1976. *The !Kung of Nyae Nyae.* Cambridge, MA: Harvard University Press.

———. 1999. *Nyae Nyae !Kung Beliefs and Rites*, Peabody Museum Monographs, No. 8. Cambridge, MA: Harvard University.

Mboti, Nyasha. 2014. To Exhibit or Be Exhibited: The Visual Art of Vetkat Regopstaan Boesman Kruiper. *Critical Arts: South-North Cultural and Media Studies* 28: 472–492.

McCall, Daniel F. 1970. Wolf Courts Girl: The Equivalence of Hunting and Mating in Bushman Thought. *Ohio University Papers in International Studies, Africa Series*, No. 7. Athens: Ohio University, Center for International Studies.

McGranaghan, Mark. 2012. Foragers on the Frontiers: The /Xam Bushmen of the Northern Cape, South Africa, in the Nineteenth Century. Unpublished Ph.D. Thesis, Oxford University.

———. 2014. 'He Who Is a Devourer of Things': Monstrosity and the Construction of Difference in /Xam Bushman Oral Literature. *Folklore* 125: 1–21.

McGranaghan, Mark, and Sam Challis. 2016. Reconfiguring Hunting Magic: Southern Bushmen (San) Perspectives on Taming and Their Implications for Understanding Rock Art. *Cambridge Archaeological Journal* 26 (4): 579–599.

McGranaghan, Mark, Sam Challis, and David Lewis-Williams. 2013. Joseph Millerd Orpen's 'A Glimpse into the Mythology of the Maluti Bushmen': A Contextual Introduction and Republished Text. *Southern African Humanities* 25: 137–166.

McNiven, Ian J. 2013. Between the Living and the Dead: Relational Ontologies and the Ritual Dimension of Dugong Hunting Across Torres Strait. In *Relational Archaeologies Humans / Animals / Things*, ed. Christopher Watts, 97–116. London: Routledge.

Merleau-Ponty, Maurice. 1962. *Phenomenology of Perception*. Trans. Colin Smith. London: Routledge and Kegan Paul.

Metzger, Fritz. 1990. *Naro und seine Sippe*. Windhoek: Namibia Wissenschaftliche Gesellschaft/Scientific Society.

Mguni, Siyakha. 2006. Iconography of Termites' Nests and Termites: Symbolic Nuances of Foundlings in Southern African Rock Art. *Cambridge Archaeological Journal* 16 (1): 53–71.

———. 2015. *Termites of the Gods: San Cosmology in Southern African Rock Art*. Johannesburg: Wits University Press.

Mitchell, Robert W., and Nicholas S. Thompson, eds. 1986. *Deception: Perspectives on Human and Nonhuman Deceit*. Albany: SUNY Press.

Moffatt, Robert. 1844. *Missionary Labours and Scenes in Southern Africa*. New York: Robert Carter.

Morris, Rosalind C. 2011. Crowds and Powerlessness: Reading //Kabbo and Canetti with Derrida in (South) Africa. In *Demenagerie: Thinking (of) Animals After Derrida*, ed. Anne Emmanuelle Berger and Marta Segarra, 167–212. Amsterdam: Brill/Rodopi.

Münzel, Mark. 2013. Warum verlassen die Geister die Insel? In *Wege im Garten der Ethnologie: Zwischen dort und hier: Festschrift für Maria Susanna Cipolleti*, ed. Harald Grauer, 35–49. Sankt Augustin: Akademia.

Northover, Richard A. 2017. Interrogating the Sacred Art of Vetkat Regopstaan Boesman Kruiper. In *Indigenous Creatures, Native Knowledge, and the Arts: Animal Studies in Modern Worlds*, ed. W. Woodward and S. McHugh, 59–84. London: Palgrave Macmillan.

Orpen, Joseph M. 1874. A Glimpse into the Mythology of the Maluti Bushmen. *Cape Monthly Magazine* [N.S.] 9: 1–13.

Orton, Jayson. 2008. Later Stone Age Ostrich Eggshell Bead Manufacture in the Northern Cap, South Africa. *Journal of Archaeological Science* 35: 1765–1775.

Osaki, Masakatsu. 1984. The Social Influence of Change on Hunting Techniques Among the Central Kalahari San. *African Studies Monograph* 5: 49–62.

Ouzman, Svend. 2008. Cosmology of the African San People. In *Encyclopedia of the History of Science, Technology, and Medicine in Non-Western Cultures*, ed. Helaine Selin, vol. 1, 2nd ed., 219–225. New York: Springer Science and Business Media.

Ovid. 1815. *Metamorphoses: Translated into English Verse Under the Direction of Sir Samuel Garth by John Dryden, Alexander Pope, Joseph Addison, William Congreve and Other Eminent Hands.* Book the First. London: R. McDermott & D. D. Arden.

Pager, Harald. 1989. *The Rock Paintings of the Upper Brandberg. Part I: Amis Gorge*, ed. R. Kuper. Cologne: Heinrich-Barth Institut.

Parkington, John. 2002a. *The Mantis, the Eland and the Hunter Follow the San*…. Cape Town: Creda Communications.

———. 2002b. Men, Women and Eland: Hunting and Gender Among the San of Southern Africa. In *In Pursuit of Gender Worldwide: Archaeological Approaches*, ed. Sarah M. Nelson and Myriam Rosen-Ayalon, 93–118. Walnut Creek: Altamira Press.

———. 2003. Eland and Therianthropes in Southern African Rock Art: When Is a Person an Animal? *African Archaeological Review* 20 (3): 135–147.

Parkington, John, and Andy Paterson. 2017. Somatogenesis: Vibrations, Undulations and the Possible Depiction of Sound in San Rock Paintings of Elephants in the Western Cape. *South African Archaeological Bulletin* 72 (206): 131–141.

Passarge, Siegfried. 1907. *Die Buschmänner der Kalahari.* Berlin: Dietrich Reimer.

Pickering, Travis R. 2013. *Rough and Tumble: Aggression, Hunting, and Human Evolution.* Berkeley: University of California Press.

Rabbethge-Schiller, Helga, ed. 2006. *Memory and Magic: Contemporary Art of the !Xun and Khwe.* Johannesburg: Jacana.

Rakitianskaia, Olga. forthcoming. Lions in the Night: Dual Unity of the Pantherine Image in San Belief and Its Possible Origins. *Southern African Humanities.*

Richters, Ferdinand. 1886. *Bericht über die senckenbergische naturforschende Gesellschaft in Frankfurt am Main.* Frankfurt: Senckenberger naturwissenshcftlich Gesellschaft.

Rifkin, Riann F., Christopher S. Henshilwood, and Magnus M. Haaland. 2015. Pleistocene Figurative *Art Mobilier* from Aollo 11 Cave, Karas Region, Southern Namibia. *South African Archaeological Bulletin* 770 (201): 113–123.

Ristau, Carolyn, ed. 1991. *Cognitive Ethology: The Mind of Other Animals (Essays in Honour of Donald Griffin).* Hillsdale: Lawrence Erlbaum.

Rudner, J. 1956/57. The Brandberg and Its Archaeological Remains. *Journal of the South West African Scientific Society* 12: 7–44.

Rusch, Neil. 2015. The Root and Tip of the //*Kwanna*: Introducing *Chiasmus* in Three /Xam Narratives. Unabridged Version of Rusch (2016a), Accessible at the Centre for Curating the Archive, 1–26. http://www.cca.uct.ac.za/publications/

———. 2016a. The Root and Tip of the //*Kwanna*: Introducing *Chiasmus* in Three /Xam Narratives. *Critical Arts* 30 (6): 877–897.

———. 2016b. Sounds and Sound Thinking in /Xam-ka !au: 'These Are Those to Which I Am Listening with All My Ears'. *Cogent Arts and Humanities*, Open Access Journal. http://dv.dororg/10:10.108012331983.2016.123615

———. 2017. Sound Artefacts: Recreating and Reconnecting the Sound of the *!goin !goin* with the Southern Bushmen and Bees. *Hunter Gatherer Research*, 2015 3 (2): 187–226.

Rusch, Neil, and John Parkington. 2010. *San Rock Engravings: Marking the Karoo Landscape*. Cape Town: Random House Struik.

Sahlins, Marshall. 1972. The Original Affluent Society. In *Stone Age Economics*, ed. Marshall Sahlins, 1–40. Chicago: Aldine.

Sbrzesny, Heide. 1974. !ko-Buschleute (Kalahari) - Der Eland Tanz. Kinder spielen das Mädchen. Initiationsritual. *Homo* 25: 233–234.

———. 1976. *Die Spiele der Ko-Buschleute unter besonderer Berücksichtigung ihrer sozialisierenden und gruppen-bindende Funktionen*. Munich: Piper.

Schapera, Isaac. 1930. *The Khoisan Peoples of South Africa: Bushmen and Hottentots.* London: Routledge & Kegan Paul.

Schatz, Ilse. 1993. *Unter Buschleuten auf der Farm Otjiguinas in Namibia.* Tsumeb: Ilse Schatz.

Scheepers, Catharina. 2014. I Had a Dream. In *Rêves de Kalahari: L'art du Kuru Art Project/Kalahari Dreams: The Art of the Kuru Art Project*, ed. Leïla Bararcchini, 20–23. Neuchâtel: L'Usage du Temps.

Schmidt, Sigrid. 1995. *Als die Tiere noch Menschen waren: Urzeit- und Trickstergeschichten der Damara und Nama in Namibia*, Afrika Erzählt. Vol. 3. Cologne: Rüdiger Köppe Verlag.

———. 2001. *Tricksters, Monsters and Clever Girls: African Folktales – Texts and Discussions*, Afrika Erzählt. Vol. 8. Cologne: Rüdiger Köppe Verlag.

———. 2013a. *A Catalogue of Khoisan Folktales of Southern Africa*, 2 vols. Research in Khoisan Studies 28.1&2. Cologne: Rüdiger Köppe Verlag.

———. 2013b. *South African /Xam Bushman Traditions and Their Relationship to Further Khoisan Folklore*, Research in Khoisan Studies, 31. Cologne: Rüdiger Köppe Verlag.

Schoeman, Peter J. 1961. *Hunters of the Desert Land.* Cape Town: Howard Timmins.

Schultze, Leonhard. 1907. *Aus Namaqualand und Kalahari.* Jena: Verlag von Gustav Fischer.

Scott, Michael. 2013. What I'm Reading: The Anthropology of Ontology (Religious Science?) *Journal of the Royal Anthropological Institute* (N.S.) 19: 859–872.

Shepard, Paul. 1998 [1978]. *Thinking Animals: Animals and the Development of Human Intelligence.* Athens: The University of Georgia Press.

Silberbauer, George B. 1963. Marriage and the Girl's Puberty Rite of the G/wi Bushmen. *Africa* 33: 12–24.

———. 1965. *Report to the Bechuanaland on the Bushman Survey*. Gaberones: Bechuanaland Government.
———. 1972. The G/wi Bushmen. In *Hunter-Gatherers Today*, ed. M.G. Bicchieri, 271–326. New York: Holt, Rinehart and Winston.
———. 1981. *Hunter and Habitat in the Central Kalahari Desert*. Cambridge: Cambridge University Press.
Simay, Philippe. 2012. *What Images Show. An Interview with Philippe Descola*. Trans. Michael C. Behrent. Online Article, April 5, 2012. http://www.booksandideas.net/What-Images-Show.html
Singer, Thea. 2017. The Evolution of Dance. *Scientific American* 317 (1): 66–71.
Skaanes, Thea. 2017a. Sounds in the Night: Ritual Bells, Therianthropes, and Eland Relations Among the Hadza. In *Human Origins: Contributions from Social Anthropology*, ed. Camilla Powers, Morna Finnegan, and Hillary Callan, 204–223. Oxford: Berghahn Books.
———. 2017b. Cosmology Matters: Power Objects, Rituals, and Meat-Sharing Among the Hadza of Tanzania. Ph.D. Dissertation, Aarhus University.
Skinner, Andrew. 2017. The Changer of Ways: Rock Art and Frontier Ideologies on the Strandberg, Northern Cape, South Africa. Unpublished Master's Thesis, University of the Witwatersrand.
Skotnes, Pippa. 1990. Rock Art: Is There Life After Trance? *De Arte* 44: 16–24.
———. 1996. The Thin Black Line: Diversity and Transformation in the Bleek and Lloyd Collection and the Paintings of the Southern San. In *Voices from the Past: /Xam Bushmen and the Bleek and Lloyd Collection*, ed. Janette Deacon and Thomas A. Dowson, 234–244. Johannesburg: Wits University Press.
———, ed. 2007. *Claims to the Country: The Archives of Lucy Lloyd and Wilhelm Bleek*. Johannesburg/Athens: Jacana Media/Ohio University Press.
Smith, Andrew B. 2016. Why Would Southern African Hunters Be Reluctant Food Producers? *Hunter-Gatherer Research* 2 (4): 1–12. (Online Open Access journal).
Solomon, Anne. 1997. The Myth of Ritual Origin? Ethnography, Mythology and the Interpretation of San Rock Art. *South African Archaeological Bulletin* 52: 3–13.
———. 1999. Meanings, Models and Minds: A Reply to Lewis-Williams. *South African Archaeological Bulletin* 54: 51–60.
———. 2000. On Different Approaches to San Rock Art. *South African Archaeological Bulletin* 55: 77–78.
———. 2007. Images, Words and Worlds: The /Xam Testimonies and the Rock Arts of the Southern San. In *Claims to the Country: The Archives of Lucy Lloyd and Wilhelm Bleek*, ed. Pippa Skotnes, 149–159. Johannesburg: Jacana Media.
———. 2008. Myths, Making and Consciousness. *Current Anthropology* 49: 59–86.

———. 2009. Broken Strings: Interdisciplinarity and /Xam Oral Literature. *Critical Arts* 23 (1): 25–41.

———. 2011. Writing San Histories: The /Xam and 'Shamanism' Revisited. *Journal of Southern African Studies* 37: 99–117.

———. 2014. Truths, Representationalism and Disciplinarity in Khoesan Researches. *Critical Arts: South-North Cultural and Media Studies* 28: 710–721.

Stephenson, Jessica. 2016. Picture the Past, Creating the Future: Art of the !Xun and Khwe Cultural Project. In *Memory and Magic: Contemporary Art of the !Xun and Khwe*, ed. Helga Rabbethge-Schiller, 17–25. Johannesburg: Jacana Media.

Stow, George W. 1905. *The Native Races of South Africa: A History of the Intrusion of the Hottentots and Bantu into the Hunting Grounds of the Bushmen, the Aborigines of the Country*. London: Swan Sonnenschein.

Straube, Helmut. 1955. *Tierverkleidungen der afrikanischen Naturvölker*. Wiesbaden: Franz Steiner Verlag.

Streck, Bernhard. 2013. *Sterbendes Heidentum: Die Rekonstruktion der ersten Weltreligion*. Leipzig: Eudora-Verlag.

Sugawara, Kazuyoshi. 2007. The Lion as the Symbol of the Other from the Perspective of the Gwi, a Hunting Tribe of the Kalahari Desert. *Biohistory Journal* (Spring): 1–2. Online Journal. https://www.brh.co.jp/en/simsishi/journal/052/research2-2.html. Accessed 19 Oct 2016.

———. 2015. On the G/ui Experiences of Being Hunted: An Analysis of Oral Discourse on the Man-Killings by Lions. Paper Presented at the 12th International Conference of Hunting-Gathering Societies (CHAGS 12), University of Vienna, Vienna, September 7–12, 2015.

Suzman, James. 2017. *Affluence Without Abundance: The Disappearing World of the Bushmen*. New York: Bloomsbury.

Szalay, Miklós, ed. *Der Mond dals Schuh: Zeichnungen der San/The Moon as Shoe: Drawings of the San*. Zurich: Scheidegger & Spiess.

Tanaka, Jiro. 1991. Egalitarianism and the Cash Economy Among Central Kalahari San. In *Cash, Commoditisation and Changing Foragers*, ed. Nicolas Peterson and Toshio Matsuyama. *Senri Ethnological Studies* 30: 117–135.

———. 1996. The World of Animals Viewed by the San. *African Studies Monographs*, Supplementary Issue, 22: 11–28.

Tanner, Adrian. 1979. *Bringing Home the Animals: Religious Ideology and Mode of Production of the Mistassini Cree Hunters*. St. John's: Institute of Social and Economic Research Memorial University of Newfoundland.

Taussig, Michael. 1993. *Mimesis and Alterity: A Particular History of the Senses*. London/New York: Routledge.

Thackeray, J. Francis. 1983. Disguises, Animal Behaviour and Concepts of Control in Relation to Rock Art of Southern Africa. In *New Approaches to Southern African Rock Art, Goodwin Series*, ed. David J. Lewsi-Willimas, vol. 4, 38–43.

———. 2005a. Eland, Hunters and Concepts of 'Sympathetic Control' in Southern African Rock Art. *Cambridge Archaeological Journal* 15 (1): 37–34.
———. 2005b. The Wounded Roan: A Contribution to the Relation of Hunting and Trance in Southern African Rock Art. *Antiquity* 79: 5–18.
Thomas, Elizabeth M. 1958. *The Harmless People*. New York: Random House.
———. 1990. The Old Way. *New Yorker*, October 15: 78–110.
———. 1993. *The Hidden Lives of Dogs*. New York: Pocket Star Books.
———. 1994. *The Tribe of Tiger: Cats and Their Culture*. New York: Pocket Books.
———. 2003. The Lion/Bushman Relationship in Nyae Nyae in the 1950s: A Relationship Crafted in the Old Way. *Anthropologica* 45: 73–78.
———. 2006. *The Old Way: A Story of the First People*. New York: Farrar Strauss Giroux.
Thomas, Roie. 2016. *Bushmen in the Tourist Imaginary*. Newcastle upon Tyne: Cambridge Scholars Publishing.
Thorp, Carolyn. 2013. 'Frog People' of the Drakensberg. *Southern African Humanities* 25: 245–262.
———. 2015. Rain's Things and Girl's Rain: Marriage, Potency and Frog Symbolism. *Southern African Humanities* 27: 165–190.
Thurner, Ingrid. 1983. *Die transzendenten und mythischen Wesen der San (Buschmänner) Eine religionsethnologische Analyse historischer Quellen*. Wien: Föhrenau.
Trenk, Oberleutnent. 2005. The Bushmen of the Namib and Their Legal and Family Conditions. In *Kalahari and Namib Bushmen in German South West Africa: Ethnographic Reports by Colonial Soldiers and Settlers*, ed. and trans. Mathias Guenther, 237–246. Cologne: Rudiger Kope Verlag.
Turner, Victor. 1970. *The Forest of Symbols*. Ithaca: Cornell University Press.
UNESCO. 2012. *Memory of the World: The Treasures that Record our History Form 1700 BC to the Present Day*. Paris/Glasgow: UNESCO/Harper/Collins.
Van der Camp, Ankie. 2012. *Bushman in the Tourist Imagery*. Newcastle-upon-Tyne: Cambridge Scholars Publishing.
Van Gennep, A. 1908. *The Rites of Passage*. Trans. M.B. Vizedom and G.C. Caffee. Chicago: Phoenix Books.
Van Vuuren, Helize. 2016. *A Necklace of Springbok Ears: /Xam Orality and South African Literature*. Stellenbosch: Sun Press.
Vedder, Heinrich. 1937. Die Buschmänner Südwestafrikas und ihre Weltanschauung. *South African Journal of Science* 34: 416–436.
Viereck, A. 1962. *Südwestafrikanische Felsmalereien*. Windhoek: South West African Scientific Society.
Viestad, Maria V. 2018. *Dress as Social Relations: An Interpretation of Bushman Dress*. Johannesburg/New York: Wits Press/NYU Press.
Vinnicombe, Patricia. 2001 [1976]. *People of the Eland: Rock Paintings of the Drakensberg Bushmen as a Reflection of Their Life and Thought*. Johannesburg: Wits Press.

Visser, Hessel. 1997. *Naro Dictionary*. 3rd ed. Gantsi: Naro Language Project.

Vivieros de Castro, Eduardo. 1998a. Cosmologies: Perspectivism. In *Cosmological Perspectivism in Amazonia and Elsewhere*, ed. Eduardo Vivieros de Castro. Four Lectures Given in the Department of Social Anthropology, Cambridge University, February–March, 1998. HAU: Masterclass Series, 1. *Journal of Ethnographic Theory*, 19 pp. Online Journal. http://www.haujournal.org/index.php/masterclass/article/view/107/135

———. 1998b. Cosmological Deixis and Amerindian Perspectivism. *The Journal of the Royal Anthropological Institute* (N.S.) 4: 469–488.

Vogelsang, Ralf, Jürgen Richter, Zenobia Jacobs, Barbara Eichorn, Veerle Linseele, and Richar G. Roberts. 2010. New Excavations of Middle Stone Age Deposits at Apollo II Rockshelter, Namibia: Stratigraphy, Archaeology, Chronology and Past Environment. *Journal of African Archaeology* 8 (2): 185–218.

Wagner-Robertz, Dagmar. 1976. Schamanismus bei den Hai//om in Südwestafrika. *Anthropos* 71: 533–554.

Weinstock, Jeffrey A. 2014. *The Ashgate Encyclopedia of Literary and Cinematic Monsters*. Burlington: Farnham & Ashgate.

Werner, H. 1906. Anthropologische, ethnologische und ethnographische Beobachtungen über die Heikum- und Kung-Buschleute. *Zeitschrift für Ethnologie* 3 (1): 241–268.

Wessels, Michael. 2010. *Bushman Letters: Interpreting /Xam Narratives*. Johannesburg: Wits Press.

———. 2012. The Khoisan Origin of the Interconnected World View in Antje Krog's *Begging to Be Black*. *Current Writing: Text and Reception in Southern Africa* 24 (2): 186–197.

———. 2013. Story of a /Xam Bushman Narrative. *Journal of Literary Studies* 29 (3): 1–22.

Whitridge, Peter. 2013. The Imbrication of Human and Animal Paths: An Arctic Case Study. In *Relational Archaeologies: Humans, Animals, Things*, ed. Christopher Watts, 228–244. London: Routledge.

Wikar, Hendrik J. 1935. *The Journal of Hendrik Jacob Wikar (1779) with an English Translation by A. W. van der Horst and the Journals of Jacobus Coetsé Jansz (1760) and Willem van Reenen (1790) with an English Translation by Dr. E. E. Mossop*. Cape Town: The van Riebeeck Society.

Wilhelm, Joachim Helmut. 2005. The !Kung Bushmen. In *Kalahari and Namib Bushmen in German South West Africa: Ethnographic Reports by Colonial Soldiers and Settlers*, ed. Mathias Guenther, 91–184. Cologne: Rüdiger Köppe Verlag.

Willerslev, Rane. 2011. Frazer Strikes Back from the Armchair: A New Search for the Animist Soul. *Journal of the Royal Anthropological Institute* (N.S.) 17: 504–526.

Wilmsen, Edwin N. 1997. *The Kalahari Ethnographies (1896–1898) of Siegfried Passarge: Nineteenth Century Khoisan- and Bantu-Speaking Peoples*. Cologne: Rudiger Koppe Verlag.

Witleson, David W. 2018. Frogs or People: Dorothea Bleek and a Genealogy of Ideas in Rock Art Research. *Azania: Archaeological Research in Africa* 53: 1–25. Online publication.

Wittenberg, Hermann. 2012. Wilhelm Bleek and the Khoisan Imagination: A Study of Censorship, Genocide and Colonial Science. *Journal of Southern African Studies* 38: 667–679.

———. 2014. The Story of //Kabbo and *Reynard the Fox*: Notes on Cultural and Textual History. In *Courage of //Kabbo: Celebrating the 100th Anniversary of the Publication of* Specimens of Bushman Folklore, ed. Janette Deacon and Pippa Skotnes, 93–98. Cape Town: UCT Press.

Woodhouse, Bert. 1984. *When Animals Were People: A-Z of Animals of Southern Africa as the Bushmen Saw and Thought Them and as the Camera Sees Them Today*. Melville: Chris van Rensburg Publications (Pty) Limited.

Wynn, Thomas, and Frederick Coolidge. 2012. *How to Think Like a Neandertal*. New York: Oxford University Press.

Young, David E., and Jean-Guy Goulet, eds. 1998. *Being Changed by Cross-Cultural Encounters: The Anthropology of Extraordinary Experience*. Peterborough: Broadview Press.

Zimmerman, Mary. 2002. *Metamorphoses: A Play Based on David R. Slavitt's Translation of* The Metamorphoses of Ovid. Evanston: Northwestern University Press.

Index[1]

[1] Note: Page numbers followed by 'n' refer to notes.

© The Author(s) 2020
M. Guenther, *Human-Animal Relationships in San and Hunter-Gatherer Cosmology, Volume I*,
https://doi.org/10.1007/978-3-030-21182-0

The manufacturer's authorised representative in the EU is Springer Nature Customer Service Centre GmbH, Europaplatz 3, 69115 Heidelberg, Germany. If you have any concerns regarding our products, please contact ProductSafety@springernature.com

Printed and bound by CPI Group (UK) Ltd, Croydon, CR0 4YY
30/09/2026
02225434-0007